Introduction to
ENVIRONMENTAL SCIENCE

Revised Printing

Kendall Hunt
publishing company

SUSANNAH SANDRIN

Kendall Hunt
publishing company

www.kendallhunt.com
Send all inquiries to:
4050 Westmark Drive
Dubuque, IA 52004-1840

CONTENTS

CHAPTER 1

Home Sweet Earth: An Introduction to Environmental Science

1.1 Introduction

The earth is the only place that humans live. We cannot at present move to anyplace else. All of us are intertwined with the earth in a symbiotic relationship where everything we do as living organisms has an effect on the earth and every other living organism on the earth. The earth's air, water, and soil, along with the sun, provide us with everything we need to survive, and with the exception of the sun, we interact with these components directly. This means that the earth is our environment, and that the quality of its air, water, and soil is what lets us survive and thrive.

The World Population

The human population is over 7 billion, and the United Nations projects 9 billion by 2050 (Figure 1.1). This growth is only sustainable due to our modern technology and the use of fossil fuels. Both of these are limited, and without them the earth's population would be a fraction of what it is today. This growth rate is fueled by a worldwide birth rate of 19.2 births/1,000 population as of 2011. The population declines by a worldwide

1

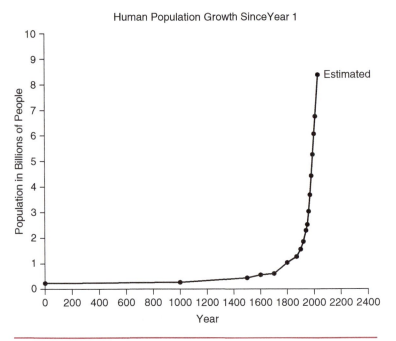

FIGURE 1.1 **World population in modern times, population growth from 7 to 9 billion people is estimated by 2050.** Figure by Lynn Burgess with data from U.S Census Bureau.

death rate of 8.12 deaths/1,000 population; therefore, there is a net gain of 11.08 people/1,000 population each year. Even if the world were to control the number of births, as China has done, and lower the birth rate to replacement level, the population would continue to grow for 50 to 70 years. The **replacement rate** is the number of births per female needed to replace her and one male plus the number of persons who die during the childbearing years, which end at 44–49 years old. It also includes those who are inhibited to procreate or have other dysfunctions that prevent childbearing and some adjustments, since the number of male births is slightly higher than female births. The replacement rate is 2.1 births per women in industrialized countries and ranges from 2.5 to 3.3 in developing countries due to a higher death rate.

China has had a one-child limit for urban couples and a two-child limit for rural couples for over 40 years. China's present birth rate is 1.09/1,000 population.

Its **population momentum** (population growth at the national level that would occur even if levels of childbearing immediately declined to replacement level) will continue to increase its population until 2030 before it stabilizes or declines. This increase will be about 9%. But it will have taken China almost 60 years to stabilize its population, and this can only be done with a strict dictatorial government.

We have a finite supply of air, water, and soil, and as the world population grows exponentially, we continue to displace other living organisms, use up the available fresh water and tillable soil, and pollute everything. Even our living space is becoming limited. Even if the human population suddenly ceased to increase, there would still be an increasing demand for the earth's resources. People in the developing countries are demanding more and more of the lifestyle enjoyed by citizens of the industrial world. People everywhere want to improve their standard of living and have more luxury items and larger homes. The size of the average home in the United States more than doubled between 1950 and 2010 (Figure 1.2). At the same time, the number of people who live together has decreased from an average of 3.37 in 1950 to 2.63 in 2010.

Let's consider a single consumer item like cellphones, since they are relatively new but are seen everywhere. Cellphone companies are offering cellphone service to 193 of the 196 countries in the world. Only three areas in the world have no cellphone service; these are the Falkland Islands, Norfolk Island, and the Western Sahara. These areas can still be covered with Internet phone service and satellite phones. China has over twice as many cellphones as the United States. So how does a cellphone affect the earth's resources? Each phone requires plastic and some scarce metals called **rare earth metals** (set of 17 chemical elements in the periodic table, specifically the 15 lanthanides plus scandium and yttrium). Energy to manufacture the phone and more energy each day to run the phone and the systems to operate it are also required. Mining of the rare earth metals is limited to a few areas and has a long history of environmental damage and illegal operations. There is limited recycling of these phones (and other electronic devices), and the disposal of electronic devices and their batteries is a significant source of pollution.

Therefore the trend is for each person on the planet to require more of our resources each year. From the soil to grow the food we need to the metals being mined, each of us will make a bigger impact on the earth than our ancestors did. Our demand for energy grows even faster than the population, and we use very little **renewable** or **green energy**. Most of our energy is derived from fossil fuels or hydroelectric dams which impact the air, water, and soil. A hydroelectric dam may be "cleaner" than a coal-fired electric plant but it is not without an environmental impact. Much of the water behind the dam is lost to evaporation and is warmed as it sits in the reservoir, the wildlife of the area is changed, the reservoir slowly fills in with silt, and soil erosion downstream is increased.

Every person on earth requires a certain amount of clear air, freshwater, and food. They need healthy soil and freshwater to grow their food. Each person contributes wastes that pollute and change the air, water, and soil. Even the most "green" person on the planet still pollutes and contributes wastes. Everyone breathes out CO_2 and creates urine and fecal wastes. The CO_2 adds to global climatic change and the bodily wastes

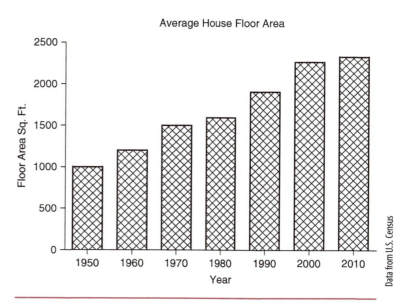

FIGURE 1.2 Increasing size of the average home in the United States by floor area over the last 60 years. Figure by Lynn Burgess.

Data from U.S. Census

must be recycled in the **ecosystem** (a system formed by the interaction of a community of organisms with their environment). This is not a problem when there are few humans spread over a large area, but as the population increases there are very few areas left where the population density is low enough that waste is not a problem. The urban areas where most people live are growing rapidly and have created a major challenge for environmental professionals, engineers, and government officials to find ways of treating and disposing of the human wastes. When the other wastes that are produced by our technology, such as plastics, burned fossil fuels, and old electronic components, are added, then the problem of disposal and recycling becomes huge, and at present we are not keeping up with the task.

Many urban areas in the United States have to reuse the water from their wastewater treatment plants as part of their domestic water supply. Many are only using the wastewater for irrigation for places like golf courses, but a few are using it as drinking water, and more will be using it for this purpose in the future as the growth of urban areas outstrips their freshwater supply. In reality, wastewater reuse is not new. Many communities with municipal wastewater treatment facilities discharge their water into a river or lake and then another community downstream takes its water from the same river or lake. Just think about the entire Mississippi River system and all of the cities upstream. Where does their wastewater go to after treatment? The exceptions are those cities that discharge into the ocean, but we then swim and eat fish caught from the ocean.

There is also the problem of disposing of or recycling solid wastes. Most of these wastes are not recycled and are either burned or buried. Both of these methods create pollution of the air, water, and soil. Some energy can be obtained from the incineration of solid wastes, but it is also a source of air pollution, and the ash has to be disposed of in a landfill, creating additional pollution problems. The burial of solid wastes is a huge loss of both energy and material resources. This process can also cause air, water, and soil pollution, as will be discussed in further chapters.

Carrying Capacity

As our population grows geometrically, the biggest question we face is, Have we reached the human carrying capacity of the earth? **Carrying capacity** is an ecological term that refers to the population that an area (in this case, the entire planet) can support without environmental deterioration. It can also be stated as the maximum numbers of a species that the environment can sustain indefinitely. If we take the second definition of carrying capacity, then we can continue to put more humans on this planet as long as we can increase our technology and keep finding either fossil fuel or alternative energy resources. If we use the first definition, then we reached the earth's human carrying capacity long ago. We need then ask the question, When does the displacement of many other species and serious pollution and degradation of the air, water, and soil become irreversible environmental deterioration? Other animal populations are controlled by predation, infective diseases, food and water availability and other resource limitations associated with space, like nesting sites or cover. Humans have lost their predators and have learned to control many of the infective diseases to such a point that chronic diseases of aging, heart disease and cancer are our biggest killers. These diseases generally occur after the childbearing age and have less effect on population growth than infectious diseases had in the past. The biggest threats to our growing population, or what is contributing to our carrying capacity, are food, freshwater, and space. Pollution-caused climatic changes are also limiting our population growth. We are building cities and houses on our farmland and therefore reducing the amount of area we have to grow food. Freshwater is in very limited supply in most areas of the world. In the sub-Saharan areas of Africa, a person may have to walk 8 or 10 miles a day to get water, and many times the water is polluted or has parasites.

As space is reduced in cities and farms are lost, there is a direct reduction in the quality of life and the standard of living. In the United States, we are now experiencing a reduction in the per capita number of home-owners and an increase in apartments and other multi-unit housing. Most of the people in multi-unit housing rent rather than own their homes. So what faces humans as we reach our maximum carrying capacity on the earth? The predictions vary between (1) increased frequency and extent of war (i.e., humans acting as their own predators), (2) new and old diseases that will cause worldwide pandemics, (3) increased areas of famine due to drought, higher temperatures, desertification, and loss of available fertilizers (phosphate supplies are very close to this today), and (4) overwhelming pollution and degradation of our air, water, and soil. With the possible exception of war, all of the predictions pertaining to a dramatic reduction in quality of life can be moderated by the environmental health profession.

ESSENTIAL EARTH SPHERES: COMPONENTS FOR LIFE

CHAPTER 2

Essential Resources: The Hydrosphere

Shutterstock/Zffoto

2.1 Introduction

When viewed from space, Earth is a watery planet. Indeed, oceans comprise about 71% of our planet's surface, while glistening glaciers cover 10% of the continents. Yet water shortages are a serious problem. This contradiction is due, in part, to the fact that the abundance of ocean water is too salty for human use, and much of the remaining fresh water is polluted, used up, or distributed unequally. As summed up by one prominent ecologist: "The wet places in the world don't need the runoff because they are wet. The dry places of the world need the runoff to irrigate the land, but they don't get much" (Pimm, 2001).

Globally, humans extract about one-sixth of the total volume of all the rivers in the world. But the amount of consumption varies from place to place. In British Columbia, for example, winter snowfall and summer rain are abundant and the population density is low. Thus, rivers carry copious amounts of water into the oceans. On the other hand, in the southwest desert of the United States, precipitation is low and the population is much denser, so the consumption of fresh water outpaces its replenishment. In addition, the Colorado River, which

7

meanders through the American southwest, no longer reaches the ocean. Instead, much of the water from the once free-flowing Colorado River is diverted and extracted to irrigate the parched landscape.

The Colorado River used to flow from the United States into Mexico and empty into the Sea of Cortez. But now, Mexico receives only a trickle of salty river water. The resulting conflict between the United States and Mexico is serious enough, but it involves two friendly allies and trading partners. More volatile circumstances surrounding water issues have become potentially explosive in the Middle East, however, where the Jordan River flows through desert and near desert regions between Israel and its enemies, Syria and Jordan. The population in this parched region is skyrocketing, and as a result the potential for armed conflict is always present.

Human water use is necessary, complicated, and political. Indeed, our very existence depends on access to fresh, clean water. Therefore, in this chapter we will look at long-term prospects for how to resolve such politically charged water shortage issues. We will also investigate examples of water conservation and management, as well as look to the future of sustainable water use.

2.2 Water Supply and the Hydrologic Cycle

*The **hydrologic** or water cycle describes the movement and storage of water between different spheres on the planet. At any point in time water may be stored in the oceans, in glaciers, ice and snow, as groundwater, in rivers, streams, lakes, ponds and wetlands, and lastly, as water vapor in the atmosphere. Water is constantly moving between these spheres through the processes of evaporation, precipitation, infiltration, melting, and groundwater flow. Only 3 percent of the Earth's water is freshwater (not salty), and close to 70 percent of that freshwater is frozen in ice caps at the poles and in glaciers. Therefore, we are ultimately dependent for our survival on a small, but critical portion of the overall planetary*

> **hydrologic cycle**: the continuous movement of water on, above, and below the surface of the Earth; also known as the water cycle.

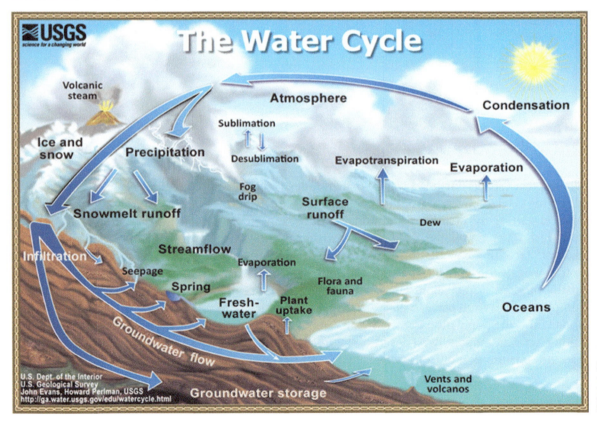

US Geological Survey

water cycle. In this Ecological Society of America (ESA) report written by ecologist Robert B. Jackson, et al we see just how massive and important the water cycle actually is.

Beyond our obvious uses of freshwater for drinking, bathing, and washing, our society makes use of water for many other purposes. Scientists break these uses down into categories known as consumptive and instream. Consumptive (or extractive) uses involve removing water from its source for drinking or other residential purposes as well as for industrial use and for irrigation of crops. Instream (or non-extractive) uses involve deriving benefits from water without removing it from where it is located. Examples of instream uses include transportation/navigation, recreation, habitat for fish and other aquatic life, hydroelectric power generation, and waste processing. The sustainable management of water supplies frequently involves tradeoffs and/or conflicts between consumptive and instream uses. This was dramatically illustrated in recent years in the Pacific Northwest when irrigation water from rivers was denied to farmers in order to maintain water levels needed for salmon populations.

*In addition to surface water management there are also concerns over how groundwater resources are managed. Hydrologists [scientists who study the movement and distribution of water] make a distinction between renewable and nonrenewable groundwater. Over three fourths of all underground water is considered nonrenewable since it is found in **aquifers** that formed tens of thousands of years ago that are not replenished. A primary example of nonrenewable groundwater or "fossil water" source is the Ogallala Aquifer in the central United States where rapid pumping of water from the Ogallala to irrigate Midwestern farmland has lowered the level of the aquifer significantly.*

> **aquifers**: an underground bed or layer of porous rock, sediment, or soil that yields water

In contrast, renewable groundwater refers to aquifers that are regularly replenished by rainfall or snowmelt. Even though groundwater resources are renewable, they can be seriously mismanaged. Examples include when water is removed faster than it is replenished and when pollutants or waste products from the surface are allowed to seep into them. Renewable groundwater deposits can be thought of as bank accounts. As long as we do not withdraw more than is going in we can maintain a positive balance. Nonrenewable groundwater is better thought of as a one-time inheritance or windfall that we can make use of but only at the expense of lower future balances.

By R.B. Jackson, et al

*L*ife on earth depends on the continuous flow of materials through the air, water, soil, and food webs of the biosphere. The movement of water through the **hydrological cycle** comprises the largest of these flows, delivering an estimated 110,000 cubic kilometers (km^3) of water to the land each year as snow and rainfall [a cubic kilometer is an area 1,000 meters wide by 1,000 meters deep by 1,000 meters high, or roughly 10 football fields across, deep, and tall]. Solar energy drives the hydrological cycle, vaporizing water from the surface of oceans, lakes, and rivers as well as from soils and plants (evapotranspiration). Water vapor rises into the atmosphere where it cools, condenses, and eventually rains down anew. This renewable freshwater supply sustains life on the land, in estuaries, and in the freshwater ecosystems of the Earth.

Renewable fresh water provides many services essential to human health and well being, including water for drinking, industrial production, and irrigation, and the production of fish, waterfowl, and shellfish. Fresh water also provides many benefits while it remains in its channels (nonextractive or instream benefits), including flood control, transportation, recreation, waste processing, hydroelectric power, and habitat for aquatic plants and animals. Some benefits, such as irrigation and hydroelectric power, can be achieved only by damming, diverting, or creating other major changes to natural water flows. Such changes often diminish or preclude other instream benefits of fresh water, such as providing habitat for aquatic life or maintaining suitable water quality for human use.

The ecological, social, and economic benefits that freshwater systems provide, and the trade-offs between consumptive and instream values, will change dramatically in the coming century. Already, over the past one hundred years, both the amount of water humans withdraw worldwide and the land area under irrigation have risen exponentially. Despite this greatly increased consumption, the basic water needs of many people in the world are not being met. Currently, 1.1 billion people lack access to safe drinking water, and 2.8 billion lack

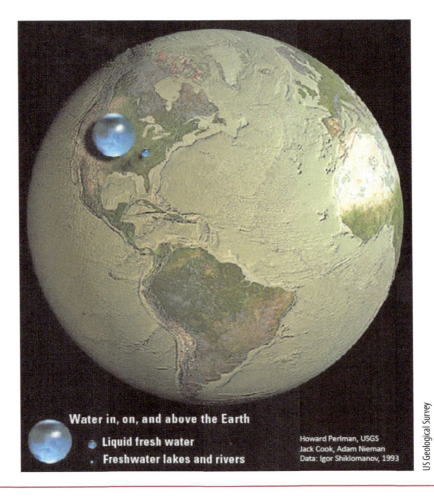

Water in, on, and above the Earth

● Liquid fresh water

● Freshwater lakes and rivers

Howard Perlman, USGS
Jack Cook, Adam Nieman
Data: Igor Shiklomanov, 1993

US Geological Survey

FIGURE 2.1   All of the World's Water.

basic sanitation services. These deprivations cause approximately 250 million cases of water-related diseases and five to ten million deaths each year. Also, current unmet needs limit our ability to adapt to future changes in water supplies and distribution. Many current systems designed to provide water in relatively stable climatic conditions may be ill prepared to adapt to future changes in climate, consumption, and population. While a global perspective on water withdrawals is important for ensuring sustainable water use, it is insufficient for regional and local stability. How fresh water is managed in particular basins and in individual watersheds is the key to sustainable water management.

The Global Water Cycle

Surface Water

Most of the earth is covered by water, more than one billion km³ of it. The vast majority of that water, however, is in forms unavailable to land-based or freshwater ecosystems. Less than 3 percent is fresh enough to drink or to irrigate crops, and of that total, more than two-thirds is locked in glaciers and ice caps. Freshwater lakes and rivers hold 100,000 km³ globally, less than one ten-thousandth of all water on earth.

Water vapor in the atmosphere exerts an important influence on climate and on the water cycle, even though only 15,000 km³ of water is typically held in the atmosphere at any time. This tiny fraction, however, is vital for the biosphere. Water vapor is the most important of the so-called greenhouse gases (others include carbon dioxide, nitrous oxide, and methane) that warm the earth by trapping heat in the atmosphere. Water vapor contributes approximately two-thirds of the total warming that greenhouse gases supply. Without these gases, the mean surface temperature of the earth would be well below freezing, and liquid water would be absent over much of the planet. Equally important for life, atmospheric water turns over every ten days or so as water vapor condenses and rains to earth and the heat of the sun evaporates new supplies of vapor from the liquid reservoirs on earth.

Solar energy typically evaporates about 425,000 km³ of ocean water each year. Most of this water rains back directly to the oceans, but approximately 10 percent falls on land. If this were the only source of rainfall, average precipitation across the earth's land surfaces would be only 25 centimeters (cm) a year, a value typical for deserts or **semi-arid regions**. Instead, a second, larger source of water is recycled from plants and the soil through **evapotranspiration**. The water vapor from this source creates a direct feedback between the land surface and regional climate. The cycling of other materials such as carbon and nitrogen (biogeochemical cycling) is strongly coupled to this water flux through the patterns of plant growth and microbial decomposition, and this coupling creates additional feedbacks between vegetation and climate. This second source of recycled water contributes two-thirds of the 70 cm of precipitation that falls over land each year. Taken together, these two sources account for the 110,000 km³ of renewable freshwater available each year for terrestrial, freshwater, and estuarine ecosystems.

Because the amount of rain that falls on land is greater than the amount of water that evaporates from it, the extra 40,000 km³ of water returns to the oceans, primarily via rivers and underground aquifers. A number of factors affect how much of this water is available for human use on its journey to the oceans. These factors include whether the precipitation falls as rain or snow, the timing of precipitation relative to patterns of seasonal temperature and sunlight, and the regional topography. For example, in many mountain regions, most precipitation falls as snow during winter, and spring snowmelt causes peak flows that flood major river systems. In some tropical regions, monsoons rather than snowmelt create seasonal flooding. In other regions, excess precipitation percolates into the soil to recharge ground water or is stored

> **semi-arid region**: a region characterized by low annual rainfall that is subject to frequent and prolonged droughts

> **evapotranspiration**: the process by which water is transferred from land to the atmosphere through evaporation; the process of converting water to water vapor, from the soil and other surfaces and by transpiration or (the process of giving off water vapor) from the leaves of plants

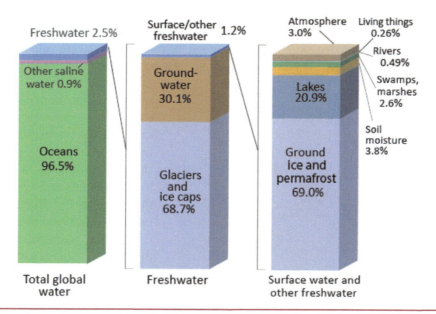

FIGURE 2.2 Where is Earth's Water?

Source: Igor Shiklomanov's chpater "World fresh water resources" in Peter H. Gleick (editor), 1993, Water in Crisis: A Guide to the World's Fresh Water Resources.

Note: Numbers are rounded, so percent summations may not add to 100.

in wetlands. Widespread loss of wetlands and floodplains, however, reduces their ability to absorb these high flows and speeds the runoff of excess nutrients and contaminants to estuaries and other coastal environments. More than half of all wetlands in the U. S. have already been drained, dredged, filled, or planted.

Available water is not evenly distributed globally. Two thirds of all precipitation falls in the tropics (between 30 degrees N and 30 degrees S latitude) due to greater solar radiation and evaporation there. Daily evaporation from the oceans ranges from 0.4 cm at the equator to less than 0.1 cm at the poles. Typically, tropical regions also have larger runoff. Roughly half of the precipitation that falls in rainforests becomes runoff, while in the deserts low rainfall and high evaporation rates combine to greatly reduce runoff. The Amazon, for example, carries 15 percent of all water returning to the global oceans. In contrast, the Colorado River drainage, which is one-tenth the size of the Amazon, has a historic annual runoff 300 times smaller. Similar variation occurs at continental scales. Average runoff in Australia is only 4 cm per year, eight times less than in North America and orders of magnitude less than in tropical South America.

As a result of these and many other disparities, freshwater availability varies dramatically worldwide.

Ground Water

Approximately 99 percent of all liquid fresh water is in underground aquifers, and at least a quarter of the world's population draws its water from these groundwater supplies. Estimates of the global water cycle generally treat rates of groundwater inflow and outflow as if they were balanced. In reality, however, this resource is being depleted globally. Ground water typically turns over more slowly than most other water pools, often in hundreds to tens of thousands of years, although the range in turnover rates is large. Indeed, a majority of ground water is not actively turning over or being recharged from the earth's surface at all. Instead, it is "fossil water," a relic of wetter ancient climatic conditions and melting Pleistocene ice sheets that accumulated over tens of thousands of years. Once used, it cannot readily be replenished.

The distinction between renewable and nonrenewable ground water is critical for water management and policy. More than three-quarters of underground water is non-renewable, meaning it has a replenishment period of centuries or more. The High Plains or Ogallala Aquifer that underlies half a million km^2 of the central United States is arguably the largest aquifer in the world. The availability of turbine pumps and relatively inexpensive energy has spurred the drilling of about 200,000 wells into the aquifer since the 1940s, making the Ogallala the primary water source for a fifth of irrigated U.S. farmland. The extent of irrigated cropland in the region peaked around 1980 at 5.6 million hectares and at pumping rates of about 6 trillion gallons of water a year. That has since declined somewhat due to groundwater depletion and socioeconomic changes in the region. However, the average thickness of the Ogallala declined by more than 5 percent across a fifth of its area in the 1980s alone.

In contrast, renewable aquifers depend on current rainfall for refilling and so are vulnerable to changes in the quantity and quality of recharge water. For example, groundwater pumping of the Edwards Aquifer, which supplies much of central Texas with drinking water, has increased four-fold since the 1930s and at times now exceeds annual recharge rates. Increased water withdrawal makes aquifers more susceptible to drought and other changes in weather and to contamination from pollutants and wastes that percolate into the ground water. Depletion of ground water can also cause land subsidence [to sink] and compaction [consolidation of sediments] of the porous sand, gravel, or rock of the aquifer, permanently reducing its capacity to store water. The Central Valley of California has lost about 25 km^3 of storage in this way, a capacity equal to more than 40 percent of the combined storage capacity of all human-made reservoirs in the state.

Shutterstock/Ingrid Curry

Irrigation on the Ogallala Aquifer is the primary water source for one fifth of irrigated U.S. farmland..

Renewable ground water and surface waters have commonly been viewed separately, both scientifically and legally. This view is changing, however, as studies in streams, rivers, reservoirs, wetlands, and estuaries show the importance of interactions between renewable surface and ground waters for water supply, water quality, and aquatic habitats. Where extraction of ground water exceeds recharge rates, the result is lower **water tables.** In summer, when a high water table is needed to sustain minimum flows in rivers and streams, low groundwater levels can decrease low-flow rates, reduce perennial stream habitat, increase summer stream temperatures, and impair water quality. Trout and salmon species select areas of groundwater upwelling in streams to moderate extreme seasonal temperatures and to keep their eggs from overheating or freezing. Dynamic exchange of surface and ground waters alters the dissolved oxygen and nutrient concentrations of streams and dilutes concentrations of dissolved contaminants such as pesticides and volatile organic compounds. Because of such links, human development of either ground water or surface water often affects the quantity and quality of the other.

> **water tables:** the uppermost level of an aquifer, below which the ground is saturated with water

The links between surface and ground waters are especially important in regions with low rainfall. Arid and semi-arid regions cover a third of the earth's lands and hold a fifth of the global population. Ground water is the primary source of water for drinking and irrigation in these regions, which possess many of the world's largest aquifers. Limited recharge makes such aquifers highly susceptible to groundwater depletion. For example, exploitation of the Northern Sahara Basin Aquifer in the 1990s was almost twice the rate of replenishment, and many springs associated with this aquifer are drying up. For non-renewable groundwater sources, discussing sustainable or appropriate rates of extraction is difficult. As with deposits of coal and oil, almost any extraction is non-sustainable. Important questions for society include at what rate groundwater pumping should be allowed, for what purpose, and who if anyone will safeguard the needs of future generations. In the Ogallala Aquifer, for example, the water may be gone in as little as a century.

Adapted from Jackson, R.B. et al. 2001. Water in a Changing World. Ecological Society of America, Issues in Ecology, Number 9. Available online at: http://www.esa.org/science_resources/issues/FileEnglish/issue9.pdf.

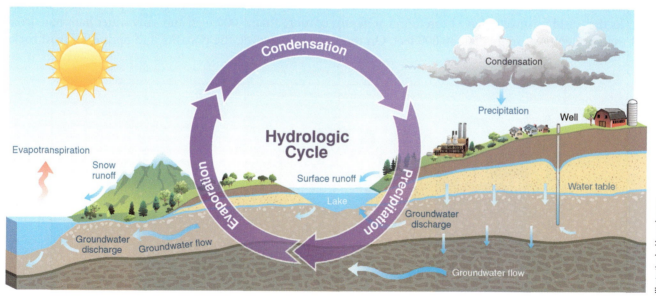

FIGURE 2.3 Interaction Between Ground Water and Surface Water. In Earth's hydrologic cycle, precipitation not only sustains ecosystems and human activity, but also recharges shallow and deep aquifers. The remaining water returns to the atmosphere via evapotranspiration. Human industrial, agricultural, and municipal users are using groundwater faster than it can be recharged, resulting in depletion of this precious natural resource. A major cause of fresh water depletion is that most groundwater is not actively recharged. Instead, it is "fossil" water—a relic of wetter ancient climatic conditions.

Aquifers

Groundwater is one of our most valuable resource—even though you probably never see it or even realize it is there. As you may have read, most of the void spaces in the rocks below the water table are filled with water. But rocks have different porosity and permeability characteristics, which means that water does not move around the same way in all rocks below ground.

When a water-bearing rock readily transmits water to wells and springs, it is called an aquifer. Wells can be drilled into the aquifers and water can be pumped out. Precipitation eventually adds water (recharge) into the porous rock of the aquifer. The rate of recharge is not the same for all aquifers, though, and that must be considered when pumping water from a well. Pumping too much water too fast draws down the water in the aquifer and eventually causes a well to yield less and less water and even run dry. In fact, pumping your well too fast can even cause your neighbor›s well to run dry if you both are pumping from the same aquifer.

In the diagram below, you can see how the ground below the water table (the blue area) is saturated with water. The unsaturated zone above the water table (the greenish area) still contains water (after all, plants roots live in this area), but it is not totally saturated with water. You can see this in the two drawings at the bottom of the diagram, which show a close-up of how water is stored in between underground rock particles.

Sometimes the porous rock layers become tilted in the earth. There might be a confining layer of less porous rock both above and below the porous layer. This is an example of a confined aquifer. In this case, the rocks surrounding the aquifer confines the pressure in the porous rock and its water. If a well is drilled into this "pressurized" aquifer, the internal pressure might (depending on the ability of the rock to transport water) be enough to push the water up the well and up to the surface without the aid of a pump, sometimes completely out of the well. This type of well is called artesian. The pressure of water from an artesian well can be quite dramatic.

A relationship does not necessarily exist between the water-bearing capacity of rocks and the depth at which they are found. A very dense granite that will yield little or no water to a well may be exposed at the land surface. Conversely, a porous sandstone, such as the Dakota Sandstone mentioned previously, may lie hundreds or thousands of feet below the land surface and may yield hundreds of gallons per minute of water. Rocks that yield freshwater have been found at depths of more than 6,000 feet, and salty water has come from oil wells at depths of more than 30,000 feet. On the average, however, the porosity and permeability of rocks

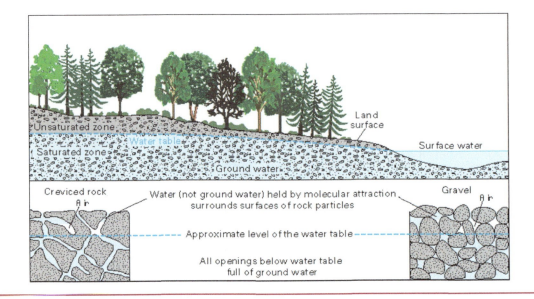

FIGURE 2.4

decrease as their depth below land surface increases; the pores and cracks in rocks at great depths are closed or greatly reduced in size because of the weight of overlying rocks.

Pumping Can Affect the Level of the Water Table

Groundwater occurs in the saturated soil and rock below the water table. If the aquifer is shallow enough and permeable enough to allow water to move through it at a rapid-enough rate, then people can drill wells into it and withdraw water. The level of the water table can naturally change over time due to changes in weather cycles and preciptiation patterns, streamflow and geologic changes, and even human-induced changes, such as the increase in impervious surfaces on the landscape.

The pumping of wells can have a great deal of influence on water levels below ground, especially in the vicinity of the well, as this diagram shows. If water is withdrawn from the ground at a faster rate that it is replenished, either by infiltration from the surface of from streams, then the water table can become lower, resulting in a "cone of depression" around the well. Depending on geologic and hydrologic conditions of the aquifer, the impact on the level of the water table can be short-lived or last for decades, and it can fall a small amount or many hundreds of feet. Excessive pumping can lower the water table so much that the wells no longer supply water—they can "go dry."

Water Movement in Aquifers

Water movement in aquifers is highly dependent of the pearmeablility of the aquifer material. Permeable material contains interconnected cracks or spaces that are both numerous enough and large enough to allow water to move freely. In some permeable materials groundwater may move several metres in a day; in other places, it moves only a few centimetres in a century. Groundwater moves very slowly through relatively impermeable materials such as clay and shale. (Source: Environment Canada)

After entering an aquifer, water moves slowly toward lower lying places and eventually is discharged from the aquifer from springs, seeps into streams, or is withdrawn from the ground by wells. Groundwater in aquifers between layers of poorly permeable rock, such as clay or shale, may be confined under pressure. If such a confined aquifer is tapped by a well, water will rise above the top of the aquifer and may even flow from the well onto the land surface. Water confined in this way is said to be under artesian pressure, and the aquifer is called an artesian aquifer.

Visualizing Artesian Pressure

Here's a little experiment to show you how artesian pressure works. Fill a plastic sandwich baggie with water, put a straw in through the opening, tape the opening around the straw closed, DON'T point the straw towards your teacher, and then squeeze the baggie. Artesian water is pushed out through the straw.

Some information on this page is from "Ground Water and the Rural Homeowner, Pamphlet", U.S. Geolgoical Survey, by Waller, Roger M.,1982

Groundwater Serves Many Purposes

Fresh groundwater was used for many important purposes, with the largest amount going toward irrigating crops, such as the delicious eggplants, squash, and rutabagas that children love to have for dinner. Local city and county water departments withdraw a lot of groundwater for public uses, such as for delivery to homes, businesses, and industries, as well as for community uses such as firefighting, water services at public buildings, and for keeping local residents happy by keeping community swimming pools full of water. Industries and mining facilities also used a lot of groundwater. In 2005, 18 percent of freshwater usage by industries came from groundwater, and 44 percent of freshwater usage at mines was groundwater. The majority of water used for self-supplied domestic and livestock purposes came from groundwater sources.

FIGURE 2.5 **The San Joaquin Valley in California has seen huge drops in the land surface due to subsidence over the past century.**

Groundwater Use in the U.S.

About 23 percent of the freshwater used in the United States in 2005 came from groundwater sources. The other 77 percent came from surface water. Groundwater is an important natural resource, especially in those parts of the country that don't have ample surface-water sources, such as the arid West. It often takes more work and costs more to access groundwater as opposed to surface water, but where there is little water on the land surface, groundwater can supply the water needs of people.

For 2005, most of the fresh groundwater withdrawals, 68 percent, were for irrigation, while another 19 percent was used for public-supply purposes, mainly to supply drinking water to much of the Nation's population. Groundwater also is crucial for those people who supply their own water (domestic use), as over 98 percent of self-supplied domestic water withdrawals came from groundwater.

Land Subsidence in California

This photo shows the approximate location of maximum subsidence in the United States, identified by research efforts of Dr. Joseph F. Poland (pictured). The site is in the San Joaquin Valley southwest of Mendota, California. Signs on pole show approximate altitude of land surface in 1925, 1955, and 1977.

In this case, excessive groundwater pumping allowed the upper soil layers to dry out and compress and compact, which is by far the single largest cause of subsidence. Soil compaction results in a reduction of the pore sizes between soil particles, resulting in essentially a permanent condition—rewetting of the underground soil and rock does not cause the land to go back up in altitude. This results in a lessening of the total storage capacity of the aquifer system. Here, the term "groundwater mining" is really true.

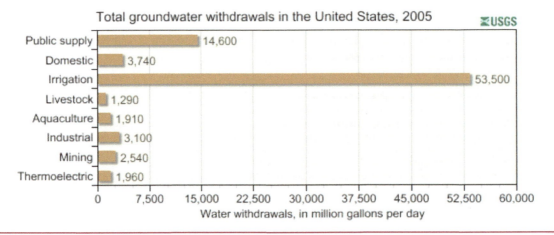

FIGURE 2.6

2.3 Water Shortages

In this article by Don Hinrichsen of the Worldwatch Institute, we see that the overuse of freshwater supplies has a significant impact on wildlife, ecosystems, and human societies. For example, diversion of freshwater for agriculture, industry, and residential uses leaves less water for other species to drink. Use of rivers and other water bodies as dumping grounds for waste reduces both the quality and quantity of water for other uses. And in some regions described in this reading lack of clean drinking water has already reached crisis proportions.

Specific areas of concern discussed in the following paper include the loss of wetlands, the destruction of aquatic habitats, and the creation of pollution that ends up in various water sources. Wetlands [an area of land with permanently or seasonally saturated soils] play a critically important role in maintaining freshwater supplies for two reasons—they slow moving water down and improve storage, and they help to remove impurities and increase water quality. Wetland loss due to development and other factors thus impacts both water quantity and quality. As was discussed in the last chapter, the loss of habitat is the major contributor to biodiversity loss, and this holds true for aquatic as well as terrestrial habitats. Large percentages of the world's fish, mussels, amphibians, and mollusks are endangered or have already been driven to extinction due to modification of water flows or large-scale diversions of water. In addition to habitat alteration and destruction, water pollution also takes a toll on biodiversity and supplies of freshwater for human use. Up to a certain point, moving water can assimilate some human waste and pollution. But the limit has been far exceeded and the consequences have been serious impairment of surface waters.

Many regions of the world are already experiencing, or will soon experience, serious challenges in meeting their water needs. China, reviewed in this article, increasingly has to grapple with the consumptive/extractive versus instream/nonextractive tradeoffs discussed in section 2.2. The Aral Sea Basin in Asia and Lake Chad in Africa offer cautionary examples of how bad the situation can get when water is mismanaged. Because increasing water supplies bumps up against the limits of the hydrological cycle, perhaps the best way to meet the water needs of a growing human population—while leaving enough water for other species—is to become much more efficient in our use. Such a "Blue Revolution" in the efficiency of water use will be touched on briefly in this article and in more detail in the next section.

By *Don Hinrichsen*

Billions of people around the world are experiencing the effects of the world's deepening freshwater crisis, often walking great distances to gather drinking water for their families.

On March 20, 2000, a group of monkeys, driven mad with thirst, clashed with desperate villagers over drinking water in a small outpost in northern Kenya near the border with Sudan. The Pan African News Agency reported that eight monkeys were killed and 10 villagers injured in what was described as a "fierce two-hour melee." The fight erupted when relief workers arrived and began dispensing water from a tanker truck. Locals claimed that a prolonged drought had forced animals to roam out of their natural habitats to seek life-giving water in human settlements. The monkeys were later identified as generally harmless vervets.

The world's deepening freshwater crisis—currently affecting 2.3 billion people—has already pitted farmers against city dwellers, industry against agriculture, water-rich state against water-poor state, county against county, neighbor against neighbor. Inter-species rivalry over water, such as the incident in northern Kenya, stands to become more commonplace in the near future.

"The water needs of wildlife are often the first to be sacrificed and last to be considered," says Karin Krchnak, population and environment program manager at the National Wildlife Federation (NWF) in Washington, D.C. "We ignore the fact that working to ensure healthy freshwater ecosystems for wildlife would mean healthy waters for all." As more and more water

Shutterstock/ventdusud

To quench civilization's thirst for fresh water and protect against shortages, humans have built thousands of large dams, which have profoundly altered aquatic ecosystems. One result is that the water needs of wildlife are often the first to be sacrificed. Glen Canyon Dam on the Colorado River has led to the endangerment of several unique species of fish, such as the humpback chub (*Gila cypha*) and the razorback sucker (*Xyrauchen texanus*).

is withdrawn from rivers, streams, lakes and aquifers to feed thirsty fields and the voracious needs of industry and escalating urban demands, there is often little left over for aquatic ecosystems and the wealth of plants and animals they support.

The mounting competition for freshwater resources is undermining development prospects in many areas of the world, while at the same time taking an increasing toll on natural systems, according to Krchnak, who co-authored an NWF report on population, wildlife, and water. In effect, humanity is waging an undeclared water war with nature.

"There will be no winners in this war, only losers," warns Krchnak. By undermining the water needs of wildlife we are not just undermining other species, we are threatening the human prospect as well.

Pulling Apart the Pipes

Currently, humans expropriate 54 percent of all available freshwater from rivers, lakes, streams, and shallow aquifers. During the 20th century water use increased at double the rate of population growth: while the global population tripled, water use per capita increased by six times. Projected levels of population growth in the next 25 years alone are expected to increase the human take of available freshwater to 70 percent, according to water expert Sandra Postel, Director of the Global Water Policy Project in Amherst, Massachusetts. And if per capita water consumption continues to rise at its current rate, by 2025 that share could significantly exceed 70 percent.

As a global average, most freshwater withdrawals—69 percent—are used for agriculture, while industry accounts for 23 percent and municipal use (drinking water, bathing and cleaning, and watering plants and grass) just 8 percent.

The past century of human development—the spread of large-scale agriculture, the rapid growth of industrial development, the construction of tens of thousands of large dams, and the growing sprawl of cities—has profoundly altered the Earth's hydrological cycle. Countless rivers, streams, floodplains, and wetlands have been dammed, diverted, polluted, and filled. These components of the hydrological cycle, which function as the Earth's plumbing system, are being disconnected and plundered, piece by piece. This fragmentation has been so extensive that freshwater ecosystems are perhaps the most severely endangered today.

Wetland Loss

Consider the plight of wetlands—swamps, marshes, fens, bogs, estuaries, and tidal flats. Globally, the world has lost half of its wetlands, with most of the destruction having taken place over the past half century. The loss of these productive ecosystems is doubly harmful to the environment: wetlands not only store water and transport nutrients, but also act as natural filters, soaking up and diluting pollutants such as nitrogen and phosphorus from agricultural runoff, heavy metals from mining and industrial spills, and raw sewage from human settlements.

In some areas of Europe, such as Germany and France, 80 percent of all wetlands have been destroyed. The United States has lost 50 percent of its wetlands since colonial times. More than 100 million hectares of U.S. wetlands (247 million acres) have been filled, dredged, or channeled—an area greater than the size of California, Nevada, and Oregon combined. In California alone, more than 90 percent of wetlands have been tilled under, paved over, or otherwise destroyed.

Biodiversity Loss

Destruction of habitat is the largest cause of biodiversity loss in almost every ecosystem, from wetlands and estuaries to prairies and forests. But biologists have found that the brunt of current plant and animal extinctions has fallen disproportionately on those species dependent on freshwater and related habitats. One fifth of the world's freshwater fish—2,000 of the 10,000 species identified so far—are endangered, vulnerable, or extinct. In North America, the continent most studied, 67 percent of all mussels, 51 percent of crayfish, 40 percent of amphibians, 37 percent of fish, and 75 percent of all freshwater mollusks are rare, imperiled, or already gone.

The global decline in amphibian populations may be the aquatic equivalent of the canary in the coal mine. Data are scarce for many species, but more than half of the amphibians studied in Western Europe, North America, and South America are in a rapid decline.

Around the world, more than 1,000 bird species are close to extinction, and many of these are particularly dependent on wetlands and other aquatic habitats. In Mexico's Sonora Desert, for instance, agriculture has siphoned off 97 percent of the region's water resources, reducing the migratory bird population by more than half, from 233,000 in 1970 to fewer than 100,000 today.

Water Pollution

Pollution is also exacting a significant toll on freshwater and marine organisms. For instance, scientists studying beluga whales swimming in the contaminated St. Lawrence Seaway, which connects the Atlantic Ocean to North America's Great Lakes, found that the cetaceans have dangerously high levels of PCBs in their blubber.

In fact the contamination is so severe that under Canadian law the whales actually qualify as toxic waste.

Waterways everywhere are used as sewers and waste receptacles. Exactly how much waste ends up in freshwater systems and coastal waters is not known. However, the UN Food and Agriculture Organization (FAO) estimates that every year roughly 450 cubic kilometers (99 million gallons) of wastewater (untreated or only partially treated) is discharged into rivers, lakes, and coastal areas. To dilute and transport this amount of waste requires at least 6,000 cubic kilometers (1.32 billion gallons) of clean water. The FAO estimates that if current trends continue, within 40 years the world's entire stable river flow would be needed just to dilute and transport humanity's wastes.

The Point of No Return?

The competition between people and wildlife for water is intensifying in many of the most biodiverse regions of the world. Of the 25 biodiversity hotspots designated by Conservation International, 10 are located in water-short regions. These regions—including Mexico, Central America, the Caribbean, the western United States, the Mediterranean Basin, southern Africa, and southwestern China—are home to an extremely high number of endemic and threatened species. Population pressures and overuse of resources, combined with critical water shortages, threaten to push these diverse and vital ecosystems over the brink. In a number of cases, the point of no return has already been reached.

China's water demands have taken a huge toll on the country's wildlife. The massive Three Gorges Dam on the Yangtze River delivers electricity and agricultural water throughout China, but has left little for species such as the Yangtze River Dolphin (*Lipotes vexillifer*), which is now believed to be extinct.

© pkujiahe/iStock/Thinkstock & Mark Carwardine/Photolibrary/Getty Images

China

China, home to 22 percent of the world's population, is already experiencing serious water shortages that threaten both people and wildlife. According to China's former environment minister, Qu Geping, China's freshwater supplies are capable of sustainably supporting no more than 650 million people—half its current population. To compensate for the tremendous shortfall, China is draining its rivers dry and mining ancient aquifers that take thousands of years to recharge.

As a result, the country has completely overwhelmed its freshwater ecosystems. Even in the water-rich Yangtze River Basin, water demands from farms, industry, and a giant population have polluted and degraded freshwater and riparian ecosystems. The Yangtze is one of the longest rivers in Asia, winding 6,300 kilometers on its way to the Yellow Sea. This massive watershed is home to around 400 million people, one-third of the total population of China. But the population density is high, averaging 200 people per square kilometer. As the river, sluggish with sediment and laced with agricultural, industrial, and municipal wastes, nears its wide delta, population densities soar to over 350 people per square kilometer.

The effects of the country's intense water demands, mostly for agriculture, can be seen in the dry lake beds on the Gianghan Plain. In 1950 this ecologically rich area supported over 1,000 lakes. Within three decades, new dams and irrigation canals had siphoned off so much water that only 300 lakes were left.

China's water demands have taken a huge toll on the country's wildlife. Studies carried out in the Yangtze's middle and lower reaches show that in natural lakes and wetlands still connected to the river, the number of fish species averages 100. In lakes and wetlands cut off and marooned from the river because of diversions and drainage, no more than 30 survive. Populations of three of the Yangtze's largest and most productive fisheries—the silver, bighead, and grass carp—have dropped by half since the 1950s.

Mammals and reptiles are in similar straits. The Yangtze's shrinking and polluted waters are home to the most endangered dolphin in the world—the Yangtze River dolphin, or Baiji. There are only around 100 of these very rare freshwater dolphins left in the wild, but biologists predict they will be gone in a decade. And if any survive, their fate will be sealed when the massive Three Gorges Dam is completed in 2013. The dam is expected to decrease water flows downstream, exacerbate the effects of pollution, and reduce the number of prey species that the dolphins eat. Likewise, the Yangtze's Chinese alligators, which live mostly in a small stretch near the river's swollen, silt-laden mouth, are not expected to survive the next 10 years. In recent years, the alligator population has dropped to between 800 and 1,000.

The Aral Sea

The most striking example of human water demands destroying an ecosystem is the nearly complete annihilation of the 64,500 square kilometer Aral Sea, located in Central Asia between Kazakhstan and Uzbekistan. Once the fourth largest inland sea in the world, it has contracted by half its size and lost three-quarters of its volume since the 1960s, when its two feeder rivers—the Amu Darya and the Syr Darya—were diverted to irrigate cotton fields and rice paddies.

The water diversions have also deprived the region's lakes and wetlands of their life source. At the Aral Sea's northern end in Kazakhstan, the lakes of the Syr Darya delta shrank from about 500 square kilometers to 40 square kilometers between 1960 and 1980. By 1995, more than 50 lakes in the Amu Darya delta had dried up and the surrounding wetlands had withered from 550,000 hectares to less than 20,000 hectares.

The unique *tugay* forests—dense thickets of small shrubs, grasses, sedges and reeds—that once covered 13,000 square kilometers around the fringes of the sea have been decimated. By 1999 less than 1,000 square kilometers of fragmented and isolated forest remained.

The habitat destruction has dramatically reduced the number of mammals that used to flourish around the Aral Sea: of 173 species found in 1960, only 38 remained in 1990. Though the ruined deltas still attract waterfowl and other wetland species, the number of migrant and nesting birds has declined from 500 species to fewer than 285 today.

Plant life has been hard hit by the increase in soil salinity, aridity, and heat. Forty years ago, botanists had identified 1,200 species of flowering plants, including 29 endemic [native only to that area] species. Today, the endemics have vanished. The number of plant species that can survive the increasingly harsh climate is a fraction of the original number.

Most experts agree that the sea itself may very well disappear entirely within two decades. But the region's freshwater habitats and related communities of plants and animals have already been consigned to oblivion.

Lake Chad

Lake Chad, too, has shrunk—to one-tenth of its former size. In 1960, with a surface area of 25,000 square kilometers, it was the second-largest lake in Africa. When last surveyed, it was down to only 2,000 square kilometers. And here, too, massive water withdrawals from the watershed to feed irrigated agriculture have reduced the amount of water flowing into the lake to a trickle, especially during the dry season.

Lake Chad is wedged between four nations: populous Nigeria to the southwest, Niger on the northwest shore, Chad to the northeast, and Cameroon on a small section of the south shore. Nigeria has the largest population in Africa, with 130 million inhabitants. Population-growth rates in these countries average 3 percent a year, enough to double human numbers in one generation. And population growth rates in the regions around the lake are even higher than the national averages. People gravitate to this area because the lake and its rivers are the only sources of surface water for agricultural production in an otherwise dry and increasingly desertified region.

Although water has been flowing into the lake from its rivers over the past decade, the lake is still in serious ecological trouble. The lake's fisheries have more or less collapsed from over-exploitation and loss of aquatic habitats as its waters have been drained away. Though some 40 commercially valuable species remain, their populations are too small to be harvested in commercial quantities. Only one species—the mudfish—remains in viable populations.

As the lake has withered, it has been unable to provide suitable habitat for a host of other species. All large carnivores, such as lions and leopards, have been exterminated by hunting and habitat loss. Other large animals, such as rhinos and hippopotamuses, are found in greatly reduced numbers in isolated, small populations. Bird life still thrives around the lake, but the variety and numbers of breeding pairs have dropped significantly over the past 40 years.

A Blue Revolution

As these examples illustrate, the challenge for the world community is to launch a "blue revolution" that will help governments and communities manage water resources on a more sustainable basis for all users. "We not only have to regulate supplies of freshwater better, we need to reduce the demand side of the equation," says Swedish hydrologist Malin Falkenmark, a senior scientist with Sweden's Natural Science Research Council. "We need to ask how much water is available and how best can we use it, not how much do we need and where do we get it." Increasingly, where we get it from is at the expense of aquatic ecosystems.

If blindly meeting demand precipitated, in large measure, the world's current water crisis, reducing demand and matching supplies with end uses will help get us back on track to a more equitable water future for everyone. While serious water initiatives were launched in the wake of the World Summit on Sustainable Development held in Johannesburg, South Africa, not one of them addressed the water needs of ecosystems.

There is an important lesson here: just as animals cannot thrive when disconnected from their habitats, neither can humanity live disconnected from the water cycle and the natural systems that have evolved to maintain it. It is not a matter of "either or" says NWF's Krchnak. "We have no real choices here. Either we as a species live within the limits of the water cycle and utilize it rationally, or we could end up in constant competition with each other and with nature over remaining supplies. Ultimately, if nature loses, we lose."

By allowing natural systems to die, we may be threatening our own future. After all, there is a growing consensus that natural ecosystems have immense, almost incalculable value. Robert Costanza, a resource economist at the University of Maryland, has estimated the global value of freshwater wetlands, including related riverine and lake systems, at close to $5 trillion a year. This figure is based on their value as flood regulators, waste treatment plants, and wildlife habitats, as well as for fisheries production and recreation.

The nightmarish scenarios envisioned for a water-starved not too distant future should be enough to compel action at all levels. The water needs of people and wildlife are inextricably bound together. Unfortunately, it will probably take more incidents like the one in northern Kenya before we learn to share water resources, balancing the needs of nature with the needs of humanity.

Adapted from Hinrichsen, D. 2003. A Human Thirst. World Watch, January/ February 2003. Available online at: http://www.worldwatch.org/system/files/EP161A.pdf

2.4 Water Conservation and Management

There is a growing consensus among water experts that the "supply-side" approaches to water management that dominated the 20th century—building dams, diversion systems, and other infrastructure to increase supply—will need to give way to a new approach in the 21st century. This new approach is more focused on the "demand-side" and emphasizes water conservation and making the most of the water supplies we already have. In the following article by Elizabeth Royte of National Geographic magazine, we see that some places, such as Albuquerque, New Mexico, have already embraced a demand-side approach with great success. Further progress in conserving water and reducing demand, especially in the agricultural sector, will depend as much or more on appropriate policy and economic incentives as on technological breakthroughs.

The situation that faced Albuquerque was similar to the bank account analogy used in the introduction to section 2.2. The city was withdrawing water from its underground aquifer faster than it was being replenished by rainfall and snowmelt, and so it was on track to run out of water from that source. Demand-side, or "soft path" approaches have reduced Albuquerque's residential per capita water use from 140 to 80 gallons per day, but it is still using water faster than it can be replenished.

In order to bring water use under control, attention has now shifted to agriculture, the largest consumer of water in many regions. In addition to helping farmers adopt water-saving technologies, such as precision irrigation, there is also a recognition that water policy needs to change, especially with regards to pricing this resource. Currently, the price of water for agriculture is heavily subsidized throughout the U.S. southwest and so farmers have relatively little incentive to conserve. Unless greater progress can be made in bringing water demand under control and improving the efficiency of water use in agriculture and other sectors, we may be forced to resort to desalination [removal of salt and other minerals from water] and/or greater reuse of wastewater. Both of these options can be expensive and energy-intensive, but in some regions they may represent the only option available to meeting water demand.

By *Elizabeth Royle*

Living in the high desert of northern New Mexico, Louise Pape bathes three times a week, military style: wet body, turn off water, soap up, rinse, get out. She reuses her drinking cup for days without washing it, and she saves her dishwater for plants and unheated shower water to flush the toilet. While most Americans use around a hundred gallons of water a day, Pape uses just about ten.

"I conserve water because I feel the planet is dying, and I don't want to be part of the problem," she says.

You don't have to be as committed an environmentalist as Pape, who edits a climate-change news service, to realize that the days of cheap and abundant water are drawing to an end. But the planet is a long way from dying of thirst. "It's inevitable that we'll solve our water problems," says Peter Gleick, president of the Pacific Institute, a non-partisan environmental think tank. "The trick is how much pain we can avoid on that path to where we want to be."

As Gleick sees it, we've got two ways to go forward. Hard-path solutions focus almost exclusively on ways to develop new supplies of water, such as supersize dams, aqueducts, and pipelines that deliver water over huge distances. Gleick leans toward the soft path: a comprehensive approach that includes conservation and efficiency, community-scale infrastructure, protection of aquatic ecosystems, management at the level of watersheds instead of political boundaries, and smart economics.

The Case of Albuquerque

Until the mid-1980s, the city of Albuquerque, some 60 miles southwest of Pape's home in Santa Fe, was blissfully unaware that it needed to follow any path at all. Hydrogeologists believed the city sat atop an underground reservoir "as big as Lake Superior," says Katherine Yuhas, conservation director of the Albuquerque Bernalillo County Water Utility Authority. The culture was geared toward greenery: Realtors attracted potential home buyers from moist regions with landscaping as verdant as Vermont; building codes required lawns. But then studies revealed startling news: Albuquerque's aquifer was nowhere near the size it once appeared to be and was being pumped out faster than rainfall and snowmelt could replenish it.

Duly alarmed, the city shifted into high gear. It revised its water-use codes, paid homeowners to take classes on reducing outdoor watering, and offered rebates to anyone who installed low-flow fixtures or a drip-irrigation system or removed a lawn. Today Albuquerque is a striving example of soft-path parsimony. Across the sprawling city, a growing number of residents and building owners funnel rainwater into barrels and underground cisterns. Almost everyone in town uses low-flow toilets and showerheads.

These efforts have shrunk Albuquerque's domestic per capita water use from 140 gallons a day to around 80. The city "anticipates another 50 years of water, economically and sustainably supplied, even with a growing population," says Yuhas. After that there's the option to desalinate brackish water nearby and new technologies such as dual plumbing: one set of pipes to deliver highly treated potable water and another to recycle less treated water for flushing toilets, watering lawns, and other nonpotable uses. Albuquerque already uses wastewater—from treatment plants and from industry—to irrigate golf courses and parks. Other municipalities have gone a step further and collect wastewater—yes, from toilets—filter and disinfect it to the nth degree, then pump it back into the local aquifer for drinking. There are similar schemes worldwide: Beijing reportedly aims to reuse 100 percent of its wastewater by 2013.

Industry, too, is adapting to less certain water supplies. Frito-Lay will soon recycle almost all its water at its plant in Casa Grande, Arizona; Gatorade and Coca-Cola remove the dust and carton lint from beverage containers using air instead of water; and Google recycles its own water to cool its giant data centers.

Water Demand for Agriculture

This is all reassuring—until you remember that irrigated agriculture accounts for 70 percent of the fresh water used by humans. Given this outsize proportion, it seems obvious that farmers have the greatest potential to conserve water.

Standing on the banks of a trickling ditch, Don Bustos—sunbaked and thickly bearded—demonstrates how he irrigates 130,000 dollars' worth of produce on 3.5 acres north of Santa Fe. "I lift this board"—he points to a plank that forms a gate in the ditch—"and I shove in a stick to hold it up." Gravity does the rest.

For 400 years farmers in the arid Southwest have relied on such acequias—networks of community-operated ditches—to irrigate their crops. The acequia diverts water from a main

Agricultural advances in water efficiency, such as drip irrigation, have helped to conserve water in the face of increasing stress on the world's fresh water supplies.

stream, then further apportions the flow through sluiceways into smaller streams and onto fields. "Without the acequia, there would be no farm," Bustos says. He's also built a water tank with drip-irrigation hoses that feed some of the acequia water directly to the plant roots—and cut his water use by two-thirds as a consequence.

Elsewhere, forward-thinking farmers have replaced flood irrigation with micro-sprinkler systems, laser leveled their fields, and installed soil-moisture monitors to better time irrigation. In California, says the Pacific Institute, such improvements could potentially conserve roughly five million acre-feet of water a year, enough to meet the household needs of 37 million people. Unfortunately, most farmers lack the incentive to install efficient but expensive irrigation systems: Government subsidies keep farm water cheap. But experts agree that more realistic water pricing and improved water management will significantly cut agricultural water use. One way or another, the developed world will get the water it needs, if not the water it wants. We can find new supplies—by desalinating water, recycling water, capturing and filtering storm water from paved surfaces, and redistributing water rights among agriculture, industry, and cities. Cheaply and quickly we can slash demand—with conservation and efficiency measures, with higher rates for water wasters, and with better management policies.

Hope for the Future?

What about the rest of the world? In places lacerated by poverty, the problem is often a lack of infrastructure—wells, pipes, pollution controls, and systems for disinfecting water. Though politically challenging to execute, the solutions are fairly straightforward: investment in appropriately scaled technology, better governance, community involvement, proper water pricing, and training water users to maintain their systems. In regions facing scarcity because of overpumped aquifers, better management and efficiency will stretch the last drops. Farmers in southern India, for example, save fuel in addition to water when they switch from flood to drip irrigation; other communities landscape their hillsides to retain rainwater and replenish aquifers.

Still, the time is coming when some farmers—the largest water users and the lowest ratepayers—may find themselves rethinking what, or if, they should plant in the first place. In the parched Murray-Darling Basin of Australia, farmers are already packing up and moving out.

It is hardly the first time that water scarcity has created environmental refugees. A thousand years ago, less than 120 miles from modern-day Santa Fe, the inhabitants of Chaco Canyon built rock-lined ditches, headgates, and dams to manage runoff from their enormous watershed. Then, starting around A.D. 1130, a prolonged drought set in. Water scarcity may not have been the only cause, but within a few decades, Chaco Canyon had been abandoned. We hardly need reminding that nature can be unforgiving: We learn to live within her increasingly unpredictable means, we move elsewhere, or we perish.

Adapted from: Royte, E. 2010. The Last Drop. National Geographic, 217(4) April: 172–177.
Available online at: http://ngm.nationalgeographic.com/print/2010/04/last-drop/royte-text

CHAPTER 3
Essential Resources: The Lithosphere

Aldo Leopold, A Sand County Almanac, 1949

"Land, then, is not merely soil; it is a fountain of energy flowing through a circuit of soils, plants, and animals."

—*Aldo Leopold, A Sand County Almanac, 1949*

3.1 Introduction to Soils & Soil Functions

Soil is an amazing, mostly natural material that covers nearly all of the land surface of the earth. Soil, along with water and air, provides the basis for human existence. It is the interface between the earth's atmosphere and bedrock or ground water. It has either formed in place or has been transported to its present location by wind,

Sections 3.1, 3.3, 3.5, and 3.10 from *Urban Soil Primer* (US Department of Agriculture)
Sections 3.2, 3.4, 3.6, 3.7, 3.8, and 3.9 Copyright © 2013 Bridgepoint Education. Reprinted by permission.

water, ice, gravity, or humans. Soils may have been deposited thousands or millions of years ago by volcanoes, glaciers, floods, or other processes or were delivered to the site by truck or other mechanical device an hour ago, a week ago, or several months ago. These facts illustrate why soils are very complex. Soil functions, as part of a natural ecosystem, are also very complex and diverse. Basic knowledge about soil allows us to use it wisely.

Soil Functions

- Soils act like sponges, soaking up rainwater and limiting runoff. Soils also impact ground-water recharge and flood-control potentials in urban areas.
- Soils act like faucets, storing and releasing water and air for plants and animals to use.
- Soils act like supermarkets, providing valuable nutrients and air and water to plants and animals. Soils also store carbon and prevent its loss into the atmosphere.
- Soils act like strainers or filters, filtering and purifying water and air that flow through them.
- Soils buffer, degrade, immobilize, detoxify, and trap pollutants, such as oil, pesticides, herbicides, and heavy metals, and keep them from entering ground-water supplies. Soils also store nutrients for future use by plants and animals above ground and by microbes within the soils.

Soil functions occur in spite of the land use. Rainwater must be dispersed or regulated in urban areas, and landscaping plant roots must have air available for growth. When areas are paved over, plans must be in place to handle rainwater. Buildings constructed on fill material must still be supported by the materials on the site. Soils perform the same or similar functions in all areas, including urban ones.

3.2 What Is Soil?

The Soil Science Society of America defines soil as "a naturally occurring surface layer formed by complex biogeochemical and physical weathering processes that contains living matter and is capable of supporting plant life." Note that there are several important parts to this definition. Soil is something we find in nature created by natural processes, it is not something that humans make. Soil is found on the surface of our planet, not at great depths. While there is no single depth that all soils go down to, in most of the United States the "bottom" of the soil is found about 1.2–1.8 m (4–6 ft) below the surface. However, the depth can be greater than 1.8 m or less than 1.2 m depending on local conditions. Soils are formed by complex bio-geochemical and physical weathering processes, meaning that soils have been significantly altered from the original materials, or parent materials, they formed in through biological, chemical, and physical processes. Chemical processes include the **dissolution** (the process of breaking a compound material into its individual elements or parts), **precipitation** (the process by which a solid substance separates, or is separated from, a liquid it is in), and alteration (changing the chemical composition of a material through the introduction of other chemical species) of materials found in the soil. Physical processes involve breaking large materials into smaller materials without changing their chemical composition, and the movement of materials within the soil profile. Biological processes involve anything driven by living organisms, which can include chemical and physical processes driven by organisms. Finally, to be considered soil, a material must be associated with living organisms.

It is important to understand that the definition of soil is not the same for everyone. Soil scientists use the definition given above, but engineers define soils as all unconsolidated materials found above bedrock. Notice that this definition is considerably different then the soil scientists' definition. Engineers are not concerned with the ability of soil to support life or with the processes that form soil. Instead, engineers are concerned with how difficult it is to move the material (soil vs. rock) and how well it will support foundations for buildings, bridges, and other structures. Geologists have another definition again, so in working with people from different professions it is important to clarify just what is meant when the term "soil" is used. Without that clarification it is possible to have two people using the same term but discussing completely different ideas.

Soils are widespread on the surface of our planet, but they are not found everywhere. When working with materials on the earth's surface, it is important to be able to tell the difference between soils and geologic materials, such as **sediments** (pieces of broken and weathered rock that have been carried by water, wind, or ice and then deposited [left behind]) or **regolith** (a layer of loose or broken up rock above solid bedrock). Soils have horizons, or layers, in them that form roughly parallel to the earth's surface (Table 3.1). These layers are created by the physical and biogeochemical weathering processes that formed the soil from parent materials (Table 3.2). Geologic sediments may have layers in them, but these layers are created by processes such as moving water, wind, or ice that left the sediments behind (Table 3.2). The properties of soils are very different than the properties of sediments and regolith, and these differences are important in some environmental health applications, such as the placement of septic systems.

TABLE 3.1 Soil horizons shown in their common order from the surface down. It is important to note that not all of these horizons are usually found in a single typical soil, but various combinations of these horizons are found in all soils. The R horizon is not an official horizon but is commonly used, the other horizons are all official horizon names

Horizon	Characteristics
O	A layer dominated by organic materials, frequently greater than 35% organic material although it can be as low as 25% given certain other conditions. Can be very thin, such as a layer of leaves on the forest floor, or several meters thick, such as the accumulation of organic material in a bog.
A	A mineral-based horizon that also contains significant amounts of organic material, this is usually less than 5% organics although it can contain as much as 35% organic material given certain conditions.
E	A light colored mineral horizon, characterized by intense leaching and the loss of clay (eluviation) to lower horizons. Common beneath forest vegetation.
B	A zone marked by a significant accumulation due to soil forming processes. These accumulations can be silicate clays (Bt horizon), pigmenting minerals marked by a color change relative to the horizons above and below (Bw horizon), or a number of other materials. The lower case letter behind the upper case B tells a soil scientist what has accumulated to create the B horizon.
C	Slightly altered parent material, represents the geologic sediment that was present prior to soil formation.
R	A layer of consolidated rock within the soil profile.

TABLE 3.2 Classification of Soil Parent Materials

Parent Material	Mode of Creation
Aeolian	Sediments that were transported and deposited by wind. A commonly known example would be sand in sand dunes.
Alluvium	Sediments that were transported and deposited by flowing water (rivers or streams).
Colluvium	Sediments that were transported and deposited by gravity, usaually found a the base of high spots such as mountains or hills.
Lacustrine	Sediments that were transported and deposited in Lakes.
Marine	Sediments that were transported and deposited in Oceans.
Organic	Created when organic materials, primarily plant debris, accumulate to thicknesses that can be several meters, thick deposits are common in wet areas (i.e., swamps).
Outwash	Sediments that were deposited by water (rivers or streams) flowing off of melting glaciers. The glacial source typically gives these sediments different properties than alluvium derived from non-glacial sources, thus the separate classification.
Residuum	Forms when solid rocks weather (break down) in place, leaving behind loose material.
Till	Sediments that were transported and deposited by glacial ice.

A horizon

B horizon

C horizon
(human
artifacts)

US Department of Agriculture

FIGURE 3.1   **Urban soil profile.**

A horizon

B horizon

C horizon

US Department of Agriculture

FIGURE 3.2   **Natural soil profile.**

3.3   Urban Soil Horizons

Horizons in urban soils may not be fully related to the natural soil-forming factors but instead may be manmade layers formed by the deposition of dredge, fill, and/or mixed materials. Human artifacts, such as bricks, bottles, pieces of concrete, plastics, glass, pesticides, petroleum products, pollutants, garbage, and disposable diapers, are often components of urban soils. Manmade materials may be added to raise a landscape to a higher level, backfill ditches or foundation walls, or construct berms. In urban areas, human activity is often the predominant activity in making soil instead of the action of the natural agents of wind, water, ice, gravity, and heat.

Urban soils differ from natural soils because they have been altered to some degree. They have been excavated, compacted, disturbed, and mixed and may no longer possess their natural soil properties and features. Many highly disturbed soils in urban areas or on construction sites have not been in place long enough for soil-forming factors to significantly change them and to form soil horizons. In areas where fill materials have been in place for a considerable time (e.g., 50 years or so), the formation of A horizons and sometimes weakly expressed B horizons has been documented. Figures 3.1 and 3.2 show soil horizons in urban and natural soil profiles.

3.4   Important Properties & Functions of Soil

Soils possess physical, chemical, and biological properties that are important to their various functions. While this chapter does not have space to cover these properties in detail, some of the most important routinely measured properties are listed in Table 3.3. An introduction to soil science textbook can be consulted for a more complete discussion of these properties.

One of the most important properties of the soil from an environmental health perspective is the organic matter content. This is because organic matter has such wide-ranging influences on other soil physical, chemical, and biological properties. In fact, most of the properties (about 80%) listed in Table 3.3 are directly or indirectly influenced by organic matter. Organic matter comes from life in the soil, such as plants (both above- and below-ground portions), earthworms, and a myriad of other organisms. Most scientists believe there is a greater diversity of life in the soil below our feet than on the earth's surface. In fact, a single tablespoon of good soil can contain billions of individual organisms. Another important property is the clay content, which is a part of the soil texture, or the distribution of sand, silt, and clay particles in the soil (Table 3.4). Organic matter and clay combine to give soils many of their most important chemical properties. This is because organic matter and clays are **colloids** (very small particles). Chemical reactions occur on surfaces, and colloids have very large surface areas relative to their volume (Table 3.5).

TABLE 3.3   Examples of some commonly measured chemical, physical, and biological properties of the soil

Chemical Properties	Physical Properties	Biological Properties
pH	Texture	Microbial biomass
Organic matter	Bulk density	Earthworm populations
Total carbon	Penetration resistance	Nematode populations
Total nitrogen	Aggregate stability	Arthropod populations
Cation exchange capacity	Water holding capacity	Mycorrihizal fungi
Major and minor nutrients	Infiltration rate	Respiration rate
Electrical conductivity	Depth to hardpan	Soil enzyme activities
Heavy metals and other plant toxins	Depth to water table	Pollutant detoxification
	Porosity	Decomposition rate
	Erosive potential	
	Aeration	

TABLE 3.4   Soil texture refers to the proportions of sand, silt, and clay found in the soil. There are different breaks between textural classes used by different professional groups so it is important to know which system is being used when discussing soil texture. Examples of the different size ranges used by some common systems for classifying mineral particle sizes are given here. In all cases gravel is the coarsest material, followed by sand and silt with clay being the finest particle size

Particle Size Name	Size Range (mm) by System				
	United States Department of Agriculture	International Society of Soil Science	American Assoc. of State Highway and Transportation Officials	Udden-Wentworth Scale	Unified System
Gravel	>2.0	>2.0	>2.0	>2.0	>5.0
Sand (total)	0.05–2.0	0.02–2.0	0.074–2.0	0.0625–2.0	0.074–5.0
Silt (total)	0.002–0.05	0.002–0.02	0.005–0.074	0.0039–0.0625	(silt and clay are combined)
Clay (total)	<0.002	<0.002	<0.005	<0.0039	<0.074

TABLE 3.5   Surface areas for different particle sizes found in soil. Note that the smaller the particle size, the larger the surface area per unit mass

Particle Size Name	Particle Size (mm)	Surface Area (m²/g)
Coarse sand	1.0–0.5	0.0023
Fine sand	0.25–0.10	0.0091
Silt	0.05–0.002	0.0454
Clay	<0.002	10 to 820

Source: Brevik, E.C. 2012. An Introduction to Soil Science Basics. In: E.C. Brevik and L.C. Burgess (Eds). Soils and Human Health. Taylor Francis Press, Boca Raton, FL. 2013.

3.5 Soil Texture

Soil is a mixture of mineral matter, organic material, air, and water. The texture of a mineral soil is based on the amounts of sand, silt, and clay in the soil. Sand, silt, and clay are defined on the basis of the size of each individual soil particle. These size relationships can be demonstrated by imagining that a sand particle is the size of a basketball, a silt particle is the size of a baseball, and a clay is the size of an aspirin tablet (Figure 3.3).

Soil texture and other soil properties vary significantly within short distances on urban or natural landscapes. This variation is caused by the movement and mixing of soil materials during construction activities or changes in any of the soil-forming factors. The combinations of different textures may improve or limit the soil for a specific use.

Soil texture affects water and air movement through the soil as particles of different sizes pack together and thus determine the size and spacing of pores and channels. Sand particles have the largest pore spaces and allow water to drain through the pores most freely. Silt particles have smaller pore spaces, so water moves through them more slowly. Clay particles have very small pores, and so they tend to adsorb and hold more water. The mixture of particle sizes affects water, nutrient, and contaminant absorption. The specific type of mineral influences engineering properties, such as shrink-swell potential and excavation difficulty, especially in expanding clays (smectite), which behave like plastics.

Measures of Water Movement

Water movement in urban soils is described in three ways (Figure 3.4).

- infiltration into the soil surface, especially from rainfall
- percolation within the soil drain lines from septic systems, which is especially important in the soil below the drain line and above a restrictive layer
- permeability within the soil from the surface to a restrictive layer

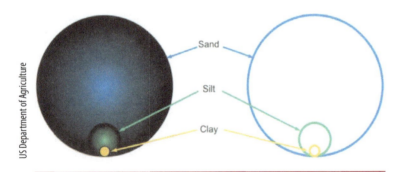

US Department of Agriculture

FIGURE 3.3 **Relative sizes of sand, silt, and clay particles.**

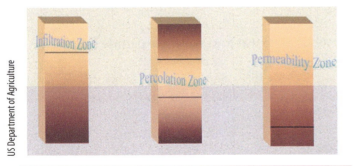

US Department of Agriculture

FIGURE 3.4 **Comparison of descriptive terms for water movement in soils.**

Key terms in understanding water movement in soils are "restrictive layer" and "water table." Restrictive layers have high density (high weight in a given volume of soil) and low porosity (limited space between particles), so that water cannot flow into or through them. Restrictive layers at the surface can cause surface sealing and limit infiltration of water into the soil. Restrictive layers within the percolation zone reduce the drainage rate of fluids in septic drain lines and can cause septic systems to backup and fail. Compaction of soil materials can occur if heavy weight is on the surface when the soil is wet, resulting in dense restrictive layers below the surface.

3.6 Chemical Properties of Soil

One of the important chemical properties for environmental health applications is **cation**

exchange capacity (the degree to which a soil can adsorb and exchange cations; cations are positively charged ions). The surfaces of most clays and organic matter have a net negative charge, giving them the ability to attract, hold, and exchange positively charged ions, or cations (Figure 3.5).

The formation of **aggregates** is another important soil property. Aggregates are masses of sand, silt, clay, and organic matter all stuck together into "clumps" in the soil (Figure 3.6). Soil aggregates are important in

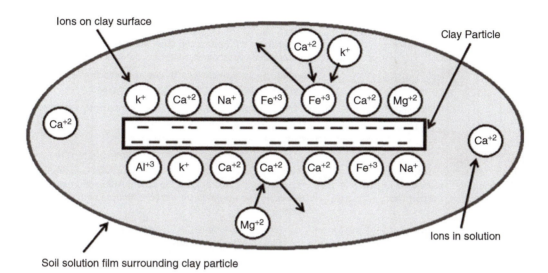

FIGURE 3.5 **The concept of cation exchange on clay surfaces. Clays have a net negative charge, which attracts cations to them.** Cations can be exchanged (note the Mg+2 exchanging for the Ca+2 and the Ca+2 and K+ exchanging for the Fe+3) between the soil solution (water plus dissolved ions) and the clay surface on a charge-for-charge basis. Figure by Eric Brevik.

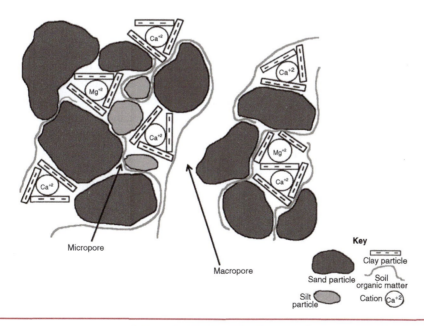

FIGURE 3.6 **Soil aggregates are composed of individual sand, silt, and clay particles that are held together by soil organic matter "glues" and electrostatic attraction between negatively charged clay particles and cations.** The large pore, or opening, between the two aggregates shown here is a macropore. Macropores serve as pathways for air, water, and many organisms to move through the soil as well as places for the penetration of plant roots. The small pores within the aggregates are micropores that store water, some of which can be accessed by plants during dry periods. Figure by Eric Brevik.

TABLE 3.6 Some common soil colors, what causes them, and what they tell us about the soil

Iron Oxide Pigmenting Agents		
Color	**Cause (Mineral Name)**	**Indicates**
yellow	goethite	Goethite forms in cool temperatures, or in moist soils in warm settings that are not water saturated. Goethite often occurs in association with hematite.
red	hematite	Hematite forms in high temperature soils that are well oxygenated.
brown	maghemite	Maghemite is formed when other iron oxides are transformed by heating in association with organic matter, i.e., in forest fires.
orange	lepidocrocite	Lepidocrocite forms in a local zone of oxidation in soils that are otherwise reduced, i.e., around the roots of wetland plants.
gray-green	hydromagnetite	Hydromagnetite forms in soils that are frequently saturated, creating reducing conditions.
Other Pigmenting Agents		
Color	**Cause**	**Indicates**
black or dark colors near the surface	coatings of organic matter on soil particles	Conditions are favorable for the accumulation of organic matter in surface horizons. This may indicate wet conditions or high levels of organic matter additions.
black in the subsurface	organic matter or manganese oxides	Organic matter—intense leaching is moving organic matter from surface horizons into the subsurface. Manganese oxides—form in wet soils that undergo frequent alterations between oxidizing and reducing conditions. Manganese oxides often occur as concretions (small hard nodules).
white	carbonates or other salts	The soils formed in an environment that is dry enough that there was not enough water to leach the salts from the soil.

Source: Brevik, E.C. 2012. An Introduction to Soil Science Basics. In: E.C. Brevik and L.C. Burgess (Eds). Soils and Human Health. Taylor Francis Press, Boca Raton, FL. 2013.

creating pores, or openings, in the soil where many soil functions, such as the physical and chemical filtration of soil water, take place. Soils that have good aggregate formation, referred to as well-aggregated soils, tend to perform better than poorly aggregated soils in providing important ecosystem services.

Soil color is a physical property that is important because of what it tells us about the chemical conditions under which the soil formed (Table 3.6). Bright colors such as yellows and reds indicate well-oxygenated environments, while dull colors such as gray-greens indicate **reducing environments** (a setting in which the primary chemical reactions involve the gaining of electrons by electron receptors, decreasing their oxidation state). Reducing environments in soil are usually associated with wet conditions, so dull colors can be used to find the position of the seasonal high water table in a soil. This is important information for things like the location of septic filter fields.

3.7 Organisms in Soil

Organisms that live in the soil also provide important contributions from an environmental health perspective. Microorganisms found in soil can neutralize the toxic effects of some chemicals, such as spilled petroleum products, and consume organic material and related contaminants released from septic systems. Because of this, humans frequently use soils as a natural disposal location for waste products, with the expectation that the soil will clean up the wastes. **Metal hyperaccumulating plants** have the ability to remove heavy metals from soils; some of these plants can accumulate metals in such high concentrations that the plant itself can be harvested and used as an ore. Properties such as these make soil organisms very important from an environmental health perspective.

3.8 Soils & Human Health

Many things are likely to come to mind when people think about their health, such as an active exercise program, wise food choices, good medical care, and proper sanitation. Few people recognize the connection between soils and human health, even though soils are actually very important to human health. Soils influence health through the nutrients taken up by plants and the animals that eat those plants, nutrients that are needed by humans for adequate nutrition for growth and development. Soils have the ability to purify water and to neutralize or sequester a wide range of contaminates. Many of our modern medicines come from soil or soil organisms. In fact, the only Nobel Prize won by a soil scientist to date was awarded to Dr. Selman Waksman in 1952 for his work in isolating antibiotic compounds from soil actinomycetes. Soils can also act to harm human health in three major ways: (1) toxic levels of substances or disease-causing organisms may enter the human food chain from the soil, (2) humans can encounter pathogenic organisms through direct contact with the soil or inhaling dust from the soil, and (3) degraded soils produce nutrient-deficient foods leading to malnutrition. Therefore, soils are an integral link in the holistic view of human health.

Soils and Water

Soil is an important part of the **hydrologic cycle** (the natural sequence through which water passes into the atmosphere as water vapor, precipitates to the earth in liquid or solid form, and ultimately returns to the oceans). As water comes in contact with soil, it has the opportunity to pick up materials from the soil that can be detrimental to human health when present in high-enough concentrations, including heavy metals, organic pollutants, and soil pathogens. These materials and organisms then have the potential to end up in potable surface or groundwater sources and cause adverse effects on human health. Contaminated water can also be introduced to the soil system by accident via sewage spills, breaching or leaching of manure pits or lagoons, or the washing of surface-applied manure into streams by rainwater. However, the physical, chemical, and biological properties of soil also allow soil to function as a natural filter, purifying water that moves through it. We make use of these capabilities in many applications, including septic systems, the use of artificial wetland systems for sewage processing, and land application of manures (Figure 3.7). Using the soil system to clean such contaminated water

Courtesy USDA

FIGURE 3.7 A manure slurry being applied to an agricultural field in Arkansas. Land application of manures is done to add nutrients to the soil, but it is also an economical means of disposing of the waste products that takes advantage of the natural filtration capacity of soil. This can be an environmentally friendly way to dispose of waste products as long as the filtration capacity of the soil is not overwhelmed.

sources works well if the application system is properly designed and sited, and the soil is not overwhelmed with an overload of contaminates, but the potential for ecosystem contamination is a constant concern.

Heavy Metals

Exposure to **heavy metals** (metal elements that have densities greater than $4500kg/m^3$) through soil contact is a major human health concern (Table 3.7). Arsenic is actually a **metalloid** (a chemical element with properties that are in-between or a mixture of those of metals and nonmetals), but is commonly grouped with the heavy metals for the purpose of human health discussions. Heavy metals can originate naturally from the weathering of rocks, but have also been introduced to soils through human activity. Heavy metals may occur as a byproduct of mining ores and are therefore present in mine spoils and in the immediate surroundings of metal processing plants. **E-wastes**, or wastes associated with electronic appliances such as computers and mobile phones, are also becoming an increasing source of heavy metals such as lead, antimony, mercury, cadmium, and nickel in the soil. Urban soils are particularly susceptible to significant accumulations of heavy metals and are frequently associated with heavy metal problems, but fertilizers, manures, and pesticides have also been sources of heavy metal additions to soils in agricultural settings. In the case of fertilizers the heavy metals typically occur as impurities in the fertilizer, while heavy metals have been used in pesticides to target and kill undesired organisms. Arsenic was frequently used in pesticides in the past, and arsenic build-up in the soils of orchards in the United States has been a particular concern. Arsenic contamination may persist in orchard soils for decades or more after the application of arsenic-containing pesticides ceases.

Organic Chemicals

Organic chemicals (molecules that possess carbon-based atoms) that end up in soil are also a major health concern. The main concern with organic chemicals comes from materials known as **persistent organic pollutants** (POPs) (Table 3.8). These are organic chemicals that resist decomposition in the environment or that **bio-accumulate** through the food chain (a process producing an increase in the concentration of chemicals [usually toxicants] in the tissues of organisms with each increase in the trophic level in the food chain), and therefore pose a risk of causing adverse effects to human health and the environment. Soils and human health

TABLE 3.7   Common anthropogenic sources of and health problems associated with selected heavy metals

Heavy Metal	Anthropogenic Sources	Health Problems
Hg	Electrical switches, fluorescent light bulbs, mercury lamps, batteries, thermometers, dental fillings, burning of coal and fuel oil, medical wastes, pesticides, mining	Central nervous system damage, coordination difficulties, eyesight problems, problems with the sense of touch, liver, heart, and kidney damage
Pb	Batteries, solder, ammunition, pigments, ceramic glaze, hair coloring, fishing equipment, leaded gasoline, mining, plumbing, burning of coal	Neurological impacts, lowers IQ and attention spans, impared hand-eye coordination, encephalopathy, deterioration of bones, hypertension
Cd	Zinc smelting, burning coal or Cd-containing garbage, rechargeable batteries, pigments, TVs, solar cells, steel, phosphorus fertilizer, metal plating, water pipes	Liver and kidney damage, carcinogenic, low bone density
As	Pesticides, mining and smelting of gold, lead, copper, and nickel, iron and steel production, burning of coal, wood preservatives	Gastrointestinal damage, skin damage, carcinogenic, heart, neurologic, and liver damage
Cr	Electroplating, corrosion protection, leather tanning, wood preservative, cooling-tower water additive	Carcinogenic, gastrointestinal disorders, hemorrhagic diathesis, convulsions

Source: Brevik, E.C. 2012. Soils and Human Health—An Overview. In: E.C. Brevik and L.C. Burgess (Eds). Soils and Human Health. Taylor Francis Press, Boca Raton, FL. 2013.

TABLE 3.8 Some organic chemical groups that are or have been commonly used and examples of some of the specific chemicals found within those groups

Chemical Group	Examples	Notes
Organochlorines[a]	para-dichlorodiphenyltrichloroethane (DDT)	Used in mosquito control
	Aldrin	Used in termite control
	Dieldrin	Protect crops from insect pests
	Endrin	Used in rodent control
	Chlordane	Used in termite control
	Heptachlor	Controls soil insects
	Hexachlorobenzene	Used in fungus control
	Mirex	Used in ant, termite control
	Toxaphene	Used in tick, mite control
	Polychlorinated biphenyls (PCBs)	Used in industrial applications
	Dioxines	Group of related compounds, by-product of manufacturing
	Furans	Created by heating of PCBs in the presence of oxygen
Organophosphates	Dichlorvos	Used in fly and flea control
	Parathion	Highly toxic to humans, banned in many western countries
	Diazinon	Insect control in homes, lawns
	Chlorpyrifos	Insect control in crops, lawns, and homes
	Malathion	Toxic to insects, fairly low toxicity to mammals
Carbamates	Carbaryl	Insect control on lawns, highly toxic to honey bees
	Aldicarb	Highly toxic to humans
Chloroacetamides	Alachlor	Weed control, carcinogen in animals
	Metolachlor	Weed control, suspected carcinogen in animals
	Acetochlor	Weed control
Glyphosate		Used in weed control, commercial name is Roundup
Phenoxy herbicides	2,4-D	Broadleaf weed control
	2,4,5-T	Used to clear brush along roads and powerlines

a—all of the organochlorines in this list are banned by the US Environmental Protection Agency due to adverse environmental impacts but may still be used in countries other than the United States.

Source: Brevik, E.C. 2012. Soils and Human Health—An Overview. In: E.C. Brevik and L.C. Burgess (Eds). Soils and Human Health. Taylor Francis Press, Boca Raton, FL. 2013.

issues arise with organic chemicals due to the widespread use of these chemicals as pesticides in agricultural situations (Figure 3.8), for lawns and households, and through the accumulation of these organic chemicals in landfills or other disposal sites due to inadequate disposal practices. Petroleum products are another major source of organic chemicals that can end up in the soil. E-wastes are a new source of POPs such as polychlorinated biphenyls (PCBs), and burning of e-wastes can generate other POPs, such as dioxins and furans, two highly toxic organic compounds. Common routes of exposure to organic chemicals include dermal contact with soil and soil ingestion. It is important to note that the complex interactions of heavy metals and the interactions and reactions that organic chemicals undergo in the soil environment and what they mean for the toxicology of these substances are not currently well understood.

FIGURE 3.8 Application of pesticides to a lettuce crop in Arizona. In this case, pesticide is being applied directly to both the crop and to exposed soil in-between the crop rows. Photo by Jeff Vanuga, U.S. Department of Agriculture, Natural Resources Conservation Service.

Radioactive Materials

Soils can be a reservoir of radioactive elements introduced through both natural and **anthropogenic** (produced by humans) sources. About 90% of human radiation exposure worldwide is from natural sources, but anthropogenic exposure sources can be significant in select locations. **Radon** (chemical element with symbol Rn and atomic number 86; it is a radioactive, colorless, odorless, tasteless noble gas, occurring naturally as a product of the decay of uranium) is a major source of natural radiation exposure, representing about half the natural radiation dose to humans (Figure 3.9). The ultimate source of radon is through the radioactive decay of uranium found in rocks, with **granites** (a coarse-grained igneous rock composed of quartz, feldspars, and micas), felsic **metamorphic rocks** (rocks created when heat and/or pressure changes previously existing rocks), organic-rich **shales** (a sedimentary rock composed of silt and clay-sized particles that are aligned in a parallel arrangement), and **phosphatic rocks** (rocks with a high phosphate content) particularly associated with high uranium content. However, most of the radon formed from the decay of uranium locked up in rocks remains trapped within the mineral grains and thus does not impact human health. Soils formed from parent materials high in uranium will also contain uranium, and significant amounts of the radon formed in these soils ends up in soil pore spaces where it can migrate via diffusion through soil gases or with water moving through the soil. When radon moves through the soil or degasses from radon-containing water sources, it can accumulate in enclosed spaces such as basements, cave dwellings, mines, and other enclosed structures in concentrations that negatively impact human health (Figure 3.10).

Positive Influences of Soils on Human Health

Much of the focus on soils to this point in the chapter has been on the negative impacts of soils on human health, but there are many positive aspects as well. The ability of soils to act as natural filters, purifying ground and surface waters if that filtration capacity is not overwhelmed, has already been discussed. Soils are also an important primary source of nutrients. Plants obtain nutrients from the soil which are then passed on to humans when the plants or animal products that were fed on the plants are consumed. Some important dietary sources of elemental

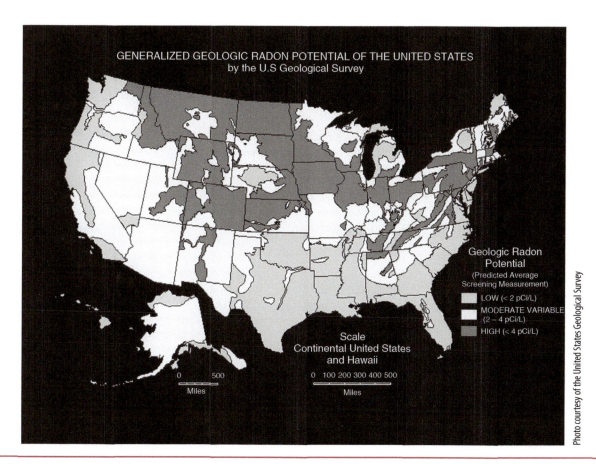

GENERALIZED GEOLOGIC RADON POTENTIAL OF THE UNITED STATES
by the U.S Geological Survey

Geologic Radon Potential
(Predicted Average Screening Measurement)

LOW (< 2 pCi/L)
MODERATE VARIABLE (2 – 4 pCi/L)
HIGH (< 4 pCi/L)

Scale
Continental United States
and Hawaii

Photo courtesy of the United States Geological Survey

FIGURE 3.9   Map showing the potential for radon gas generation and transport through geologic materials and soils in the United States.

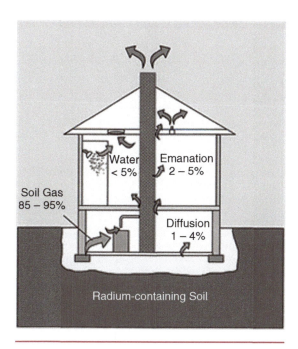

Water < 5%
Emanation 2 – 5%
Soil Gas 85 – 95%
Diffusion 1 – 4%
Radium-containing Soil

FIGURE 3.10   Diagram showing common points where radon enters houses.

nutrients essential to human life that originate in the soil are given in Table 3.9. Over 97% of the calories consumed by the average human have a land origin (Table 3.10), which means humans are consuming food loaded with nutrients from soil. Most of our current clinically relevant antibiotics come from soil organisms, and soils are rich sources of other medicines as well. About 40% of all prescription drugs have their origin in the soil, including many of our most recent cancer drugs. Antidiarrheal medicines and treatments for conditions such as diaper rash and poison ivy/oak/sumac have been derived from soil clays. Soil clays have been used in toothpaste formulas and to treat poisoning by the herbicides paraquat and diquat. And finally, there is mounting evidence to suggest that human contact with healthy soils, such as during gardening, can provide human health benefits in much the same way that exposure to green landscapes (areas displaying vegetative growth) has been shown to reduce stress, shorten recovery times from illnesses or surgery, and provide other health benefits (Figure 3.11). So while soil can cause human health problems, there are also many potential health benefits that can be derived from soil.

TABLE 3.9 Some important sources of elemental nutrients essential to human life. Ultimately all these elements come from the soil, either directly from soil to plant or from soil to plant to animal

Element	Important Plant Sources	Important Animal-Product Sources
Ca	Kale, collards, mustard greens, broccoli	Dairy products
Cl		Dairy products, meats, eggs
Cu	Beans, peas, lentils, whole grains, nuts, peanuts, mushrooms, chocolate	Organ meats
Fe		Meats, especially red meat
I	Vegetables, cereals, fruit	
K	Fruits, cereals, vegetables, beans, peas, lentils	Dairy products, meats
Mg	Seeds, nuts, beans, peas, lentils, whole grains, dark green vegetables	
Mn	Whole grains, beans, peas, lentils, nuts, tea	
Mo	Beans, peas, lentils, dark green leafy vegetables	Organ meats
Na		Dairy products, meats, eggs
P	Nuts, beans, peas, lentils, grains	Meats, eggs, dairy products
Se	Grain products, nuts, garlic, broccoli (if grown on high-Se soils)	Meats from Se-fed livestock
Zn	Nuts, whole grains, beans, peas, lentils	Meats, organ meats

Source: Brevik, E.C. 2009. Soil, Food Security, and Human Health. In: W. Verheye (Ed.). Soils, Plant Growth and Crop Production. Encyclopedia of Life Support Systems (EOLSS), Developed under the Auspices of the UNESCO, EOLSS Publishers, Oxford, UK. http://www.eolss.net.

TABLE 3.10 Daily per capita food intake as a worldwide average, 2001–2003 (Table based on information from the United Nations Food and Agriculture Organization)

Food Source	Calories[a]	Percent of Calories
Rice	557	25.5
Wheat	521	23.9
Maize	147	6.7
Sorgum	33	1.5
Potatoes	60	2.7
Cassava	42	1.9
Sugar	202	9.3
Soybean Oil	87	4.0
Palm Oil	50	2.3
Milk	122	5.6
Animal Fats (raw and butter)	62	2.8
Eggs	33	1.5
Meat (pig)	117	5.4
Meat (poultry)	46	2.1
Meat (bovine)	40	1.8
Meat (sheep and goats)	11	0.5
Fish and other aquatic products[b]	52	2.4
TOTAL	2182	

a—Aquatic products data from 2003. All other data from 2001–03.

b—Includes both marine and freshwater products.

FIGURE 3.11   Studies have shown that humans tend to respond positively to views of attractive landscapes, such as this view from Olympic National Park in Washington (left). Preliminary results indicate contact with healthy soils (right) may have similar effects to views of attractive landscapes. Photos by Eric Brevik.

3.9 Soil Health Management

Soil works for you if you work for the soil by using management practices that improve soil health and increase productivity and profitability immediately and into the future. A fully functioning soil produces the maximum amount of products at the least cost. Maximizing soil health is essential to maximizing profitability. Soil will not work for you if you abuse it.

Managing for soil health (improved soil function) is mostly a matter of maintaining suitable habitat for the myriad of creatures that comprise the soil food web. This can be accomplished by disturbing the soil as little as possible, growing as many different species of plants as practical, keeping living plants in the soil as often as possible, and keeping the soil covered all the time.

Manage More by Disturbing Soil Less

Soil disturbance can be the result of physical, chemical or biological activities. Physical soil disturbance, such as tillage, results in bare and/or compacted soil that is destructive and disruptive to soil microbes, and it creates a hostile environment for them to live. Misapplication of farm inputs can disrupt the symbiotic relationships between fungi, other microorganisms, and plant roots. Overgrazing, a form of biological disturbance, reduces root mass, increases runoff, and increases soil temperature. All forms of soil disturbance diminish habitat for soil microbes and result in a diminished soil food web.

Diversify Soil Biota with Plant Diversity

Plants use sunlight to convert carbon dioxide and water into carbohydrates that serve as the building blocks for roots, stems, leaves, and seeds. They also interact with specific soil microbes by releasing carbohydrates (sugars) through their roots into the soil to feed the microbes in exchange for nutrients and water. A diversity of plant carbohydrates is required to support the diversity of soil microorganisms in the soil. In order to achieve a high level of diversity, different plants must be grown. The key to improving soil health is ensuring that food and energy chains and webs consist of several types of plants or animals, not just one or two.

Biodiversity is ultimately the key to the success of any agricultural system. Lack of biodiversity severely limits the potential of any cropping system and increases disease and pest problems. A diverse and fully functioning soil food web provides for nutrient, energy, and water cycling that allows a soil to express its full potential. Increasing the diversity of a crop rotation and cover crops increases soil health and soil function, reduces input costs, and increases profitability.

Keep a Living Root Growing Throughout the Year

Living plants maintain a rhizosphere, an area of concentrated microbial activity close to the root. The rhizosphere is the most active part of the soil ecosystem because it is where the most readily available food is, and where peak nutrient and water cycling occurs. Microbial food is exuded by plant roots to attract and feed microbes that provide nutrients (and other compounds) to the plant at the root-soil interface where the plants can take them up. Since living roots provide the easiest source of food for soil microbes, growing long-season crops or a cover crop following a short-season crop, feeds the foundation species of the soil food web as much as possible during the growing season.

Healthy soil is dependent upon how well the soil food web is fed. Providing plenty of easily accessible food to soil microbes helps them cycle nutrients that plants need to grow. Sugars from living plant roots, recently dead plant roots, crop residues, and soil organic matter all feed the many and varied members of the soil food web.

Keep the Soil Covered as Much as Possible

Soil cover conserves moisture, reduces temperature, intercepts raindrops (to reduce their destructive impact), suppresses weed growth, and provides habitat for members of the soil food web that spend at least some of their time above ground. This is true regardless of land use (cropland, hayland, pasture, or range). Keeping the soil covered while allowing crop residues to decompose (so their nutrients can be cycled back into the soil) can be a bit of a balancing act. Producers must give careful consideration to their crop rotation (including any cover crops) and residue management if they are to keep the soil covered and fed at the same time.

Reduce Wind Erosion

Wind erosion is the movement of soil particles by wind. It occurs when land surfaces lack vegetation and the soil dries out. Windspeeds must reach a certain velocity (in most cases more than 12 miles per hour at 1 foot above the land surface) to move soil particles, depending on the size of the particles. The smaller soil particles (silt and clay) require lower windspeeds, and individual particles of organic matter move most easily and with the lowest windspeeds because of their low weight. "PM-2.5" dust refers to soil particles less than 2.5 microns in size. It can enter human lungs and cause respiratory problems. These small particles form when construction vehicles pulverize soil under dry and windy conditions. There is a high potential for dust blowing on large construction sites and in other disturbed areas.

There are three kinds of wind erosion based on particle size and weight. "Soil creep" occurs when very high wind moves coarse sand particles by rolling them along the soil surface. "Saltation" occurs when wind moves soil particles by bouncing them along the soil surface. Medium-sized sand particles usually are moved by this process. A "dust storm" occurs when wind detaches small soil particles from the land surface and suspends them in the air. Wind erosion is most visible during the suspension stage of dust storms.

Flat areas in dry climates are likely to have serious wind erosion problems. Certain areas of the United States are more prone to high-velocity winds than others. Construction activities usually disturb the land surface by removing plants and pulverizing soil aggregates, making the site more likely to dry out. Reducing traffic over the land surface, keeping the surface rough by maintaining soil clods or aggregates, watering construction site surfaces, applying mulch to disturbed sites, and maintaining windbreaks or barriers reduce the risk of urban wind erosion. Establishing grasses and other plants as soon as possible after construction is completed also helps to control wind erosion.

CHAPTER 4

Essential Resources: The Atmosphere

Shutterstock/Vadim Sadowski

4.1 Introduction to the Atmosphere

The air that surrounds us is a complex mixture of gases. It is active and changes dramatically as you go higher up away from the surface of the earth. This air is called the atmosphere, and it is composed of 78.09% nitrogen (N_2) and 20.95% oxygen (O_2). The remaining 1% is composed mostly of argon (Ar) at 0.93%, and carbon dioxide (CO_2) at 0.039%, with lower concentrations of methane and other minor gases. Carbon dioxide and methane concentrations are increasing due to the actions of humans. The concentration of water vapor varies widely between locations and time of day, and is, on average, about 1%.

The atmosphere is held by the earth's gravity and gets less dense as the altitude increases. It becomes thinner and thinner as the distance between molecules of air increases. In all, the atmosphere is only a thin layer around the earth and is only about 1/100 of the earth's diameter. About three-quarters of the atmosphere's mass is within 11 kilometers (km) (36,000 ft) of the surface. Half of the oxygen is gone within the lower 5.5 km (18,000 ft). The atmosphere has no definite boundary with space, but at 120 km the atmosphere starts to show effects on the reentry of spacecraft and at 100 km is the **Kármán line**, which is generally regarded as the edge of space. The atmosphere is commonly divided into four layers (Figure 4.1).

Troposphere

The **troposphere** is the atmospheric layer closest to the surface and extends up to about 10 km (32,000 ft), but this can vary from 8 to 18 km (26,000–59,000 ft) depending on the position of the earth and the season of the year. This is the region of the atmosphere that is heated by the surface of the earth and can harbor life. It contains most of the water vapor in the atmosphere and is where weather occurs. The weather is due to the uneven heating of the surface of the earth and the mixing of air masses with different temperatures and water contents. This mixing is also influenced by the rotation of the earth. The temperature of the air declines with increasing altitude until it reaches roughly –52°C at the top of the troposphere and then ceases to decline with altitude. This region is called the tropopause and is the beginning of the stratosphere. At this point the mixing of the air is limited between the troposphere and the stratosphere.

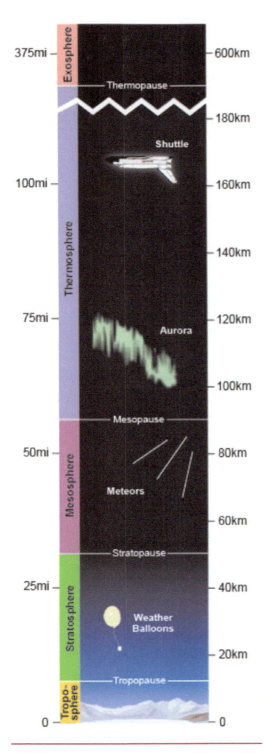

FIGURE 4.1 **Layers of the atmosphere and approximate heights in kilometers.** Figure courtesy of NOAA.

Stratosphere

The **stratosphere** extends from about 11 km (36,000 ft) to 51 km (170,000 ft). In the stratosphere the temperature increases with increasing altitude. This region contains the ozone layer, which is high in O_3 or **ozone**. This region absorbs ultraviolet light radiation from the sun, thereby not only protecting life on the surface, but also warming the air in the stratosphere. Since this area has very limited mixing, the air closest to the troposphere is the coldest at about –52 to –60°C, while the top of the stratosphere may be near freezing (0°C). The top of the stratosphere is called the **stratopause** and has about 1/1000 the air pressure of sea level. It is the boundary between the stratosphere and the mesosphere.

Mesosphere

The **mesosphere** extends up to about 80–85 km (260,000–280,000 ft). In this sphere the temperature again declines with altitude and may be as cold as –100°C, with an average of –85°C. Even though the oxygen level is very low, there is still enough oxygen for meteors to burn up as they enter the atmosphere. Since the temperatures are so low, all of the water vapor is frozen, forming ice clouds called **noctilucent clouds**. They are so faint that they are usually only seen at dawn or dusk when the sun is still behind the horizon.

Thermosphere

The **thermosphere** is the outermost layer of the atmosphere, extending from 85 to 800 km (280,000 to 2,600,000 ft) depending on solar activity and the pressure of the solar wind. The air in this area is so rarified that a molecule may have to travel a kilometer before hitting another molecule. The temperature actually increases with altitude, but since there are so few air molecules, temperature in the usual sense is not well defined. The air is poorly mixed, and usually the composition is constant. The thermosphere is where spacecraft like the International Space Station are in low orbit and the **aurora** is seen (a luminous atmospheric phenomenon appearing as streamers or bands of light sometimes visible in the night sky in northern or southern regions of the earth; it is thought to be caused by charged particles from the sun entering the earth's magnetic field and stimulating molecules in the atmosphere).

Weather and Climate

The sun sends an enormous amount of energy to the earth. Every day over 1,000 watts/m² hits the earth; 70% is absorbed and the rest is reflected into space. This solar energy heats the air that is closest to the surface, with most of the heating occurring in the troposphere. This heating is uneven due to the varied absorption and reflective properties of the surface. For example, water will absorb more heat from the sun than soil, but it returns the heat more slowly than soil, so it will lose heat over the night. This solar energy causes the air to expand and rise, evaporating water and adding water vapor to the air. As the air rises to higher altitudes, it cools and moves back toward the surface (Figure 4.2). The rotation of the planet also adds to the mixing and turbulence of the air by spinning this mixture. Air flows toward areas where the air is less dense, as from water to land during the day (this reverses at night) (Figure 4.3). These factors

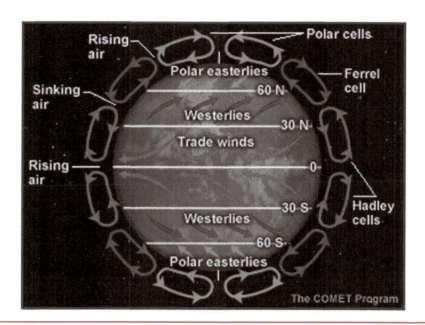

FIGURE 4.2   Mixing of the atmosphere due to uneven heating of the Earth due to water and land heat absorption differences and differences in the angle of the sun's rays at the equator and poles. Figure courtesy of NOAA.

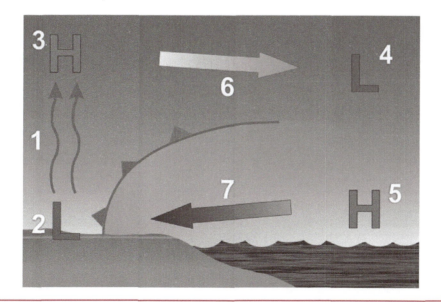

FIGURE 4.3   Land heats more rapidly than water during the day, creating a low-pressure center with rising air over the land. Higher-pressure air over the water moves in to replace the rising air over the land. As the low-pressure air over the land rises, it cools and moves offshore to replace the sea air that is moving onto land at lower altitudes. At night land cools more rapidly than water, and this process reverses. Figure courtesy of NOAA.

produce the weather and varied climates on the planet. Since the air is the most mixed component of the earth, any pollution added to the air will at some time be distributed throughout the entire planet. This dilutes the air pollution greatly but still causes pollution that can be detected everywhere. For example, the lead compound tetraethyl lead, which was added to gasoline to stop engines from pinging, was generally banned around the world in 2005. There are still six countries that allow the use of leaded gasoline, and it is still used in aviation gasoline for piston-driven aircraft in the United States. This lead has been found in the ice in both the Arctic and Antarctica.

The atmosphere is the protective bubble we live in. This protective bubble consists of several gases (Table 4.1) with the top four making up 99.998% of all gases. Of the dry composition of the atmosphere **nitrogen**, by far, is the most common. Nitrogen dilutes oxygen and prevents rapid burning at the Earth's surface. Living things need it to make proteins. **Oxygen** is used by all living things and is essential for respiration.

It is also necessary for combustion or burning.

Argon is used in light bulbs, in double-pane windows, and to preserve the original Declaration of Independence and the Constitution. Plants use **carbon dioxide** to make oxygen. Carbon dioxide also acts as a blanket that prevents the escape of heat into outer space.

These percentages of atmospheric gases are for a completely dry atmosphere. The atmosphere is rarely, if ever, dry. **Water vapor** (water in a 'gas' state) is nearly always present up to about 4% of the total volume (Table 4.2). In the Earth's desert regions (30°N/S) when dry winds are blowing, the water vapor contribution to the composition of the atmosphere will be near zero.

Water vapor contribution climbs to near 3% on extremely hot/humid days. The upper limit, approaching 4%, is found in tropical climates. The table below shows the changes in atmospheric composition with the inclusion of different amounts of water vapor.

TABLE 4.1 Dry Composition of Air

Gas	Symbol	Content
Nitrogen	N_2	78.084%
Oxygen	O_2	20.947%
Argon	Ar	0.934% ⎫
Carbon dioxide	CO_2	0.033% ⎬ 99.998%
Neon	Ne	18.20 parts per million
Helium	He	5.20 parts per million
Krypton	Kr	1.10 parts per million
Sulfur dioxide	SO_2	1.00 parts per million
Methane	CH_4	2.00 parts per million
Hydrogen	H2	0.50 parts per million
Nitrous oxide	N_2O	0.50 parts per million
Xenon	Xe	0.09 parts per million
Ozone	O_3	0.07 parts per million
Nitrogen dioxide	NO_2	0.02 parts per million
Iodine	I_2	0.01 parts per million
Carbon monoxide	CO	trace
Ammonia	NH_3	trace

TABLE 4.2   **Water Vapor in the Atmosphere**

Gas	Symbol	Content				
Nitrogen	N_2	78.084%	77.30%	76.52%	75.74%	74.96%
Oxygen	O_2	20.947%	20.74%	20.53%	20.32%	20.11%
Water Vapor	**H_2O**	**0%**	**1%**	**2%**	**3%**	**4%**
Argon	Ar	0.934%	0.92%	0.91%	0.90%	0.89%

4.2 Air Pressure

The atoms and molecules that make up the various layers in the atmosphere are constantly moving in random directions. Despite their tiny size, when they strike a surface they exert a force on that surface in what we observe as pressure.

Each molecule is too small to feel and only exerts a tiny bit of force. However, when we sum the total forces from the large number of molecules that strike a surface each moment, then the total observed pressure can be considerable.

Air pressure can be increased (or decreased) one of two ways. First, simply adding molecules to any particular container will increase the pressure. A larger number of molecules in any particular container will increase the number of collisions with the container's boundary which is observed as an increase in pressure.

A good example of this is adding (or subtracting) air in an automobile tire. By adding air, the number of molecules increase as well a the total of the collisions with the tire's inner boundary. The increased number of collisions forces the tire to expand and pressure increase.

Courtesy of the NOAA

The second way of increasing (or decreasing) is by the addition (or subtraction) of heat. Adding heat to any particular container can transfer energy to air molecules. The molecules therefore move with increased velocity striking the container's boundary with greater force and is observed as an increase in pressure.

Since molecules move in all directions, they can even exert air pressure upwards as they smash into object from underneath. In the atmosphere, air pressure can be exerted in all directions.

In the International Space Station, the density of the air is maintained so that it is similar to the density at the earth's surface. Therefore, the air pressure is the same in the space station as the earth's surface (14.7 pounds per square inch).

The Pascal

The scientific unit of pressure is the Pascal (Pa) named after after Blaise Pascal (1623–1662). One pascal equals 0.01 millibar or 0.00001 bar. Meteorology has used the millibar for air pressure since 1929.

When the change to scientific unit occurred in the 1960's many meteorologists preferred to keep using the magnitude they are used to and use a prefix "hecto" (h), meaning 100.

Therefore, 1 hectopascal (hPa) equals 100 Pa which equals 1 millibar. 100,000 Pa equals 1000 hPa which equals 1000 millibars. The end result is although the units we refer to in meteorology may be different, their numerical value remains the same. For example the standard pressure at sea-level is 1013.25 millibars and 1013.25 hPa.

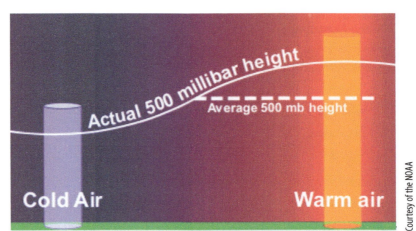

Courtesy of the NOAA

Back on Earth, as elevation increases, the number of molecules decreases and the density of air therefore is less, meaning a decrease in air pressure. In fact, while the atmosphere extends more than 15 miles (24 km) up, one half of the air molecules in the atmosphere are contained within the first 18,000 feet (5.6 km).

Because of this decrease in pressure with height, it makes it very hard to compare the air pressure at one location to another, especially when the elevations of each site differ. Therefore, to give meaning to the pressure values observed at each station, we need to convert the station air pressures reading to a value with a common denominator.

The common denominator we use is the sea-level. At observation stations around the world, through a series of calculations, the air pressure reading, regardless of the station elevation, is converted to a value that *would* be observed if that instrument were located at sea level.

The two most common units in the United States to measure the pressure are "Inches of Mercury" and "Millibars". Inches of mercury refers to the height of a column of mercury measured in hundredths of inches. This is what you will usually hear from the NOAA Weather Radio or from your favorite weather or news source. At sea level, standard air pressure in inches of mercury is 29.92.

Millibars comes from the original term for pressure "bar". Bar is from the Greek "báros" meaning weight. A millibar is 1/1000th of a bar and is approximately equal to 1000 dynes (one dyne is the amount of force it takes to accelerate an object with a mass of one gram at the rate of one centimeter per second squared). Millibar values used in meteorology range from about 100 to 1050. At sea level, standard air pressure in millibars is 1013.2. Weather maps showing the pressure at the surface are drawn using millibars.

Although the changes are usually too slow to observe directly, air pressure is almost always changing. This change in pressure is caused by changes in air density, and air density is related to temperature.

Courtesy of the NOAA

Warm air is less dense than cooler air because the gas molecules in warm air have a greater velocity and are farther apart than in cooler air. So, while the average altitude of the 500 millibar level is around 18,000 feet (5,600 meters) the actual elevation will be higher in warm air than in cold air.

The most basic change in pressure is the twice daily rise and fall in due to the heating from the sun. Each day, around 4 a.m./p.m. the pressure is at its lowest and near its peak around 10 a.m./p.m. The magnitude of the daily cycle is greatest near the equator decreasing toward the poles.

On top of the daily fluctuations are the larger pressure changes as a result of the migrating weather systems. These weather systems are identified by the blue H's and red L's seen on weather maps. The H's represent the location of the area of highest pressure. The L's represent the position of the lowest pressure.

4.3 The Transfer of Heat Energy

The heat source for our planet is the sun. Energy from the sun is transferred through space and through the earth's atmosphere to the earth's surface. Since this energy warms the earth's surface and atmosphere, some of it is or becomes heat energy. There are three ways heat is transferred into and through the atmosphere:

- radiation
- conduction
- convection

Radiation

If you have stood in front of a fireplace or near a campfire, you have felt the heat transfer known as radiation. The side of you nearest the fire warms, while your other side remains unaffected by the heat. Although you are surrounded by air, the air has nothing to do with this transfer of heat. Heat lamps, that keep food warm, work in the same way. Radiation is the transfer of heat energy through space by electromagnetic radiation.

Most of the electromagnetic radiation that comes to the earth from the sun is in the form of visible light. Light is made of waves of different frequencies. The frequency is the number of instances that a repeated event occurs, over a set time. In electromagnetic radiation, the frequency is the number of times an electromagnetic wave moves past a point each second.

Our brains interpret these different frequencies into colors, including red, orange, yellow, green, blue, indigo, and violet. When the eye views all these different colors at the same time, it is interpreted as white. Waves from the sun which we cannot see are infrared, which have lower frequencies than red, and ultraviolet, which have higher frequencies than violet light.

Most of the solar radiation is absorbed by the atmosphere and much of what reaches the earth's surface is radiated back into the atmosphere to become heat energy. Dark colored objects such as asphalt absorb more of the radiant energy and warm faster that light colored objects. Dark objects also radiate their energy faster than lighter colored objects.

Brad Thompson/Shutterstock

Conduction

Conduction is the transfer of heat energy from one substance to another or within a substance. Have you ever left a metal spoon in a pot of soup being heated on a stove? After a short time the handle of the spoon will become hot.

pan_kung/Shutterstock

This is due to transfer of heat energy from molecule to molecule or from atom to atom. Also, when objects are welded together, the metal becomes hot (the orange-red glow) by the transfer of heat from an arc. This is called conduction and is a very effective method of heat transfer in metals. However, air conducts heat poorly.

Convection

Convection is the transfer of heat energy in a fluid. This type of heating is most commonly seen in the kitchen when you see liquid boiling.

Air in the atmosphere acts as a fluid. The sun's radiation strikes the ground, thus warming the rocks. As the rock's temperature rises due to conduction, heat energy is released into the atmosphere, forming a bubble of air which is warmer than the surrounding air. This bubble of air rises into the atmosphere. As it rises, the bubble cools with the heat contained in the bubble moving into the atmosphere.

As the hot air mass rises, the air is replaced by the surrounding cooler, more dense air, what we feel as wind. These movements of air masses can be small in a certain region, such as local cumulus clouds, or large cycles in the troposphere, covering large sections of the earth. Convection currents are responsible for many weather patterns in the troposphere.

4.4 The Earth-Atmosphere Energy Balance

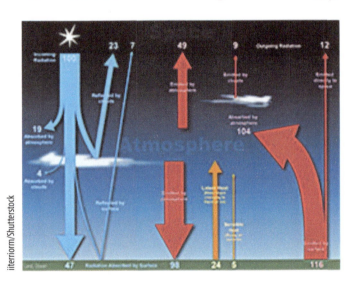

The earth-atmosphere energy balance is the balance between incoming energy from the Sun and outgoing energy from the Earth. Energy released from the Sun is emitted as shortwave light and ultraviolet energy. When it reaches the Earth, some is reflected back to space by clouds, some is absorbed by the atmosphere, and some is absorbed at the Earth's surface.

However, since the Earth is much cooler than the Sun, its radiating energy is much weaker (long wavelength) infrared energy. We can indirectly see this energy radiate into the atmosphere as heat, rising from a hot road, creating shimmers on hot sunny days. The earth-atmosphere energy balance is achieved as the energy received from the Sun *balances* the energy lost by the Earth back into space.

In this way, the Earth maintains a stable average temperature and therefore a stable climate.

The absorption of infrared radiation trying to escape from the Earth back to space is particularly important to the global energy balance. Energy absorption by the atmosphere stores more energy near its surface than it would if there was no atmosphere. The average surface temperature of the moon, which has no atmosphere, is 0°F (−18°C). By contrast, the average surface temperature of the Earth is 59°F (15°C). This heating effect is called the greenhouse effect.

Greenhouse warming is enhanced during nights when the sky is overcast. Heat energy from the earth can be trapped by clouds leading to higher temperatures as compared to nights with clear skies. The air is not

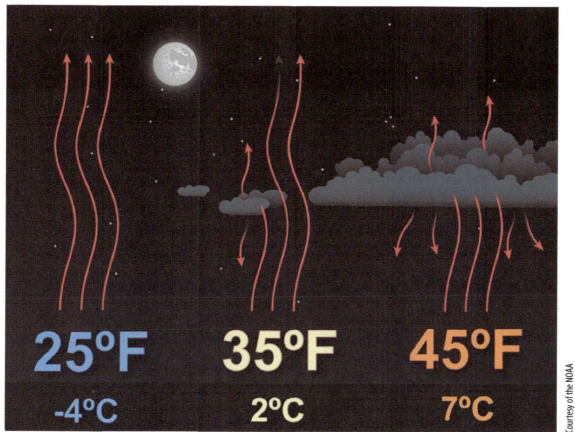

Courtesy of the NOAA

allowed to cool as much with overcast skies. Under partly cloudy skies, some heat is allowed to escape and some remains trapped. Clear skies allow for the most cooling to take place.

4.5 Air and Life

Everyone understands that we need the oxygen in the air to survive and that plants use the carbon dioxide and produce oxygen. But the atmosphere did not have the same composition in the early days of the earth. Today's atmosphere is the third type that the earth has had. The first atmosphere was constantly being stripped away due to the solar wind until enough gases were released to stabilize the atmosphere. By using volcanic evidence, scientists have found that this first atmosphere would have contained 60% hydrogen, 20% oxygen (mostly in the form of water vapor), 10% carbon dioxide, 5% to 7% hydrogen sulfide, and traces of other gases. During this time the oceans developed due to rainfall, and this water reduced the carbon dioxide content because it was absorbed by the water. The second atmosphere had little oxygen because it was tied up due to the oxidation of iron and other compounds. Life developed during this time, but it was **anaerobic** (did not use oxygen). Anaerobic life is still present on the earth in areas that have little to no oxygen. As primitive bacteria found a way to use the sun's energy in photosynthesis, oxygen was released as a toxic byproduct. Other bacteria evolved a way of using the oxygen to help them produce energy from the breakdown of carbohydrates. These changes led to the third atmosphere, the one we have today. The oxygen concentration has gone up and down since about 1.7 billion years ago when it was produced in abundance. Oxygen content was as high as 30% at its peak 280 million years ago. Carbon dioxide, oxygen, and sulfur are in constant flux due to the action of volcanoes, plants, animals and the shifting of the continents.

Shutterstock/Mihai Simonia

5.1 Global Weather

Introduction

In the previous chapter, we have seen the sun as the source for our weather through the transfer of heat energy to the earth. The equatorial region receives the bulk of the heat energy but not always directly. Relative to the sun, the earth's axis is tilted approximately 23½°. The amount of radiation any one place receives each year varies throughout the year.

In the northern hemispheric winter, the southern hemisphere received the majority of the solar radiation. The day which the daylight hours are the shortest in the northern hemisphere is December 22. Conversely, the southern hemisphere daylight hours are the longest. Six months later, on June 22, the earth has completed one half of its orbit with the northern hemisphere receiving the majority of the radiation.

Twice a year, March 21 and September 23, both hemispheres receive the same amount of radiation. The days are called the equinox meaning equal night. Both hemispheres have 12 hours of daylight and darkness.

Courtesy of National Oceanic and Atmospheric Administration

51

All sections courtesy of National Oceanic and Atmospheric Administration

5.2 Global Circulations

Global Circulations explain how air and storm systems travel over the Earth's surface. The global circulation would be simple (and the weather boring) if the Earth did not rotate, the rotation was not tilted relative to the sun, and had no water.

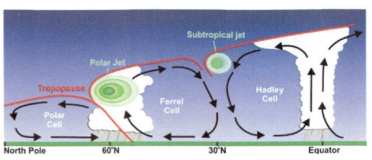

In a situation such as this, the sun heats the entire surface, but where the sun is more directly overhead it heats the ground and atmosphere more. The result would be the equator becomes very hot with the hot air rising into the upper atmosphere.

That air would then move toward the poles where it would become very cold and sink, then return to the equator (above right). One large area of high pressure would be at each of the poles with a large belt of low pressure around the equator.

However, since the earth rotates, the axis is tilted, and there is more land mass in the northern hemisphere than in the southern hemisphere, the actual global pattern is much more complicated.

Instead of one large circulation between the poles and the equator, there are three circulations...

1. **Hadley cell** – Low latitude air movement toward the equator that with heating, rises vertically, with poleward movement in the upper atmosphere. This forms a convection cell that dominates tropical and sub-tropical climates.
2. **Ferrel cell** – A mid-latitude mean atmospheric circulation cell for weather named by Ferrel in the 19th century. In this cell the air flows poleward and eastward near the surface and equatorward and westward at higher levels.
3. **Polar cell** – Air rises, diverges, and travels toward the poles. Once over the poles, the air sinks, forming the polar highs. At the surface air diverges outward from the polar highs. Surface winds in the polar cell are easterly (polar easterlies).

Between each of these circulation cells are bands of high and low pressure at the surface. The high pressure band is located about 30° N/S latitude and at each pole. Low pressure bands are found at the equator and 50°–60° N/S.

Usually, fair and dry/hot weather is associated with high pressure, with rainy and stormy weather associated with low pressure. You can see the results of these circulations on a globe. Look at the number of deserts located along the 30°N/S latitude around the world. Now, look at the region between 50°–60° N/S latitude. These areas, especially the west coast of continents, tend to have more precipitation due to more storms moving around the earth at these latitudes.

5.3 The Jet Stream

Jet streams are relatively narrow bands of strong wind in the upper levels of the atmosphere. The winds blows from west to east in jet streams but the flow often shifts to the north and south. Jet streams follow the boundaries between hot and cold air. Since these hot and cold air boundaries are most pronounced in winter, jet streams are the strongest for both the northern and southern hemisphere winters.

Why does the jet stream winds blow from west to east? Recall from the previous section what the global wind patterns would be like if the earth was not rotating. (The warm air rising at the equator will move toward both poles.) We saw that the earth's rotation divided this circulation into three cells. The earth's rotation is responsible for the jet stream as well.

The motion of the air is not directly north and south but is affected by the momentum the air has as it moves away from the equator. The reason has to do with momentum and how fast a location on or above the Earth moves relative to the Earth's axis.

Your speed relative to the Earth's axis depends on your location. Someone standing on the equator is moving much faster than someone standing on a 45° latitude line. In the graphic (Figure 5.1) the person at the position on the equator arrives at the yellow line sooner than the other two. Someone standing on a pole is not moving at all (except that he or she would be slowly spinning). The speed of the rotation is great enough to cause you to weigh one pound less at the equator than you would at the north or south pole.

The momentum the air has as it travels around the earth is conserved, which means as the air that's over the equator starts moving toward one of the poles, it keeps its eastward motion constant. The Earth below the air, however, moves slower as that air travels toward the poles. The result is that the air moves faster and faster in an easterly direction (relative to the Earth's surface below) the farther it moves from the equator.

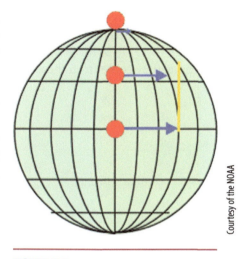

FIGURE 5.1

In addition, with the three-cell circulations mentioned previously, the regions around 30° N/S and 50°–60° N/S are areas where temperature changes are the greatest. As the difference in temperature between the two locations increase, the strength of the wind increases. Therefore, the regions around 30° N/S and 50°–60° N/S are also regions where the wind, in the upper atmosphere, is the strongest.

The 50°–60° N/S region is where the polar jet located with the subtropical jet located around 30°N. Jet streams vary in height of four to eight miles and can reach speeds of more than 275 mph (239 kts/442 kp/h).

The actual appearance of jet streams result from the complex interaction between many variables—such as the location of high and low pressure systems, warm and cold air, and seasonal changes. They meander around the globe, dipping and rising in altitude/latitude, splitting at times and forming eddies, and even disappearing altogether to appear somewhere else.

Jet streams also "follow the sun" in that as the sun's elevation increases each day in the spring, the average latitude of the jet stream shifts poleward. (By Summer in the Northern Hemisphere, it is typically found near the U.S. Canadian border.) As Autumn approaches and the sun's elevation decreases, the jet stream's average latitude moves toward the equator.

Also, the jet stream is often indicated by a line on maps and by television meteorologist. The line generally points to the location of the strongest wind. Jet streams are typically wider and not as distinct but a region where the wind increase toward a core of strongest wind.

One way of visualizing this is to consider a river. The river's current is generally the strongest in the center with decreasing strength as one approaches the river's bank. It can be said that jet streams are "rivers of air".

5.4 Climate

The earth's tilt, rotation and land/sea distribution affect the global weather patterns we observe. While the weather varies from day-to-day at any particular location, over the years, the same type of weather will reoccur. The reoccurring "average weather" found in any particular place is called climate.

German climatologist and amateur botanist Wladimir Köppen (1846–1940) divided the world's climates into several major categories based upon general temperature profile related to latitude. These categories are as follows:

The classical length of record to determine the climate for any particular place is 30 years, as defined by the World Meteorological Organization (WMO). The quantities most often observed are temperature, precipitation, and wind.

The "normals" are computed once every 10 years which helps to smooth out year-to-year variations. For example, the current 30-year normals were calculated from the actual weather data that occurred during the 30 years of 1981–2010. So, when you hear what the normal high and low temperature for your location, for example, they come from these 30-year averages.

A—Tropical Climates   Tropical moist climates extend north and south from the equator to about 15° to 25° latitude. In these climates all months have average temperatures greater than 64°F (18°C) and annual precipitation greater than 59″.

B—Dry Climates   The most obvious climatic feature of this climate is that potential evaporation and transpiration exceed precipitation. These climates extend from 20°–35° North and South of the equator and in large continental regions of the mid-latitudes often surrounded by mountains.

C—Moist Subtropical Mid-Latitude Climates   This climate generally has warm and humid summers with mild winters. Its extent is from 30°50° of latitude mainly on the eastern and western borders of most continents. During the winter, the main weather feature is the mid-latitude cyclone. Convective thunderstorms dominate summer months.

D—Moist Continental Mid-latitude Climates   Moist continental mid-latitude climates have warm to cool summers and cold winters. The location of these climates is poleward of the C climates. The average temperature of the warmest month is greater than 50°F (10°C), while the coldest month is less than −22°F (−30°C). Winters are severe with snowstorms, strong winds, and bitter cold from Continental Polar or Arctic air masses.

FIGURE 5.2   Climate Categories (*Continued on next page*)

E—Polar Climates Polar climates have year-round cold temperatures with the warmest month less than 50°F (10°C). Polar climates are found on the northern coastal areas of North America, Europe, Asia, and on the land masses of Greenland and Antarctica.

H—Highlands Unique climates based on their elevation. Highland climates occur in mountainous terrain where rapid elevation changes cause rapid climatic changes over short distances.

FIGURE 5.2 Climate Categories.

The map (below) shows where these major categories occur in the mainland United States.

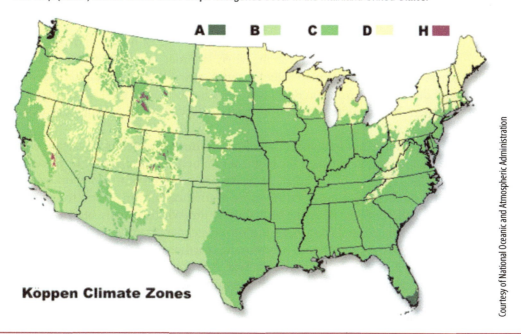

Köppen Climate Zones

A   B   C   D   H

FIGURE 5.3 Koppen Climate Zones.

5.5 Tropical Rain Belt

This section is an article from the National Oceanic and Atmospheric Administration [US] (also known as NOAA) website.

Annual Migration of Tropical Rain Belt

Rebecca Lindsey
Caitlyn Kennedy

Tuesday, May 3, 2011

Near the Earth's equator, solar heating is intense year round. The Sun heats the ocean, evaporating tremendous amounts of water. The sun-warmed ocean heats the overlying air, which rises like a hot air balloon. As air rises, it cools, and water vapor condenses into rain. This daily cycle of heating, evaporation, and convection creates a persistent band of showers and storms around Earth's middle.

The rising air near the equator is replaced by air from north and south of the equator. The Earth rotates beneath this moving air, causing it to turn to the west and creating the easterly trade winds. Because these trade winds converge near the equator, the tropical rain belt is known to meteorologists as the Inter-tropical Convergence Zone.

Visit this website (https://www.climate.gov/news-features/understanding-climate/annual-migration-tropical-rain-belt) to view an animation of monthly average global rain rates. The sequence begins in March, when summer is drawing to a close in the Southern Hemisphere. The highest rain rates on the map (blue) occur south of the equator, and places like Madagascar, northern Australia, and the southern Amazon Rainforest are near the end of their rainy seasons. By June, those places are dry, and Northern Hemisphere tropical locations, including India, Southeast Asia, and Africa south of the Sahara are receiving rain.

Centered on the Pacific Ocean, this animation of monthly average rain rates shows how the tropical rain belt follows the Sun. The sequence begins in March, when summer is drawing to a close in the Southern Hemisphere. The highest rain rates on the map (blue) occur south of the equator, and places like Madagascar, northern Australia, and the southern Amazon Rainforest are near the end of their rainy seasons. By June, those places are dry, and Northern Hemisphere tropical locations, including India, Southeast Asia, and Africa south of the Sahara are receiving rain.

The tropical rain belt stretches out in a virtually unbroken line across the open ocean. Over land, the line is distorted because of the interaction between the tropical rain belt and monsoons. In places with a monsoon climate, the prevailing winds over land and adjacent ocean areas reverse directions on a seasonal basis. The winds reverse because the land changes temperature more dramatically from season to season than water; the differences in heating alter the patterns of rising and sinking air. Winds usually blow offshore in winter and onshore in summer—like a daily coastal sea breeze, but on a much larger scale.

For instance, the shifting of the tropical rain belt into the Northern Hemisphere in May and June coincides with the onset of the summer monsoon in Southeast Asia. As summer days lengthen and heating increases, land surfaces on the continent of Asia get hotter than surrounding oceans. The land heats the air, and the air rises. Warm, moist air from the Indian Ocean flows onshore, bringing rain.

The interaction between the monsoon and the Inter-tropical Convergence Zone produces heavy rain over India, Southeast Asia, and southeastern China throughout much of the Northern Hemisphere summer. Similar interactions between monsoons and the north-south migration of tropical rains occurs in Australia, in the Amazon basin of northern South America, West Africa, and in Mexico and the U.S. Southwest.

Highlights

- Near the Earth's equator, solar heating is intense year round. The Sun heats the ocean, evaporating tremendous amounts of water. This daily cycle of heating, evaporation, and convection creates a persistent band of showers and storms around Earth's middle.

- This tropical rain belt stretches out in a virtually unbroken line across the open ocean. Over land, the line is distorted because of the interaction between the tropical rain belt and monsoon circulations.
- The tropical rain belt migrates north and south of the equator as the seasons change, leading to pronounced wet and dry seasons in many tropical countries.

5.6 Rain Shadow Effect

This section is an article from the NOAA website. The rain shadow effect occurs in many regions across the world that are bordered by mountains, creating many of Earth's deserts.

Rain Shadows on the Summits of Hawaii

Tuesday, January 4, 2011

On Hawaii's Big Island, prevailing Pacific trade winds from the northeast bring higher levels of rainfall to northern mountain slopes, producing dramatic changes in vegetation as you move from northeast to west across the island.

This pattern is mostly due to large mountains that block passage of precipitation systems, casting a "shadow" of dryness behind them. Ancient eruptions from the extinct 5,480-foot Kohala and the 13,796-foot dormant Mauna Kea Volcanoes shaped the mountainous northwest portion of the island, seen above in this Landsat 7 satellite image from January 2001. Clouds hover over the lush, dark green forests on the rainier eastern slopes, while drier western slopes appear mostly earthy-brown. The sharp contrast in vegetation is called the rain shadow effect.

On the Hawaiian Islands, rain shadowing occurs when coastal winds from the northeast push air up and over the mountains. As the air rises, it chills in the cool, high-altitude temperatures. Moisture condenses and falls as rain or snow on the windward slopes. The air that crests the mountaintop warms and dries as it heads down the leeward slopes, leaving them dry and making vegetation sparse.

When drought occurs on Hawaii, the most severe conditions tend to be on the rain-shadowed western slopes, away from the prevailing winds. This was obvious in 2010, when the islands were hit by an intense drought that impacted agriculture yields and water availability.

Courtesy of National Oceanic and Atmospheric Administration

FIGURE 5.4   Satellite image of Hawaii's Big Island.

5.7 Monsoons

What Is a Monsoon?

The word monsoon is derived from the Arabic word mausim, which means season. Traders plying the waters off the Arabian and Indian coasts noted for centuries that dry northeast winds in the winter suddenly turn to the southwest during the summer, and bring beneficial yet torrential rains to the Asian subcontinent. We now know that these large scale wind shifts, from dry desert areas to moist tropical areas, occur in other parts of the Earth, including the Oceanic subcontinent, Southeast Asia, Australia, North America, Africa and South America.

These wind shifts, and the dramatic change in weather they bring, are all more or less driven by a similar mechanism. For much of the year, low level winds in dry subtropical regions tend to blow from the land toward the sea (Figure 5.5). However by late spring, strong solar heating causes temperatures to soar over these land areas. The intense heat causes surface air pressure to fall, forming an area of low pressure known as a thermal low. Adjacent large bodies of water are also warmed, but not as quickly. Thus air pressures remain high relative to the land. Eventually, the pressure difference increases to the point that the cooler and much more humid air over the ocean is drawn toward the hot, dry air over land (Figure 5.6). This moist air moving onto the hot land eventually becomes unstable and develops into thunderstorms. Once this occurs and rain begins to fall, humidity levels increase over land, which only triggers more thunderstorms. This cycle will continue until land areas begin to cool in the early fall and water temperatures reach their peak in early fall. This reduces the pressure difference, which in turn causes the moist onshore flow to diminish, and the monsoon gradually ends.

Monsoon patterns also share a similar upper level flow characteristic. As surface low pressure forms over the hot land areas, the air in the upper levels of the atmosphere also sinks and warms. The sinking air aloft forms high pressure at jet stream level and causes upper level winds to weaken. The jet stream, which blows from west-to-east around the globe, is forced toward the poles as the upper level high expands (Figure 5.7). As the upper high migrates north, upper level temperatures south of the high cool slightly, while winds aloft over a monsoon region turn around to the east (Figure 5.8). These easterly winds aloft import considerable moisture off nearby oceans. When combined with the low level moisture, a favorable environment for thunderstorm

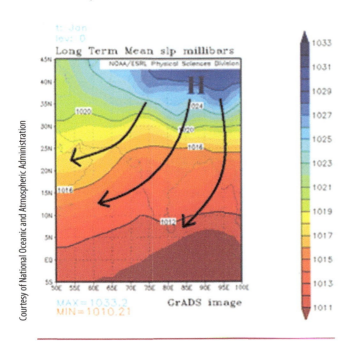

FIGURE 5.5

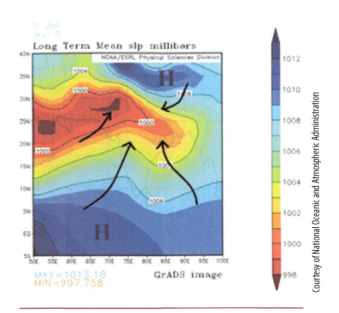

FIGURE 5.6

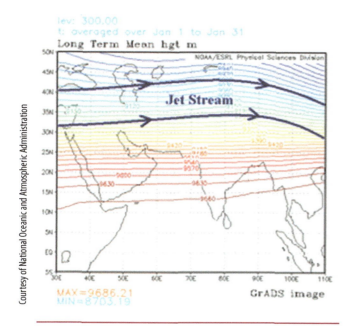

FIGURE 5.7

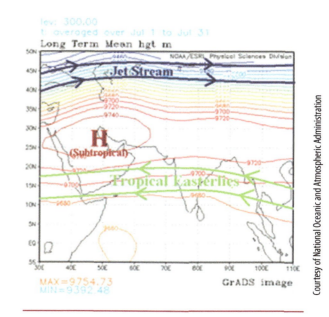

FIGURE 5.8

development is created over areas that are typically dry for much of the year. As rain begins to fall, humidity levels increase over land, triggering more thunderstorms. This cycle continues until land areas cool in early fall and ocean water temperatures reach their peak. This reduces the pressure difference and the moist onshore flow, which in turn ends the monsoon.

Monsoons typically occur in areas with a large, elevated landmass which further enhances temperature and pressure contrasts between land and ocean, enhances moisture transport, and supports stronger subtropical highs. All of these, in turn, enhance rainfall in monsoon regions. This explains why the Indian Monsoon is the strongest and largest. The presence of the Tibetan Plateau, which resides to the north of the Indian subcontinent, is the largest and highest elevated landmass on Earth.

The North American Monsoon

Until the late 1970s, there was serious debate about whether a monsoon truly existed in North America. However, considerable research, which culminated in the Southwest Arizona Monsoon Project (SWAMP) in 1990 and 1993, established the fact that a bonafide monsoon, characterized by large-scale wind and rainfall shifts in the summer, develops over much of Mexico and the intermountain region of the U.S. Published papers at the time called this pattern by different names, including the "Summer Thunderstorm Season," "The Mexican Monsoon," "The Southwest Monsoon," and the "Arizona Monsoon."

In 2004, a major multinational research project was conducted in northwest Mexico and the southwest U.S. The North American Monsoon Experiment (NAME) sought to better describe the monsoon in North America, and increase our ability to predict it on a daily, weekly and seasonal basis. NAME showed that despite its many names, the weather pattern we see during the summer is not only a true monsoon, but it also affects the weather over a large portion of North America. Thus the generally accepted name is now "North American Monsoon."

The North American Monsoon is not as strong or persistent as its Indian counterpart, mainly because the Mexican Pleateau is not as high or as large as the Tibetan Plateau in Asia. However, the North American Monsoon shares most of the basic characteristics of its Indian counterpart. There is a shift in wind patterns in summer which occurs as Mexico and the southwest U.S. warm under intense solar heating. As this happens, the flow reverses from dry land areas to moist ocean areas. In the North American Monsoon, the low level moisture is transported primarily from the Gulf of California and eastern Pacific. The Gulf of California, a narrow body of water surrounded by mountains, is particularly important for low-level moisture transport into Arizona and Sonora. Upper level moisture is also transported into the region, mainly from the Gulf of Mexico by easterly winds aloft. Once the forests of the Sierra Madre Occidental green up from the initial monsoon rains, evaporation and plant transpiration can add additional moisture to the atmosphere which will then flow into Arizona. Finally, if the southern Plains of the U.S. are unusually wet and green during the early summer months, that area can also serve as a moisture source. This combination causes a distinct rainy season over large portions of western North America, which develops rather quickly and sometimes dramatically. Figure 5.9 shows the general moisture sources for the North American Monsoon.

FIGURE 5.9 Moisture sources for the North American Monsoon.

Courtesy of National Oceanic and Atmospheric Administration

Rainfall during the monsoon is not continuous. It varies considerably, depending on a variety of factors. There are usually distinct "burst" periods of heavy rain during the monsoon, and "break" periods with little or no rain. Monsoon precipitation, however, accounts for a substantial portion of annual precipitation in northwest Mexico and the Southwest U.S. Most of these areas receive over half their annual precipitation from the monsoon.

The North American Monsoon circulation pattern typically develops in late May or early June over southwest Mexico. By mid to late summer, thunderstorms increase over the "core" region of the southwest U.S. and northwest Mexico, including the U.S. and Mexican states of Arizona, New Mexico, Sonora, Chihuahua, Sinaloa and Durango. The monsoon typically arrives in mid to late June over northwest Mexico, and early July over the southwest U.S. Once the monsoon is underway, mountain ranges, including the Sierra Madre Occidental and the Mogollon Rim provide a focusing mechanism for the daily development of thunderstorms. Thus much of the monsoon rainfall occurs in mountainous terrain. For example, monsoon rainfall in the Sierra Madre Occidental typically ranges from 10 to 15 inches. Since the southwest U.S. is at the northern fringe of the monsoon, precipitation is less and tends to be more variable. Areas further west of the core monsoon region, namely California and Baja California, typically receive only spotty monsoon-related rainfall. In those areas, the intense solar heating isn't strong enough to overcome a continual supply of cold water from the North Pacific Ocean moving down the west coast of North America. Winds do turn toward the land in these areas, but the cool moist air actually stabilizes the atmosphere.

In addition to the lower level monsoon circulation, an upper level monsoon (or subtropical) ridge develops over the southern High Plains and northern Mexico. In June, this ridge is too far south over Mexico and actually blocks deep moisture from moving north into Arizona (Figure 5.10).

However by late June or early July, this ridge shifts north into the southern Plains or southern Rockies (Figure 5.11). As this shift takes place, mid and upper level moisture streams into Arizona, and low level moisture surges from Mexico meet less resistance.

This monsoon ridge is almost as strong as the one which develops over Asia during the summer. However, since the lower level moisture flow is not as persistent as in the Indian monsoon, the upper level steering pattern and disturbances around the ridge are critical for influencing where thunderstorms develop on any given day. The exact strength and position of the subtropical ridge also governs how far north the tropical easterly

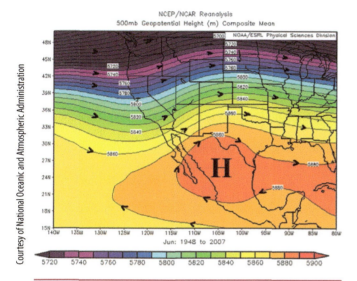

FIGURE 5.10 Mean 500mb height pattern, June. Subtropical high is strengthening over northern Mexico.

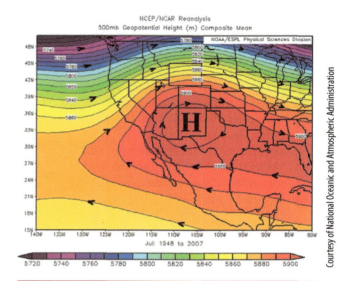

FIGURE 5.11 Mean 500mb height pattern, July. Subtropical high is near maximum seasonal strength over New Mexico.

winds aloft can spread. If the ridge is too close to a particular area, the sinking air at its center suppresses thunderstorms and can result in a significant monsoon "break." If the ridge is too far away or too weak, the east winds around the high are inadequate to bring tropical moisture into the mountains of Mexico and southwest U.S. However, if the ridge sets up in a few key locations, widespread and potentially severe thunderstorms can develop.

It is important to note that the monsoon is not an individual thunderstorm. While the word "monsoon" accurately conjures up images of torrential rains and flooding, calling a single thunderstorm a "monsoon" is incorrect. A monsoon is a large scale weather pattern which causes our summer thunderstorms.

Monsoon Inter-Annual Variability?

Monsoon variability from one summer to the next is substantial, and exceeds the normal monsoon seasonal precipitation at most locations. For example, the normal monsoon precipitation at Tucson, AZ is 6.06 inches. The driest monsoon season measured 1.59 inches, and the wettest measured 13.84 inches. Therefore a variation between seasons of 12.25 inches exists, which is over twice the normal monsoon precipitation at Tucson. Understanding the causes for this huge variation is the first step in developing an ability to forecast an upcoming monsoon season.

Research within the past decade or so has investigated the possible causes behind North American Monsoon variability. Specific factors examined include:

Sea surface temperature anomalies.
Large-scale circulation patterns.
Land surface conditions.
Tropical convergence zones.
Moisture transport mechanisms.

All of these research factors uncovered important details affecting the monsoon in the Southwest, but none of them provided a perfectly clear picture of all conditions affecting its variability. These factors are related to each other and are not independent. For example, sea surface temperatures affect all the other factors to some extent. In additional to inter-annual monsoon variability, multi-decadal variability has been observed. In other words, data from 1963 through 1996 may show results in one aspect, while larger data sets from 1900 through 1963 uncover results not seen in the later time periods.

References

Data Discovery Hurricane Science Center. The Role of the ITCZ in Generating Tropical Depressions. Accessed April 19, 2011.

Marshall, J., and Plumb, R. Alan. (2008). Atmosphere, Ocean, and Climate Dynamics, An Introductory Text. London: Elsevier Academic Press.

National Weather Service JetStream – Online School for Weather. Inter-tropical Convergence Zone.

Vanichkajorn, T, and Archevarahuprok, B. The National Climate Center of Thailand, Thailand Meteorological Department (pdf). APEC Climate Center Website.

Maps by Ned Gardiner and Hunter Allen, based on NASA Global Precipitation Climatology Projectdata provided by the NOAA/ESRL Climate Attribution Team. Caption by Rebecca Lindsey and Caityln Kennedy.

CHAPTER 6

Natural Hazards: Weather-Related & Geologic

6.1 General Weather-Related Hazards

We often think of dramatic weather events when we think of weather-related natural hazards. These may include a tornado that tears up an unsuspecting town, or a hurricane that floods coastal areas. These are indeed terrible events! But we also tend to worry about them much more than we should, given how many deaths they cause every year in the United States. If you look at the 10-year average for hurricane deaths, as an example, about as many people die annually from hurricanes as bee stings. Additionally, we often don't think as much about the less dramatic weather-related hazards. Are you surprised to see in the chart of weather-related fatalities that heat is one of the most hazardous weather conditions for human health?

The U.S. Natural Hazard Statistics provide statistical information on fatalities, injuries and damages caused by weather related hazards. These statistics are compiled by the Office of Services and the National Climatic Data Center from information contained in *Storm Data*, a report comprising data from NWS forecast offices in the 50 states, Puerto Rico, Guam and the Virgin Islands (Figure 6.1).

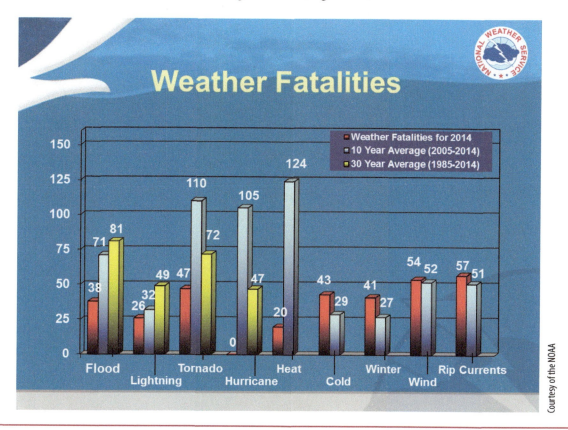

Courtesy of the NOAA

FIGURE 6.1 Weather Fatalities in the U.S. as of 2014.

Sections 6.1, 6.2, 6.3, 6.4, 6.5, and 6.6 courtesy of the National Oceanic and Atmospheric Administration
Sections 6.8, 6.10, 6.13, and 6.14 courtesy of the United States Ready Project, www.ready.gov
Sections 6.9 and 6.12 courtesy of United States Geological Survey

6.2　Heat Index

The following article from the U.S. National Weather Service's Jet Stream - Online School for Weather explains the hazards associated with hot weather. (http://www.srh.noaa.gov/jetstream/global/hi.htm). With basic climate regions around the world, there are some places where the weather is considered hot. But just heat alone does not make high temperatures a threat. There is an old saying stating **"It's not the heat, it's the humidity"**. Well, **actually it's both**.

Heat waves are not easily photographed, like the destruction of tornadoes, hurricanes and floods and therefore tend to not have the same visual impact as these other disasters. Yet, heat waves kill *more* people in the United States than all of the other weather related disasters *combined*. The 10-year average (1997–2006) for heat related deaths in the U.S. is 170 in a typical year.

Heat waves form when high pressure aloft (from 10,000 – 25,000 feet / 3,000 – 7,600 meters) strengthens and remains over a region for several days up to several weeks (Figure 6.3). This is common in summer (in both Northern and Southern Hemispheres) as the jet stream 'follows the sun'. On the equator side of the jet stream, in the middle layers of the atmosphere, is the high pressure area.

Summertime weather patterns are generally slower to change than in winter. As a result, this mid-level high pressure also moves slowly. Under high pressure, the air subsides (sinks) toward the surface. This sinking air acts as a dome capping the atmosphere.

This cap helps to trap heat instead of allowing it to lift. Without the lift there is little or no convection and therefore little or no convective clouds (cumulus clouds) with minimal chances for rain. The end result is a continual build-up of heat at the surface that we experience as a heat wave.

Our bodies dissipate heat by varying the rate and depth of blood circulation, by losing water through the skin and sweat glands, and, as the last extremity is reached, by panting. As the body heats up, the heart begins to pump more blood, blood vessels **dilate** to accommodate the increased flow, and the tiny capillaries in the upper layers of skin are put into operation.

Idambies/Shutterstock

FIGURE 6.2　Desert plants are adapted to hot and dry conditions.

The body's blood is circulated closer to the skin's surface, and excess heat drains off into the cooler atmosphere by one or a combination of three ways...

■ radiation,
■ convection, and
■ evaporation.

At lower temperatures, radiation and convection are efficient methods of removing heat. However, once the air temperature reaches 95°F (35°C), heat loss by radiation and convection ceases. It is at this point that heat loss by sweating becomes all-important. But sweating, by itself, does nothing to cool the body, unless the water is removed by evaporation (sweat changing to water vapor). The downside of this method of cooling is that **high relative humidity retards evaporation**.

Relative humidity is a measure of the amount of water vapor contained in the air, divided by the maximum amount the air can hold, expressed as a percent. A relative humidity of 50% means the air contains ½ of the water vapor it can actually hold. **The maximum amount of water vapor the air can hold is dependent upon the temperature** (the "relative" in relative humidity).

The higher the temperature, the more water (actually water vapor) the air can hold. For example, air with a temperature of 32°F (0°C) can hold about 0.16 ounce of water. Air with a temperature of 80°F (27°C) can hold about an ounce of water.

So, what does this all mean? Sweat is evaporated (changes from a liquid to a gas, i.e. water vapor) when heat is added. The heat is supplied by your body. The results are summed up in Table 6.1.

We, at the National Weather Service, as part of our mission for protecting life and property, have a measure of how the hot weather "feels" to the body. The **Heat Index** is based on work by R.G. Steadman and published in 1979 under the title "The Assessment of Sultriness, Parts 1 and 2." In this work, Steadman constructed a

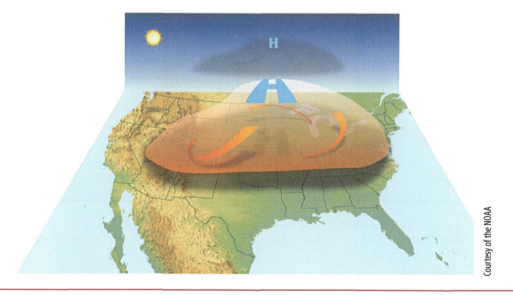

Courtesy of the NOAA

FIGURE 6.3 High pressure in the middle layers of the atmosphere acts as a dome or cap allowing heat to build up at the earth's surface.

TABLE 6.1 Relative humidity and the potential for heat stroke

Relative Humidity	Capacity for air to hold water	Amount of Evaporation	HEAT removed from the body
low	LARGER	HIGHER	MORE
HIGH	smaller	lower	less

table which uses **relative humidity** and **dry bulb temperature** to produce the **"apparent temperature"** or the **temperature the body "feels"**.

We use this table to provide you with **Heat Index** values. These values are for **shady locations only**. Exposure to **full sunshine** can **increase** heat index values by up to **15°F (8°C)**. Also, strong winds, particularly with very hot, dry air, can be extremely hazardous as the wind **adds heat** to the body, as shown by the Heat Index Chart (Figure 6.4).

Here's how to read the chart. Follow the temperature line until it intersects the relative humidity line. Then read the Heat Index on the curved line. For example, an air temperature of 100°F (38°C) and Relative Humidity of 40%. Follow the 100°F (38°C) temperature line until it intersects the 40% relative humidity line. Then curved line that also intersects is the Heat Index of 110°F (43°C), or **Very Hot**.

That is the temperature the body thinks it is and attempts to compensate for that level of heat. **Remember, these values are in the SHADE.** You can add up to 15°F (8°C) to these values if you are in direct sunlight.

Table 6.2 tells you the risk to the body from continued exposure to the excessive heat.

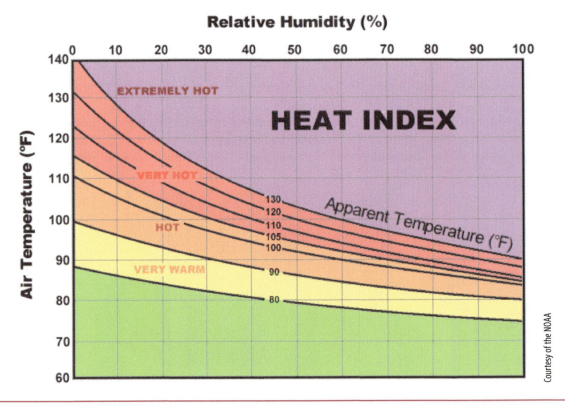

Courtesy of the NOAA

FIGURE 6.4 Heat Index Chart.

TABLE 6.2 Potential for heat risk due to heat exposure

Category	Classification	Heat Index/Apparent Temperature	General Affect on People in High Risk Groups
I	Extremely Hot	130°F or Higher (54°C or Higher)	Heat/Sunstroke **HIGHLY LIKELY** with continued exposure
II	Very Hot	105°F - 130°F (41°C - 54°C)	Sunstroke, heat cramps, or heat exhaustion **LIKELY**, and heat stroke **POSSIBLE** with prolonged exposure and/or physical activity
III	Hot	90°F - 105°F (32°C - 41°C)	Sunstroke, heat cramps, or heat exhaustion **POSSIBLE** with prolonged exposure and/or physical activity
IV	Very Warm	80°F - 90°F (27°C - 32°C)	Fatigue **POSSIBLE** with prolonged exposure and/or physical activity

6.3 Thunderstorms

Introduction

It is estimated that there are as many as 40,000 thunderstorm occurrences each day world-wide. This translates into an astounding 14.6 million occurrences annually! The United States certainly experiences its share of thunderstorm occurrences.

Figure 6.5 shows the average number of thunderstorm days each year throughout the U.S. The most frequent occurrence is in the southeastern states, with Florida having the highest number 'thunder' days (80 to 100+ days per year).

It is in this part of the country that warm, moist air from the Gulf of Mexico and Atlantic Ocean (which we will see later are necessary ingredients for thunderstorm development) is most readily available to fuel thunderstorm development.

The Necessary Ingredients for Thunderstorms

All thunderstorms require three ingredients for their formation:

- moisture,
- instability, and
- a lifting mechanism.

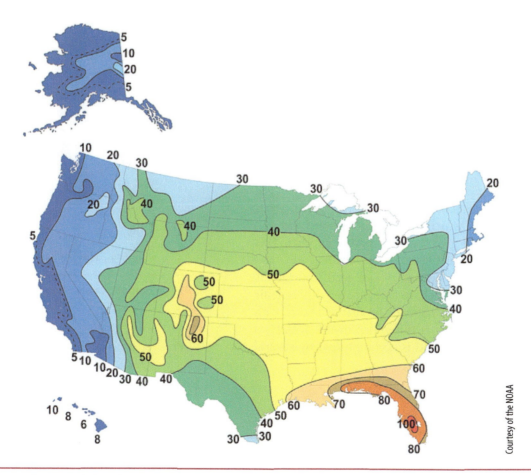

Courtesy of the NOAA

FIGURE 6.5 Distribution of Thunderstorms in the U.S.

Sources of Moisture

Typical sources of moisture are large bodies of water such as the Atlantic and Pacific oceans as well as the Gulf of Mexico (Figure 6.6).

Water temperature also plays a large role in how much moisture is in the atmosphere. Warm ocean currents occur along east coasts of continents and cool ocean currents occur along *westcoasts*. Evaporation is higher in warm ocean currents and therefore puts more moisture into the atmosphere than with cold ocean currents at the same latitude.

Therefore, in the southeastern U.S. the warm water from the two moisture sources (Atlantic Ocean and Gulf of Mexico) helps explain why there is much more rain in that region as compared to the same latitude in Southern California.

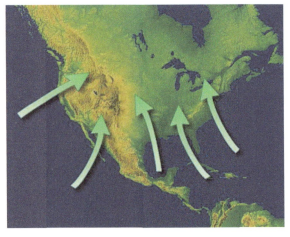

FIGURE 6.6 In the U.S., moisture that feeds thunderstorms comes primarily from the south, including the Atlantic Ocean, the Gulf of Mexico and the Pacific Ocean.

Instability

Air is considered *unstable* if it continues to rise when given a nudge upward (or continues to sink if given a nudge downward). An unstable air mass is characterized by warm *moist* air near the surface and cold *dry* air aloft (Figure 6.7).

In these situations, if a bubble or parcel of air is forced upward it will continue to rise on its own. As this parcel rises it cools and some of the water vapor will condense forming the familiar tall cumulonimbus cloud that is the thunderstorm.

Sources of Lift (upward)

Typically, for a thunderstorm to develop, there needs to be a mechanism which initiates the upward motion, something that will give the air a nudge upward. This upward nudge is a direct result of air density.

Some of the sun's heating of the earth's surface is transferred to the air which, in turn, creates different air densities. The propensity for air to rise increases with decreasing density. This is difference in air density is the main source for lift and is accomplished by several methods.

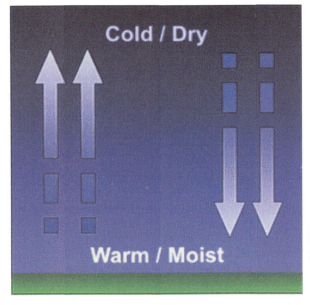

FIGURE 6.7 Schematic of an unstable air mass.

Differential Heating

The sun's heating of the earth's surface is not uniform. For example, a grassy field will heat at a slower rate than a paved street. A body of water will heat slower than the nearby landmass.

This will create two adjacent areas where the air is of different densities. The cooler air sinks, pulled toward the surface by gravity, forcing up the warmer, less dense air, creating thermals.

Sharon Day/Shutterstock

FIGURE 6.8 Grassy areas heat up at a slower rate than paved areas, causing differential heating.

Fronts, Drylines and Outflow Boundaries

Fronts are the boundary between two air masses of different temperatures and therefore different air densities. The colder, more dense air behind the front lift warmer, less dense air abruptly. If the air is moist thunderstorms will often form along the cold front.

Drylines are the boundary between two air masses of different moisture content and separates warm, moist air from hot, dry air. Moist air is less dense then dry air. Drylines therefore, act similar to fronts in that a boundary exists between the two air masses of different densities.

The air temperature behind a dryline is often much higher due to the lack of moisture. That alone will make the air less dense but the moist air ahead of the dryline has an even lower density making it more buoyant. The end result is air lifted along the dryline forming thunderstorms. This is common over the plains in the spring and early summer.

Outflow boundaries are a result of the rush of cold air as a thunderstorm moves overhead. The rain-cooled, more dense, air acts as a "mini cold front", called an outflow boundary. Like fronts, this boundary lifts warm moist air and can cause new thunderstorms to form.

Terrain

As air encounters a mountain it is forced up because of the terrain. Upslope thunderstorms are common in the Rocky Mountain west during the summer.

Life Cycle of a Thunderstorm

The building block of all thunderstorms is the thunderstorm cell. The thunderstorm cell has a distinct life-cycle that lasts about 30 minutes and includes 3 stages.

The Towering Cumulus Stage

A cumulus cloud begins to grow vertically, perhaps to a height of 20,000 feet (6 km). Air within the cloud is dominated by updraft with some turbulent eddies around the edges (Figure 6.9).

The Mature Cumulus Stage

The storm has considerable depth, often reaching 40,000 to 60,000 feet (12 to 18 km). Strong updrafts and downdrafts coexist. This is the most dangerous stage when large hail, damaging winds, and flash flooding may occur (Figure 6.10a & b).

The Dissipating Stage

The downdraft cuts off the updraft. The storm no longer has a supply of warm moist air to maintain itself and therefore it dissipates. Light rain and weak outflow winds may remain for a while during this stage, before leaving behind just a remnant anvil top (Figure 6.11).

Thunderstorm Hazards - Hail

Hail is precipitation that is formed when updrafts in thunderstorms carry raindrops upward into extremely cold areas of the atmosphere. Hail can damage aircraft, homes and cars, and can be deadly to livestock and people. One of the people killed during the March 28, 2000 tornado in Fort Worth was killed when struck by grapefruit-size hail.

While Florida has the most thunderstorms, New Mexico, Colorado, and Wyoming usually have the most hail storms. Why? The freezing level in the Florida thunderstorms is so high, the hail often melts before reaching the ground.

Hailstones grow by collision with supercooled water drops. (Supercooled drops are liquid drops surrounded by air that is below freezing which is a common occurrence in thunderstorms.) There are two methods by which the hailstone grows, wet growth and dry growth, and which produce the "layered look" of hail.

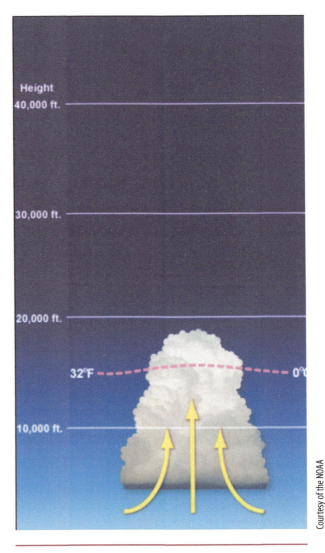

FIGURE 6.9 Towering cumulus stage.

In wet growth, the hailstone nucleus (a tiny piece of ice) is in a region where the air temperature is below freezing, but not super cold. Upon colliding with a supercooled drop the water does not immediately freeze around the nucleus.

Instead liquid water spreads across tumbling hailstones and slowly freezes. Since the process is slow, air bubbles can escape resulting in a layer of clear ice.

With dry growth, the air temperature is well below freezing and the water droplet immediately freezes as it collides with the nucleus. The air bubbles are "frozen" in place, leaving cloudy ice.

Strong updrafts create a rain-free area in supercell thunderstorms (above right). We call this area a **WER** which stands for "weak echo region".

This term, WER, comes from an apparently rain free region of a thunderstorm which is bounded on one side AND above by very intense precipitation indicted by a strong echo on radar.

This rain-free region is produced by the updraft and is what suspends rain and hail aloft producing the strong radar echo. (right)

1. The hail nucleus, buoyed by the updraft is carried aloft by the updraft and begins to grow in size as it collides with supercooled raindrops and other small pieces of hail.
2. Sometimes the hailstone is blown out of the main updraft and begins to fall to the earth.

FIGURE 6.10a A mature cumulus cloud.

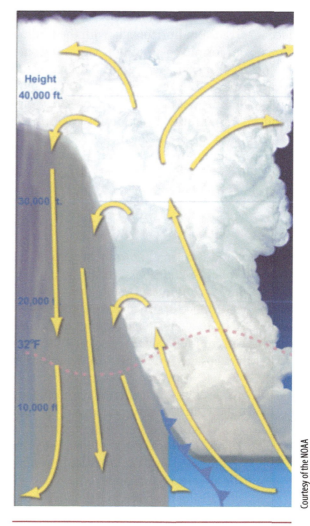

FIGURE 6.10b Formation of the mature cumulus stage.

3. If the updraft is strong enough it will move the hailstone back into the cloud where it once again collides with water and hail and grows. This process may be repeated several times.

4. In all cases, when the hailstone can no longer be supported by the updraft it falls to the earth. The stronger the updraft, the larger the hailstones that can be produced by the thunderstorm.

Multi-cell thunderstorms produce many hail storms but usually not the largest hailstones.

However, the sustained updraft in supercell thunderstorms support large hail formation by repeatedly lifting the hailstones into the very cold air at the top of the thunderstorm cloud.

In all cases, the hail falls when the thunderstorm's updraft can no longer support the weight of the ice. The stronger the updraft the larger the hailstone can grow.

Thunderstorm Hazards - Damaging Wind

Damaging wind from thunderstorms is much more common than damage from tornadoes. In fact, many confuse damage produced by "straight-line" winds and often erroneously attribute it to tornadoes.

The source for damaging winds is well understood and it begins with the **downdraft**. As air rises, it will cool to the point of condensation where water vapor forms tiny water droplets, comprising the cumulus cloud we see. As the air continues to rise further condensation occurs and the cloud grows. Near the center of the updraft, the particle begin to collide and coalescence forming larger droplets. This continues until the rising air can no longer support the ever increasing size of water drops.

Once the rain drops begin to fall friction causes the rising air to begin to fall towards the surface itself. Also, some of the falling rain will evaporate. Through evaporation heat energy is removed from the atmosphere cooling the air associated with the precipitation.

As a result the cooling, the density of the air increases causing it to sink toward the earth. The downdraft also signifies the end of the convection with the thunderstorm and it subsequent decrease.

When this dense rained-cooled air reaches the surface it spreads out horizontally with the leading edged of the cool air forming a gust front. The gust front marks the boundary of a sharp temperature decrease and increase in wind speed. The gust front can act as a point of lift for the development of new thunderstorm cells or cut off the supply of moist unstable air for older cells.

Downbursts are defined as strong winds produced by a downdraft over a horizontal area up to 6 miles (10 kilometers). Downbursts are further subdivided into **microbursts** and **macrobursts**.

Microbursts and Macrobursts

A microburst is a small downburst with an outflow less than 2½ miles (4 kilometers) in horizontal diameter and last for only 2-5 minutes. Despite their small size, microbursts can produce destructive winds up to 168 mph (270 km/h). Also, they create hazardous conditions for pilots and have been responsible for several disasters. For example see the steps below and the corresponding Figures 6.14.

1. As aircraft descend (right) into the airport they follow an imagery line called the "glide slope" (solid light blue line) to the runway.
2. Upon entering the microburst, the plane encounters a "headwind", an increase in wind speed over the aircraft. The stronger wind creates additional lift causing the plane to rise above the glide slope. To return the plane to the proper position, the pilot lowers the throttle to decrease the plane's speed thereby causing the plane to descend.
3. As the plane flies through to the other side of the microburst, the wind direction shifts and is now a "tailwind" as it is from behind the aircraft. This decreases the wind over the wing reducing lift. The plane sinks below the glide slope.
4. However, the "tailwind" remains strong and even with the pilot applying full throttle trying to increase lift again, there may be little, if any, room to recover from the rapid descent causing the plane to crash short of the runway.

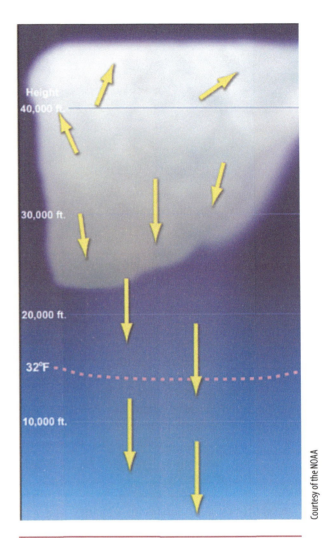

Courtesy of the NOAA

FIGURE 6.11 Dissipating Stage.

Since the discovery of this effect in the early to mid 1980's, pilots are now trained to recognize this event and take appropriate actions to prevent accidents. Also, many airports are now equipped with equipment to detect microbursts and warn aircraft of their occurrences.

A **macroburst** is larger than a microburst with a horizontal extent more than 2½ miles (4 km) in diameter. Also a macroburst is not quite a strong as a microburst but still can produce winds as high as 130 mph (210 km/h). Damaging winds generally last longer, from 5 to 20 minutes, and produce tornado-like damage up to an EF-3 scale.

In wet, humid environments, macrobursts and microbursts will be accompanied by intense rainfall at the ground. If the storm forms in a relatively dry environment, however, the rain may evaporate before it reaches the ground and these downbursts will be without precipitation, known as **dry microbursts**.

In the desert southwest, dust storms are a rather frequent occurrence due to downbursts (Figure 6.15). The city of Phoenix, AZ typically has 1-3 dust storms each summer due to the cooler dense air spreading out from thunderstorms.

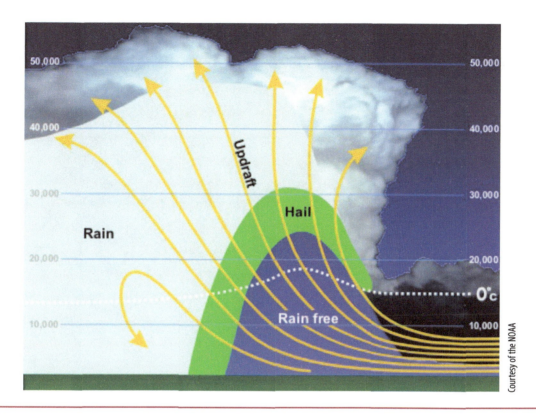

FIGURE 6.12   Location of hail, rain and no rain in a hail storm.

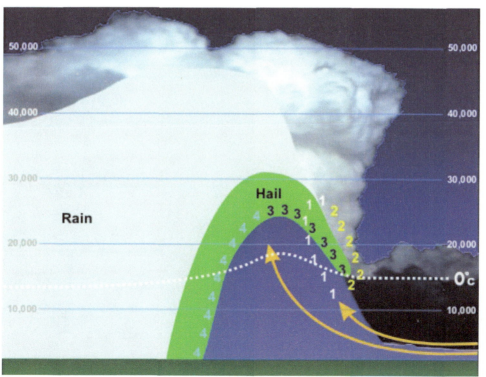

FIGURE 6.13   Steps 1-4 in the formation of hail.

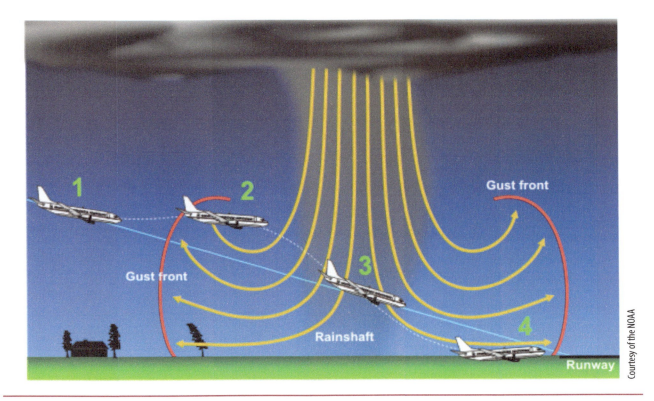

Courtesy of the NOAA

FIGURE 6.14   Microburst hazards to airplanes.

cholder/Shutterstock

FIGURE 6.15   Dust storms are caused by dry micro- or macrobursts.

On July 5, 2011, a massive dust storm resulted in widespread areas of zero or near zero visibility in Phoenix. The wind that produced this storm was generated by downbursts from thunderstorms with winds up to 70 mph (110 kp/h).

Heat Bursts

Dry heatbursts are responsible for a rare weather event called "Heat Bursts". Heat bursts usually occur at night, are associated with decaying thunderstorms, and are marked by gusty, sometimes damaging, winds, a sharp increase in temperature and a sharp decrease in dewpoint.

While not fully understood, it is thought that the process of creating a dry microburst begins higher in the atmosphere for heat bursts. A pocket of cool air aloft forms during the evaporation process since heat energy is required. In heat bursts, all the precipitation has evaporated and this cooled air, being more dense than the surrounding environment, begins to sink due to gravity.

As the air sinks it compresses and with no more water to evaporate the result is the air rapidly warms. In fact, it can become quite hot and very dry. Temperatures generally rise 10 to 20 degrees in a few minutes and have been known to rise to over 120°F (49°C) and remain in place for several hours before returning to normal.

Derechos

If the atmospheric conditions are right, widespread and long-lived windstorms, associated with a band of rapidly moving showers or thunderstorms, can result. The word "derecho" is of Spanish origin, and means straight ahead. A derecho is made up of a "family of *downburst* clusters" and by definition must be at least 240 miles in length.

6.4 Tornadoes

Thunderstorm Hazards - Tornadoes

A tornado is a violently rotating (usually counterclockwise in the northern hemisphere) column of air descending from a thunderstorm and in contact with the ground. Although tornadoes are usually brief, lasting only a few minutes, they can sometimes last for more than an hour and travel several miles causing considerable damage.

The United States experiences more tornadoes by far than any other country. In a typical year about 1300 tornadoes will strike the United States. The peak of the tornado season is April through June and more tornadoes strike the central United States than any other place in the world. This area has been nicknamed "tornado alley (Figure 6.16)."

Most tornadoes are spawned from supercell thunderstorms. Supercell thunderstorms are characterized by a persistent rotating updraft and form in environments of strong vertical wind shear.

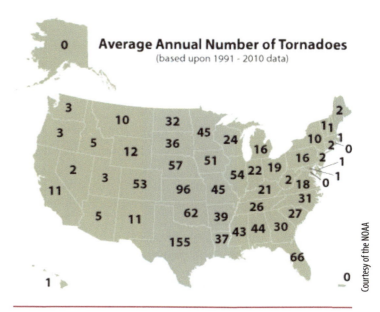

FIGURE 6.16 Distribution of tornadoes in the U.S. by state. The U.S. experiences more tornadoes than any other country in the world.

Wind shear is the change in wind speed and/or direction with height (Figure 6.17).

The updraft lifts the rotating column of air created by the speed shear. This provides two different rotations to the supercell; cyclonic or counter clockwise rotation and an anti-cyclonic of clockwise rotation.

The directional shear amplifies the cyclonic rotation and diminishes the anti-cyclonic rotation (the rotation on the right side of the of the updraft in Figure 6.18). All that remains is the cyclonic rotation called a mesocyclone. By definition a supercell is a rotating thunderstorm.

When viewed from the top (Figure 6.19), the counter-clockwise rotation of the mesocyclone gives the supercell its classic "hook" appearance when seen by radar. As the air rises in the storm, it becomes stretched and more narrow with time.

The image (Figure 6.20) is from the Doppler radar in Springfield, Missouri, May 22, 2011. This image was taken at 5:43 p.m., as an EF-5 strength tornado was moving through Joplin, Missouri.

The colors indicate the intensity of the rain with green representing light rain, the yellow and orange for moderate rain and reds and fuchsia for the heaviest rain and hail. The classic "hook" pattern of the supercell from which a tornado was observed can be clearly seen.

See if you can spot the hook. Hint: look where the rain is most intense (fuchsia color).

The exact processes for the formation of a funnel are not known yet. Recent theories suggest that once a mesocyclone is underway, tornado development is related to the temperature differences across the edge of downdraft air wrapping around the mesocyclone.

However, mathematical modeling studies of tornado formation also indicate that it can happen without such temperature patterns; and in fact, very little temperature variation was observed near some of the most destructive tornadoes in history on May 3, 1999 in Oklahoma.

The funnel cloud of a tornado consists of moist air. As the funnel descends the water vapor within it condenses into liquid droplets. The liquid droplets are identical to cloud droplets yet are not considered part of the cloud since they form within the funnel (Figure 6.21).

The descending funnel is made visible because of the water droplets. The funnel takes on the color of the cloud droplets, which is white.

Due to the air movement, dust and debris on the ground will begin rotating, often becoming several feet high and hundreds of yards wide (Figure 6.22).

After the funnel touches the ground and becomes a tornado, the color of the funnel will change. The color often depends upon the type of dirt and debris is moves over (red dirt produces a red tornado, black dirt a black tornado, etc.).

Tornadoes can last from several seconds to more than an hour but most last less than 10 minutes. The size and/or shape of a tornado is no measure of its strength.

FIGURE 6.17 Wind shear diagram.

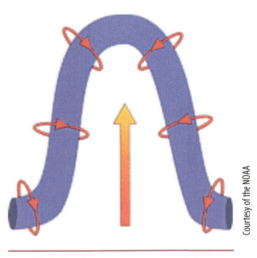

FIGURE 6.18 Effect of directional shear on a cyclonic rotation.

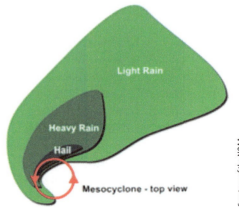

FIGURE 6.19 Mesocyclone.

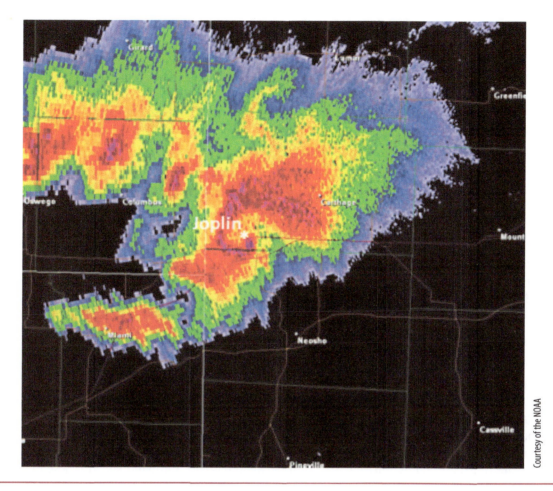

Courtesy of the NOAA

FIGURE 6.20   Doppler radar image of an EF-5 tornado in Joplin, MO on May 22, 2011.

Jon Bilous/Shutterstock

FIGURE 6.21   A descending funnel cloud.

Courtesy of the NOAA

FIGURE 6.22   A funnel cloud touches ground.

TABLE 6.3   The Enhanced F-Scale

EF scale	Class	Wind speed		Description
		mph	km/h	
EF0	weak	65-85	105-137	Gale
EF1	weak	86-110	138-177	Moderate
EF2	strong	111-135	178-217	Significant
EF3	strong	136-165	218-266	Severe
EF4	violent	166-200	267-322	Devastating
EF5	violent	> 200	> 322	Incredible

Occasionally, small tornadoes do major damage and some very large tornadoes, over a quarter-mile wide, have produced only light damage.

The tornado will gradually lose intensity. The condensation funnel decreases in size, the tornado becomes tilted with height, and it takes on a contorted, rope-like appearance before it completely dissipates.

The Enhanced F-Scale

The Fujita (F) Scale was originally developed by Dr. Tetsuya Theodore Fujita to estimate tornado wind speeds based on damage left behind by a tornado. An Enhanced Fujita (EF) Scale, developed by a forum of nationally renowned meteorologists and wind engineers, makes improvements to the original F scale. This EF Scale has replaced the original F scale, which has been used to assign tornado ratings since 1971 (Table 6.3).

The original F scale had limitations, such as a lack of damage indicators, no account for construction quality and variability, and no definitive correlation between damage and wind speed. These limitations may have led to some tornadoes being rated in an inconsistent manner and, in some cases, an overestimate of tornado wind speeds.

The EF Scale takes into account more variables than the original F Scale did when assigning a wind speed rating to a tornado. The EF Scale incorporates 28 damage indicators (Dis) such as building type, structures, and trees. For each damage indicator, there are 8 degrees of damage (DOD) ranging from the beginning of visible damage to complete destruction of the damage indicator. The original F Scale did not take these details into account.

For example, with the EF Scale, an F3 tornado will have estimated wind speeds between 136 and 165 mph (218 and 266 km/h), whereas with the original F Scale, an F3 tornado has winds estimated between 162-209 mph (254-332 km/h).

The wind speeds necessary to cause "F3" damage are not as high as once thought and this may have led to an overestimation of some tornado wind speeds.

There is still some uncertainty as to the upper limits of the strongest tornadoes so F5 ratings do not have a wind speed range. Wind speed estimations for F5 tornadoes are left open ended and assigned wind speeds greater than 200 mph (322 km/h).

6.5 Lightning

Introduction

- Lightning is one of the oldest observed natural phenomena on earth. At the same time, it also is one of the least understood. While lightning is simply a gigantic spark of static electricity (the same kind of electricity that sometimes shocks you when you touch a doorknob), scientists do not have a complete grasp on how it works, or how it interacts with solar flares impacting the upper atmosphere or the earth's electromagnetic field.

■ Lightning has been seen in volcanic eruptions, extremely intense forest fires, surface nuclear detonations, heavy snowstorms, and in large hurricanes. However, it is most often seen in thunderstorms. In fact, lightning (and the thunder that results) is what makes a thunderstorms.

■ At any given moment, there can be as many as 2,000 thunderstorms occurring across the globe. This translates to more than 14.5 MILLION storms each year. NASA satellite research indicated these storms produce lightning flashes about 40 times a second worldwide (Figure 6.23).

■ This is a change from the commonly accepted value of 100 flashes per second which was an estimate from 1925. Whether it is 40, 100, or somewhere in between, we live on an electrified planet.

How Lightning Is Created

The conditions needed to produce lightning have been known for some time. However, exactly how lightning forms has never been verified so there is room for debate. Leading theories focus around separation of electric charge and generation of an electric field within a thunderstorm. Recent studies also indicate that ice, hail, and semi-frozen water drops known as graupel are essential to lightning development. Storms that fail to produce large quantities of ice usually fail to produce lightning.

Forecasting when and where lightning will strike is not yet possible and most likely never will be. But by educating yourself about lightning and learning some basic safety rules, you, your family, and your friends can avoid needless exposure to the dangers of one of the most capricious and unpredictable forces of nature.

Charge Separation

Thunderstorms have very turbulent environments. Strong updrafts and downdrafts occur with regularity and within close proximity to each other. The updrafts transport small liquid water droplets from the lower regions of the storm to heights between 35,000 and 70,000 feet, miles above the freezing level.

Meanwhile, downdrafts transport hail and ice from the frozen upper regions of the storm. When these collide, the water droplets freeze and release heat. This heat in turn keeps the surface of the hail and ice slightly warmer than its surrounding environment, and a "soft hail", or "graupel" forms.

When this graupel collides with additional water droplets and ice particles, *a critical phenomenon occurs*: Electrons are sheared off of the ascending particles and collect on the descending particles. Because electrons carry a negative charge, the result is a storm cloud with a negatively charged base and a positively charged top (Figure 6.25).

Annual Flash Rate Maps (NASA)

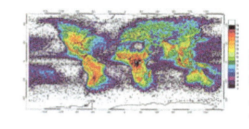

Courtesy of the NOAA

FIGURE 6.23 Satellite image of the frequency of lightning strikes.

Oleksii Natykach/Shutterstock

FIGURE 6.24 A cloud to ground lightning strike.

Courtesy of the NOAA

FIGURE 6.25 Charge separation in a storm cloud.

Field Generation

In the world of electricity, opposites attract and insulators inhibit. As positive and negative charges begin to separate within the cloud, an electric field is generated between its top and base. Further separation of these charges into pools of positive and negative regions results in a strengthening of the electric field (Figure 6.26).

However, the atmosphere is a very good insulator that inhibits electric flow, so a TREMENDOUS amount of charge has to build up before lightning can occur. When that charge threshold is reached, the strength of the electric field overpowers the atmosphere's insulating properties, and lightning results.

The electric field within the storm is not the only one that develops. Below the negatively charged storm base, positive charge begins to pool within the surface of the earth (Figure 6.27). This positive charge will shadow the storm wherever it goes, and is responsible for cloud-to-ground lightning. However, the electric field within the storm is much stronger than the one between the storm base and the earth's surface, so most lightning (~75-80%) occurs within the storm cloud itself.

FIGURE 6.26 Positive charges begin to collect at the ground surface as a result of the negatively charged storm cloud above the ground.

How Lightning Develops Between the Cloud and the Ground

A moving thunderstorm gathers another pool of positively charged particles along the ground that travel with the storm (Figure 6.27, image 1). As the differences in charges continue to increase, positively charged particles rise up taller objects such as trees, houses, and telephone poles.

A channel of negative charge, called a "stepped leader" will descend from the bottom of the storm toward the ground (Figure 6.27, image 2). It is invisible to the human eye, and shoots to the ground in a series of rapid steps, each occurring in less time than it takes to blink your eye. As the negative leader approaches the ground, positive charge collects in the ground and in objects on the ground.

This positive charge "reaches" out to the approaching negative charge with its own channel, called a "streamer" (Figure 6.27, image 3). When these channels connect, the resulting electrical transfer is what we see

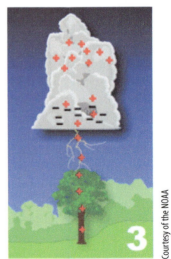

FIGURE 6.27 The formation of cloud to ground lightning strikes.

as lightning. After the initial lightning stroke, if enough charge is leftover, additional lightning strokes will use the same channel and will give the bolt its flickering appearance.

Tall objects such as trees and skyscrapers are commonly struck by lightning. Mountains also make good targets. The reason for this is their tops are closer to the base of the storm cloud. Remember, the atmosphere is a good electrical insulator. The less insulation the lightning has to burn through, the easier it is for it to strike. However, this does not always mean tall objects will be struck. It all depends on where the charges accumulate. Lighting can strike the ground in an open field even if the tree line is close by.

The Sound of Thunder

Thunder is the acoustic shock wave resulting from the extreme heat generated by a lightning flash. Lightning can be as hot as 54,000°F (30,000°C), a temperature that is five times the surface of the sun! When lightning occurs, it heats the air surrounding its channel to that same incredible temperature in a fraction of a second.

Like all gases, when air molecules are heated, they expand. The faster they are heated, the faster their rate of expansion. But when air is heated to 54,000°F in a fraction of a second, a phenomenon known as "explosive expansion" occurs. This is where air expands so rapidly that it compresses the air in front of it, forming a shock wave similar to a sonic boom. Exploding fireworks produce a similar result.

6.6 Typhoons

Tropical Cyclone Introduction

A tropical cyclone is a warm-core, low pressure system without any "front" attached, that develops over the tropical or subtropical waters, and has an organized circulation. Depending upon location, tropical cyclones have different names around the world. In the:

- Atlantic/Eastern Pacific Oceans - *hurricanes*
- Western Pacific - *typhoons*
- Indian Ocean - *cyclones*

Regardless of what they are called, there are several favorable environmental conditions that must be in place before a tropical cyclone can form. They are:

- Warm ocean waters (at least 80°F / 27°C) throughout a depth of about 150 ft. (46 m).
- An atmosphere which cools fast enough with height such that it is potentially unstable to moist convection.
- Relatively moist air near the mid-level of the troposphere (16,000 ft. / 4,900 m).
- Generally a minimum distance of at least 300 miles (480 km) from the equator.
- A pre-existing near-surface disturbance.
- Low values (less than about 23 mph / 37 km/h) of vertical wind shear between the surface and the upper troposphere. Vertical wind shear is the change in wind speed with height.

FIGURE 6.28 Satellite image of a tropical cyclone.

pio3/Shutterstock

Tropical Cyclone Formation Basin

Given that sea surface temperatures need to be at least 80°F (27°C) for tropical cyclones to form, it is natural that they form near the equator. However, with only the rarest of occasions, these storms do not form within 5° latitude of the equator. This is due to the lack of sufficient **Coriolis Force**, the force that causes the cyclone to spin. However, tropical cyclones form in seven regions around the world (Figure 6.29).

FIGURE 6.29 The 7 regions in which tropical cyclones form.

One rare exception to the lack of tropical cyclones near the equator was Typhoon Vamei which formed near Singapore on December 27, 2001. Since tropical cyclone observations started in 1886 in the North Atlantic and 1945 in the western North Pacific, the previous recorded lowest latitude for a tropical cyclone was 3.3°N for Typhoon Sarah in 1956. With its circulation center at 1.5°N Typhoon Vamei's circulation was on both sides of the equator. U.S. Naval ships reported maximum sustained surface wind of 87 mph and gust wind of up to 120 mph.

The seedlings of tropical cyclones, called "disturbances", can come from:

■ **Easterly Waves:** Also called tropical waves, this is an inverted trough of low pressure moving generally westward in the tropical easterlies. A trough is defined as a region of relative low pressure. The majority of tropical cyclones form from easterly waves.

■ **West African Disturbance Line (WADL):** This is a line of convection (similar to a squall line) which forms over West Africa and moves into the Atlantic Ocean. WADL's usually move faster than tropical waves.

■ **TUTT:** A TUTT (Tropical Upper Tropospheric Trough) is a trough, or cold core low in the upper atmosphere, which produces convection. On occasion, one of these develops into a warm-core tropical cyclone.

■ **Old Frontal Boundary:** Remnants of a polar front can become lines of convection and occasionally generate a tropical cyclone. In the Atlantic Ocean storms, this will occur early or late in the hurricane season in the Gulf of Mexico or Caribbean Sea.

Once a disturbance forms and sustained convection develops, it can become more organized under certain conditions. If the disturbance moves or stays over warm water (at least 80°F), and upper level winds remain weak, the disturbance can become more organized, forming a depression.

The warm water is one of the most important keys as it is water that powers the tropical cyclone (Figure 6.30). As water vapor (water in the gaseous state) rises, it cools. This cooling causes the water vapor to condense into a liquid we see as clouds. In

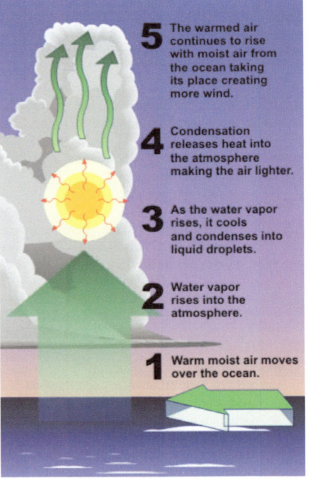

5 The warmed air continues to rise with moist air from the ocean taking its place creating more wind.

4 Condensation releases heat into the atmosphere making the air lighter.

3 As the water vapor rises, it cools and condenses into liquid droplets.

2 Water vapor rises into the atmosphere.

1 Warm moist air moves over the ocean.

FIGURE 6.30 How a disturbance over warm water turns into a cyclone.

the process of condensation, heat is released. This heat warms the atmosphere making the air lighter still which then continues to rise into the atmosphere. As it does, more air moves in near the surface to take its place which is the strong wind we feel from these storms.

Therefore, once the eye of the storm moves over land will begin to weaken rapidly, not because of friction, but because the storm lacks the moisture and heat sources that the ocean provided. This depletion of moisture and heat hurts the tropical cyclone's ability to produce thunderstorms near the storm center. Without this convection, the storm rapidly diminishes.

The NASA image (Figure 6.31) is Hurricane Wilma in October 2005. The color of the ocean represents sea surface temperature with orange and red colors indicating temperatures of 82°F or greater.

As Wilma moves northwest, then eventually northeast, the water temperature decreases (indicated by the change to light blue color) after the storm passes a particular location. This is the result of the heat that is removed from the ocean and provided to the storm.

Therein shows the purpose of tropical cyclones. Their role is to take heat, stored in the ocean, and transfer it to the upper atmosphere where the upper level winds carry that heat to the poles. This keeps the polar regions from being as cold as they could be and helps keep the tropics from overheating.

There are many suggestions for the mitigation of tropical cyclones such as "seeding" storms with chemicals to decrease their intensity, dropping water absorbing material into the storm to soak-up some of the moisture, to even using nuclear weapons to disrupt their circulation thereby decreasing their intensity. While well meaning, the ones making the suggestions vastly underestimate the amount of energy generated and released by tropical cyclones.

Even if we could disrupt these storms, it would not be advisable. Since tropical cyclones help regulate the earth's temperature, any decrease in tropical cyclone intensity means the oceans retain more heat. Over time, the build-up of heat could enhance subsequent storms and lead to more numerous and/or stronger events.

Tropical Cyclone Classification

Tropical cyclones with an organized system of clouds and thunderstorms with a defined circulation, and maximum sustained winds of **38 mph (61 km/h) or less** are called "tropical depressions". Once the tropical cyclone reaches winds of at least 39 mph (63 km/h) they are typically called a "tropical storm" and assigned a name.

If maximum sustained winds reach 74 mph (119 km/h), the cyclone is called:

- A **hurricane** in the North Atlantic Ocean, the Northeast Pacific Ocean east of the dateline, and the South Pacific Ocean east of 160°E, (The word hurricane comes from the Carib Indians of the West Indies, who called this storm a *huracan*. Supposedly, the ancient Tainos tribe of Central America called their god of evil "Huracan". Spanish colonists modified the word to hurricane.),
- A **typhoon** in the Northwest Pacific Ocean west of the dateline (**super typhoon** if the maximum sustained winds are at least 150 mph / 241 km/h),
- A **severe tropical cyclone** in the Southwest Pacific Ocean west of 160°E or Southeast Indian Ocean east of 90°E,
- A **severe cyclonic storm** in the North Indian Ocean, and
- Just a **tropical cyclone** in the Southwest Indian Ocean.

Hurricanes are further classified according to their wind speed. The **Saffir-Simpson Hurricane Wind Scale** is a 1-5 rating based on the hurricane's present intensity. This scale only addresses the wind speed and does not

Oct 21 2005

Courtesy of the NOAA

FIGURE 6.31 Hurricane Wilma making landfall in October 2005.

take into account the potential for other hurricane-related impacts, such as storm surge, rainfall-induced floods, and tornadoes.

Earlier versions of this scale – known as the Saffir-Simpson Hurricane Scale – incorporated central pressure and storm surge as components of the categories. However, hurricane size (extent of hurricane-force winds), local bathymetry (depth of near-shore waters), topography, the hurricane's forward speed and angle to the coast also affect the surge that is produced.

For example, the very large Hurricane Ike (with hurricane force winds extending as much as 125 miles (200 kilometers) from the center) in 2008 made landfall in Texas as a Category 2 hurricane and had peak storm surge values of about 20 feet (6 meters). In contrast, tiny Hurricane Charley (with hurricane force winds extending at most 25 miles (40 kilometers) from the center) struck Florida in 2004 as a Category 4 hurricane and produced a peak storm surge of only about 7 feet (2.1 meters). These storm surge values were substantially outside of the ranges suggested in the original scale.

To help reduce public confusion about the impacts associated with the various hurricane categories as well as to provide a more scientifically defensible scale, the storm surge ranges, flooding impact and central pressure statements were removed from the scale and only peak winds are now employed.

Tropical Cyclone Structure

The main parts of a tropical cyclone are the rainbands, the eye, and the eyewall (Figure 6.32). Air spirals in toward the center in a counter-clockwise pattern in the northern hemisphere (clockwise in the southern hemisphere), and out the top in the opposite direction. In the very center of the storm, air sinks, forming an "eye" that is mostly cloud-free.

The Eye

The hurricane's center is a relatively calm, generally clear area of sinking air and light winds that usually do not exceed 15 mph (24 km/h) and is typically 20–40 miles (32–64 km) across. An eye will usually develop when the maximum sustained wind speeds go above 74 mph (119 km/h) and is the calmest part of the storm.

But why does an eye form? The cause of eye formation is still not fully understood. It probably has to do with the combination of "the conservation of angular momentum" and centrifugal force. The conservation

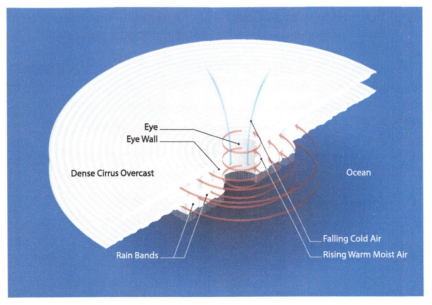

FIGURE 6.32 A cross section of a cyclone.

of angular momentum means is objects will spin faster as they move toward the center of circulation. So air increases it speed as it heads toward the center of the tropical cyclone. One way of looking at this is watching figure skaters spin. The closer they hold their hands to the body, the faster they spin. Conversely, the farther the hands are from the body the slower they spin. In tropical cyclone, as the air moves toward the center, the speed must increase.

However, as the speed increases, an outward-directed force, called the centrifugal force, occurs because the wind's momentum wants to carry the wind in a straight line. Since the wind is turning about the center of the tropical cyclone, there is a pull outward. The sharper the curvature, and/or the faster the rotation, the stronger is the centrifugal force.

Around 74 mph (119 km/h) the strong rotation of air around the cyclone balances inflow to the center, causing air to ascend about 10-20 miles (16-32 km) from the center forming the eyewall. This strong rotation also creates a vacuum of air at the center, causing some of the air flowing out the top of the eyewall to turn inward and sink to replace the loss of air mass near the center.

This sinking air suppresses cloud formation, creating a pocket of generally clear air in the center. People experiencing an eye passage at night often see stars. Trapped birds are sometimes seen circling in the eye, and ships trapped in a hurricane report hundreds of exhausted birds resting on their decks. The landfall of hurricane Gloria (1985) on southern New England was accompanied by thousands of birds in the eye.

The sudden change of very strong winds to a near calm state is a dangerous situation for people ignorant about a hurricane's structure. Some people experiencing the light wind and fair weather of an eye may think the hurricane has passed, when in fact the storm is only half over with dangerous eyewall winds returning, this time from the opposite direction within a few minutes.

The Eyewall

The eyewall consists of a ring of tall thunderstorms that produce heavy rains and usually the strongest winds. Changes in the structure of the eye and eyewall can cause changes in the wind speed, which is an indicator of the storm's intensity. The eye can grow or shrink in size, and double (concentric) eyewalls can form.

Rainbands

Curved bands of clouds and thunderstorms that trail away from the eye wall in a spiral fashion. These bands are capable of producing heavy bursts of rain and wind, as well as tornadoes. There are sometimes gaps in between spiral rain bands where no rain or wind is found.

In fact, if one were to travel between the outer edge of a hurricane to its center, one would normally progress from light rain and wind, to dry and weak breeze, then back to increasingly heavier rainfall and stronger wind, over and over again with each period of rainfall and wind being more intense and lasting longer (Figure 6.33).

Tropical Cyclone Size

Typical hurricane strength tropical cyclones are about 300 miles (483 km) wide although they can vary considerably. as shown in the two enhanced satellite images below. Size is not necessarily an indication of hurricane intensity. Hurricane Andrew (1992), the second most devastating hurricane to hit the United States, next to Katrina in 2005, was a relatively small hurricane (Figure 6.33).

On record, **Typhoon Tip** (1979) was the largest storms with gale force winds (39 mph/63 km/h) that extended out for 675 miles (1087 km) in radius in the Northwest Pacific on 12 October,

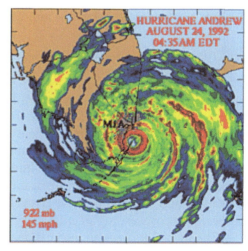

FIGURE 6.33 Rainbands (shown in red) in a cyclone.

Courtesy of the NOAA

1979 (Figure 6.34). The smallest storm was **Tropical Cyclone Tracy** with gale force winds that only extended 30 miles (48 km) radius when it struck Darwin, Australia, on December 24, 1974.

However, the hurricane's destructive winds and rains cover a wide swath. Hurricane-force winds can extend outward more than 150 miles (242 km) for a large one. The area over which tropical storm-force winds occur is even greater, ranging as far out as almost 300 miles (483 km) from the eye of a large hurricane.

The strongest hurricane on record for the Atlantic Basin is Hurricane Wilma (2005). With a central pressure of 882 mb (26.05"), Wilma produced sustained winds of 175 mph (280 km/h).

Courtesy of the NOAA

The relative sizes of the largest and smallest tropical cyclones on record as compared to the United States.

FIGURE 6.34 Comparing sizes of 2 tropical cyclones - their size can very a lot.

Tropical Cyclone Names

For several hundred years, many hurricanes in the West Indies were named after the particular saint's day on which the hurricane occurred. Ivan R. Tannehill describes in his book "Hurricanes" the major tropical storms of recorded history and mentions many hurricanes named after saints. For example, there was "Hurricane Santa Ana" which struck Puerto Rico with exceptional violence on July 26, 1825, and "San Felipe" (the first) and "San Felipe" (the second) which hit Puerto Rico on September 13 in both 1876 and 1928.

The first known meteorologist to assign names to tropical cyclones was Clement Wragge, an Australian meteorologist. Before the end of the 19th century, he began by using letters of the Greek alphabet, then from Greek and Roman mythology and progressed to the use of feminine names. In the United states, an early example of the use of a woman's name for a storm was in the novel "Storm" by George R. Stewart, published by Random House in 1941. During World War II, this practice became widespread in weather map discussions among forecasters, especially Air Force and Navy meteorologists who plotted the movements of storms over the wide expanses of the Pacific Ocean.

In 1953, the United States abandoned a confusing a two-year old plan to name storms by a phonetic alphabet (Able, Baker, Charlie, etc.). That year, this Nation's weather services began using female names for storms. The practice of naming hurricanes solely after women came to an end in 1978 when men's and women's names were included in the Eastern North Pacific storm lists. In 1979, male and female names were included in lists for the Atlantic and Gulf of Mexico.

Why Tropical Cyclones Are Named

Experience shows that the use of short, distinctive given names in written as well as spoken communications is quicker and less subject to error than the older more cumbersome latitude-longitude identification methods. These advantages are especially important in exchanging detailed storm information between hundreds of widely scattered stations, airports, coastal bases, and ships at sea.

The use of easily remembered names greatly reduces confusion when two or more tropical storms occur at the same time. For example, one hurricane can be moving slowly westward in the Gulf of Mexico, while at exactly the same time another hurricane can be moving rapidly northward along the Atlantic coast. In the past, confusion and false rumors have arisen when storm advisories broadcast from one radio station were mistaken for warnings concerning an entirely different storm located hundreds of miles away.

The name lists have an international flavor because hurricanes affect other nations and are tracked by the public and weather services of countries other than the United States. Names for these lists agreed upon by the nations involved during international meetings of the World Meteorological Organization.

The only time that there is a change in the list is if a storm is so deadly or costly that the future use of its name on a different storm would be inappropriate for reasons of sensitivity. If that occurs, then at an annual meeting by the WMO committee (called primarily to discuss many other issues) the offending name is stricken from the list and another name is selected to replace it.

The National Hurricane Center (RSMC Miami, FL), is responsible for the Atlantic basin west of 30°W. If a disturbance intensifies into a tropical storm the Center will give the storm a name from one of the six lists below.

A separate set is used each year beginning with the first name in the set. After the sets have all been used, they will be used again. The 2013 set, for example, will be used again to name storms in the year 2018.

The letters Q, U, X, Y, and Z are not included because of the scarcity of names beginning with those letters. If over 21 named tropical cyclones occur in a year, the Greek alphabet will be used following the "W" name.

In addition, after major land-falling storms having major economic impact, the names are retired.

Tropical Cyclone Hazards

Each year beginning around June 1st, the Gulf and East Coast states are at great risk for tropical cyclones. While most people know that tropical cyclones can contain damaging wind, many do not realize that they also produce several other hazards, both directly and indirectly.

Storm Surge

Storm surge is simply water that is pushed toward the shore by the force of the winds swirling around the storm. This advancing surge combines with the normal tides to create the hurricane storm tide, which can increase the average water level 15 feet (4.5 m) or more (Figure 6.35).

In addition, wind driven waves are superimposed on the storm tide. This rise in water level can cause severe flooding in coastal areas, particularly when the storm tide coincides with the normal high tides. Because much of the United States' densely populated Atlantic and Gulf Coast coastlines lie less than 10 feet above mean sea level, the danger from storm tides is tremendous.

The level of surge in a particular area is also determined by the slope of the continental shelf. A shallow slope off the coast will allow a greater surge to inundate coastal communities. Communities with a steeper continental shelf will not see as much surge inundation, although large breaking waves can still present major problems. Storm tides, waves, and currents in confined harbors severely damage ships, marinas, and pleasure boats.

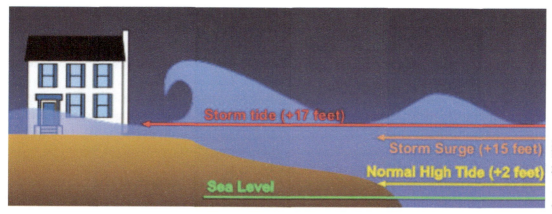

FIGURE 6.35 Storm surge.

6.7 Floods

Floods are one of the most common hazards in the United States, however not all floods are alike. Some floods develop slowly, while others such as flash floods, can develop in just a few minutes and without visible signs of rain. Additionally, floods can be local, impacting a neighborhood or community, or very large, affecting entire river basins and multiple states.

Flash floods can occur within a few minutes or hours of excessive rainfall, a dam or levee failure, or a sudden release of water held by an ice jam. Flash floods often have a dangerous wall of roaring water carrying rocks, mud and other debris. Overland flooding, the most common type of flooding event typically occurs when waterways such as rivers or streams overflow their

FIGURE 6.36 A car is submerged in flood waters.

banks as a result of rainwater or a possible levee breach and cause flooding in surrounding areas. It can also occur when rainfall or snowmelt exceeds the capacity of underground pipes, or the capacity of streets and drains designed to carry flood water away from urban areas.

Be aware of flood hazards no matter where you live or work, but especially if you are in low-lying areas, near water, behind a levee or downstream from a dam. Even very small streams, gullies, creeks, culverts, dry streambeds or low-lying ground that appear harmless in dry weather can flood.

Causes of Flooding

Tropical Storms and Hurricanes - Hurricanes pack a triple punch: high winds, soaking rain, and flying debris. They can cause storm surges to coastal areas, as well as create heavy rainfall which in turn causes flooding hundreds of miles inland. While all coastal areas are at risk, certain cities are particularly vulnerable and could have losses similar to or even greater than those caused by the 2005 hurricane, Katrina, in New Orleans and Mississippi.

When hurricanes weaken into tropical storms, they generate rainfall and flooding that can be especially damaging since the rain collects in one place. In 2001, Tropical Storm Allison produced more than 30 inches of rainfall in Houston in just a few days, flooding over 70,000 houses and destroying 2,744 homes.

Spring Thaw - During the spring, frozen land prevents melting snow or rainfall from seeping into the ground. Each cubic foot of compacted snow contains gallons of water and once the snow melts, it can result in the overflow of streams, rivers, and lakes. Add spring storms to that and the result is often serious spring flooding.

Heavy Rains - Several areas of the country are at heightened risk for flooding due to heavy rains. The Northwest is at high risk due to La Niña conditions, which include snowmelts and heavy rains. And the Northeast is at high risk due to heavy rains produced from Nor'easters. This excessive amount of rainfall can happen throughout the year, putting your property at risk.

West Coast Threats - Although floods can occur throughout the year, the West Coast rainy season usually lasts from November to April. This window increases the chance of heavy flooding and flash flood risks.

Wildfires have dramatically changed the landscape and ground conditions on the West Coast, causing fire-scorched land to develop in to mudflows under heavy rain. Experts believe it will take years for the vegetation to be fully restored, which in turn will help stabilize these areas.

In addition to the heavy rains and wildfires, the West Coast has thousands of miles of levees, which were constructed to help protect homes and land in case of a flood. However, levees are not fail-proof and can, weaken, or overtop when waters rise, often causing catastrophic results.

Levees & Dams - Levees (Figure 6.37) are designed to protect hold back a certain level of water. However, levees can and do fail; and when they fail, they can fail catastrophically. Weakening of levees over time, or as a result of weather events exceeding the levee's level of support, can cause the levee to be overtopped or breached, thus increasing the chance for flooding. Homeowners and renters insurance policies usually do not cover flood loss, therefore FEMA strongly encourages those who live and work behind levees to consider flood insurance as a dependable financial security from a flood event.

Flash Floods - Flash floods are the #1 weather-related killer in the U.S. since they can roll boulders, tear out trees, and destroy buildings and bridges. A flash flood is a rapid flooding of low-lying areas in less than six hours, which is caused by intense rainfall from a thunderstorm or several thunderstorms. Flash floods can also occur from the collapse of a man-made structure or ice dam.

New Development - Construction and development can change the natural drainage and create brand new flood risks. That's because new buildings, parking lots, and roads mean less land to absorb excess precipitation from heavy rains, hurricanes, and tropical storms.

Thunderstorm Hazards - Flash Floods

Except for heat related fatalities, more deaths occur from flooding than any other hazard. Why? Most people fail to realize the power of water. For example, six inches of fast-moving flood water can knock you off your feet.

While the number of fatalities can vary dramatically with weather conditions from year to year, the national 30-year average for flood deaths is 127. That compares with a 30-year average of 73 deaths for lightning, 68 for tornadoes and 16 for hurricanes.

National Weather Service data also shows:

- Nearly half of all flash flood fatalities are vehicle-related,
- The majority of victims are males, and
- Flood deaths affect all age groups.

Most flash floods are caused by slow moving thunderstorms, thunderstorms that move repeatedly over the same area or heavy rains from tropical storms and hurricanes. These floods can develop within minutes or hours depending on the intensity and duration of the rain, the topography, soil conditions and ground cover.

Sky Light Pictures/Shutterstock

FIGURE 6.37 Kananaskis Dam, Alberta, Canada.

Flash floods can roll boulders, tear out trees, destroy buildings and bridges, and scour out new channels. Rapidly rising water can reach heights of 30 feet or more. Furthermore, flash flood-producing rains can also trigger catastrophic mud slides.

Occasionally, floating debris or ice can accumulate at a natural or man-made obstruction and restrict the flow of water. Water held back by the ice jam or debris dam can cause flooding upstream. Subsequent flash flooding can occur downstream if the obstruction should suddenly release.

Flood Hazard Terms

Familiarize yourself with these terms to help identify a flood hazard:

Flood Watch - Flooding is possible. Tune in to NOAA Weather Radio, commercial radio or television for information.

Flash Flood Watch - Flash flooding is possible. Be prepared to move to higher ground; listen to NOAA Weather Radio, commercial radio or television for information.

Flood Warning - Flooding is occurring or will occur soon; if advised to evacuate, do so immediately.

Flash Flood Warning - A flash flood is occurring; seek higher ground immediately.

Charlie Edward/Shutterstock

FIGURE 6.38 Streets in York, England flooded by inland flooding.

BestPhotoPlus/Shutterstock

FIGURE 6.39 A snowstorm blankets cars.

Driving: Flood Facts

The following are important points to remember when driving in flood conditions:

- Six inches of water will reach the bottom of most passenger cars causing loss of control and possible stalling.
- A foot of water will float many vehicles.
- Two feet of rushing water can carry away most vehicles including sport utility vehicles (SUV's) and pick-ups.
- Do not attempt to drive through a flooded road. The depth of water is not always obvious. The road bed may be washed out under the water, and you could be stranded or trapped.
- Do not drive around a barricade. Barricades are there for your protection. Turn around and go the other way.
- Do not try to take short cuts. They may be blocked. Stick to designated evacuation routes.
- Be especially cautious driving at night when it is harder to recognize flood dangers.

Inland Flooding

In addition to the storm surge and high winds, tropical cyclones threaten the United States with their torrential rains and flooding. Even after the wind has diminished, the flooding potential of these storms remains for several days.

Since 1970, nearly 60% of the 600 deaths due to floods associated with tropical cyclones occurred inland from the storm's landfall. Of that 60%, almost a fourth (23%) of U.S. tropical cyclone deaths occur to people who drown in, or attempting to abandon, their cars.

Also, over three-fourths (78%) of children killed by tropical cyclones drowned in freshwater floods. In fact, fatalities occur because people underestimate the power of moving water and *purposely walk or drive into flooding conditions.*

It is common to think the stronger the storm the greater the potential for flooding. However, this is not always the case. A weak, slow moving tropical storm can cause more damage due to flooding than a more powerful fast moving hurricane. This was very evident with Tropical Storm Allison in June 2001.

Allison, the first named storm of the 2001 Atlantic Hurricane Season, devastated portions of Southeast Texas, including the Houston Metro area and surrounding communities, with severe flooding. Allison spent five days over Southeast and East Texas and dumped record amounts of rainfall across the area. Allison deposited **up to three feet of rain** to the east and northeast of Houston, Texas during a 5-day period.

6.8 Dangerous Winds

Hurricanes are known for their damaging wind. They are rated in strength by their wind also. However, when the NWS's National Hurricane Center issues a statement concerning the wind and catagory, that value is for *sustained* wind only. This hurricane scale does not include gusts or squalls.

Gusts are short but rapid bursts in wind speed and are primarily caused by turbulence over land mixing faster air aloft to the surface. Squalls, on the other hand, are longer periods of increased wind speeds and are generally associated with the bands of thunderstorms which make-up the spiral bands around the hurricane.

A tropical cyclone's wind damages and destroys structures two ways. First, many homes are damaged or destroyed when the high wind simply lifts the roof off of the dwellings. The process involved is called Bernoulli's Principle which implies the faster the air moves the lower the pressure within the air becomes. The high wind moving over the top of the roof creates lower pressure on the exposed side of the roof relative to the attic side.

The higher pressure in the attic helps lift the roof. Once lifted, the roof acts as a sail and is blown clear of the dwelling. With the roof gone, the walls are much easier to be blown down by the hurricane's wind.

The second way the wind destroys buildings can also be a result of the roof becoming airborne. The wind picks up the debris (i.e. wood, metal siding, toys, trash cans, tree branches, etc.) and sends them hurling at high speeds into other structures. Based on observations made during damage investigations conducted by the Wind Science and Engineering Research Center at Texas Tech University, researchers realized that much of the damage in windstorms is caused by flying debris.

They found, based on damage investigations, sections of wooden planks are the most typical type of debris observed due to tornado. A 15-lb 2×4 timber plank in a 250 mph (400 km/h) wind would travel at 100 mph (161 km/h). While 250 mph (400 km/h) is considerably more than even the strongest hurricane's sustained wind, the wind in squalls and tornadoes, could easily reach that speed.

6.9 Winter Storms & Wind Chill

While the danger from winter weather varies across the country, nearly all Americans, regardless of where they live, are likely to face some type of severe winter weather at some point in their lives. Winter storms can range from a moderate snow over a few hours to a blizzard with blinding, wind-driven snow that lasts for several days. Many winter storms are accompanied by dangerously low temperatures and sometimes by strong winds, icing, sleet and freezing rain.

One of the primary concerns is the winter weather's ability to knock out heat, power and communications services to your home or office, sometimes for days at a time. Heavy snowfall and extreme cold can immobilize an entire region.

The National Weather Service refers to winter storms as the "Deceptive Killers" because most deaths are indirectly related to the storm. Instead, people die in traffic accidents on icy roads and of hypothermia from prolonged exposure to cold. It is important to be prepared for winter weather before it strikes.

Know the Terms

It's a good idea to know the terms used to describe changing winter weather conditions and what actions to take. These terms can be used to determine the timeline and severity of an approaching storm. (Advisory / Watch / Warning). The U.S. National Weather Service (NWS) also issues advisories and warnings for other winter weather, including blizzards, freezes, wind chill, lake effect snow, and dense fog. Be alert to weather reports and tune in for specific guidance when these conditions develop.

Freezing Rain - Rain that freezes when it hits the ground, creating a coating of ice on roads, walkways, trees and power lines.

Sleet - Rain that turns to ice pellets before reaching the ground. Sleet also causes moisture on roads to freeze and become slippery.

Wind Chill- Windchill is the temperature it "feels like" when you are outside. The NWS provides a Windchill Chart to show the difference between air temperature and the perceived temperature and the amount of time until frostbite occurs.

Winter Weather Advisory - Winter weather conditions are expected to cause significant inconveniences and may be hazardous. When caution is used, these situations should not be life threatening. The NWS issues a winter weather advisory when conditions are expected to cause significant inconveniences that may be hazardous. If caution is used, these situations should not be life-threatening.

Winter Storm Watch - A winter storm is possible in your area. Tune in to NOAA Weather Radio, commercial radio, or television for more information. The NWS issues a winter storm watch when severe winter conditions, such as heavy snow and/or ice, may affect your area but the location and timing are still uncertain. A winter storm watch is issued 12 to 36 hours in advance of a potential severe storm. Tune in to NOAA Weather Radio, local radio, TV, or other news sources for more information. Monitor alerts, check your emergency supplies, and gather any items you may need if you lose power.

Winter Storm Warning - A winter storm is occurring or will soon occur in your area.

Blizzard Warning - Sustained winds or frequent gusts to 35 miles per hour or greater and considerable amounts of falling or blowing snow (reducing visibility to less than a quarter mile) are expected to prevail for a period of three hours or longer.

Frost/Freeze Warning - Below freezing temperatures are expected.

Wind Chill

Just as there are persistent hot places around the world, there are persistent cold places. The cold air alone can be deadly but when the air is moving if feels much colder. The **wind chill** is the effect of the wind on people and animals. The wind chill temperature is based on the rate of heat loss from exposed skin caused by wind and cold and is to give you an approximation of how cold the air feels on your body.

As the wind increases, it removes heat from the body, driving down skin temperature and eventually the internal body temperature. If the temperature is 0°F (-18°C) and the wind is blowing at 15 mph (13 kts / 24 kp/h), the wind chill temperature is -19°F (-28°C). At this level, exposed skin can freeze in just a few minutes.

The only effect wind chill has on inanimate objects, such as car radiators and water pipes, is to shorten the amount of time for the object to cool. The inanimate object *will not* cool below the actual air temperature. For example, if the temperature outside is -5°F (-21°C) and the wind chill temperature is -31°F (-35°C), then your car's radiator temperature will be no lower than the air temperature of -5°F (-21°C).

What is important about the wind chill besides feeling colder than the actual air temperature? The lower the wind chill temperature, the greater you are at risk for developing frost bite and/or hypothermia.

Frostbite occurs when your body tissue freezes. The most susceptible parts of the body are fingers, toes, ear lobes, and the tip of the nose. Hypothermia occurs when body core temperature, normally around 98.6°F (37°C) falls below 95°F (35°C).

The best way to avoid hypothermia and frostbite is to stay warm and dry indoors. When you must go outside, dress appropriately. Wear several layers of loose-fitting, lightweight, warm clothing. Trapped air between the layers will insulate you. Remove layers to avoid sweating and subsequent chill.

Outer garments should be tightly woven, water repellant, and hooded. Wear a hat, because half of your body heat can be lost from your head. Cover your mouth to protect your lungs from extreme cold. Mittens, snug at the wrist, are better than gloves. Try to stay dry and out of the wind.

6.10 Wildfires

Wildfire Hazards

A Mounting Threat

Wildfires are a growing natural hazard in most regions of the United States, posing a threat to life and property, particularly where native ecosystems meet developed areas.

TABLE 6.4   **Wildfire Impacts**

Wildfire Impacts
The Federal Government annually spends billions of dollars to suppress wildfires.
Increase the potential for flooding, debris flows, and landslides.
Smoke and other emissions contain pollutants that can cause significant health problems.
Short-term effects: destruction of timber, forage, wildlife habitats, scenic vistas, and watersheds
Long-term effects: reduced access to recreational areas; destruction of community infrastructure and cultural and economic resources

However, because fire is a natural (and often beneficial) process, fire suppression can lead to more severe fires due to the buildup of vegetation, which creates more fuel.

In addition, the secondary effects of wildfires, including erosion, landslides, introduction of invasive species, and changes in water quality, are often more disastrous than the fire itself.

Science Can Meet the Challenge

The U.S. Geological Survey (USGS) provides tools and information by identifying wildfire risks, ways to reduce wildfire hazards, providing real-time firefighting support, and assessing the aftermath of wildfires. The goal is to build more resilient communities and ecosystems.

The USGS conducts vegetation and fuels mapping to support firefighting readiness, reduce wildfire hazards in the wildland-urban interface, and assess wildfire effects on ecosystems.

To determine how current conditions differ from natural wildlfire circumstances, the USGS studies historical fire patterns—their size, how they started, how hot they burned, and what time of year they occurred.

The USGS is also developing methods to monitor the effectiveness of treatments to reduce fuel buildup, the effects of wildfire on wildlife, and the ecological effects of fuel-reduction measures and postfire rehabilitation treatments.

Land managers use this information to determine fire risk, plan fuel treatments, and develop emergency response plans.

After the Flames, the Risk Remains

The less obvious but equally devastating effects of wildfires occur after the fire is extinguished. These aftereffects include erosion, landslides, debris flows, and altered water quality.

The risk of floods and debris flows increases due to the exposure of bare ground and the loss of vegetation. Sediment, burned debris, and chemicals affect water quality as well.

USGS Rapid-Deployment Data-Collection Networks provide critical information for postfire flood and debrisflow warnings and on the response of eroded, burned areas. This information helps emergency management officials with emergency response, postfire mitigation, and rehabilitation planning.

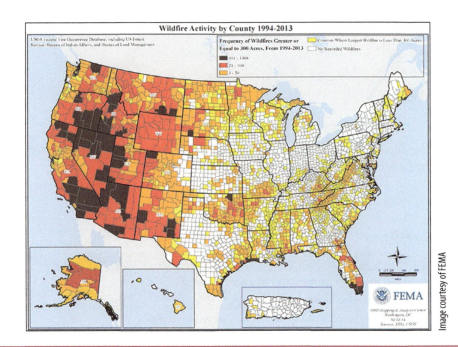

FIGURE 6.40 This map shows that wildfires are more prevalent in the Western U.S., where conditions are drier.

Fire Ecology

Wildland fires are an important ecosystem process throughout the western United States. Coniferous forests have long been subject to a frequent fire regime of low-intensity fires, which played an important role in reducing hazardous fuels and in rejuvenating the forests. In chaparral shrublands of California, high-intensity crown fires have been a strong force guiding the evolution of plant life, and regulator of ecological communities. In many desert habitats, fires have been far less frequent, and often are a more severe disturbance. Today the natural role of fire in these ecosystems is complicated by the fact that fire potentially favors plant invasions and these aliens in turn may alter fire regimes. To restore more normal fire dynamics to a particular region, managers need to know how fire has historically affected the local system, and how it functions today.

6.11 Landslides: Weather or Geologic Hazard

On March 22, 2014 a massive landslide near Oso, Washington killed 43 people. An entire hillside sloughed into the valley below, burying people and houses in its path. Landslides are one of those hazards that have both geologic and weather factors that influence when and where they take place. These factors include the slope of the ground, the amount of water in the ground, the amount of weight on top of the slope, and geologic events such as earthquakes that can trigger a slide.

Wildfire Facts
More land has been affected by wildfires in recent years than at any time since the 1960s. In 2004, wildfires burned more than 8 million acres in 40 States.
The greater Yellowstone National Park fire of 1988 burned more than 1.2 million acres.
Wildfire severity has increased and fire frequency has decreased during the past 200 years.
Many species depend on wildfires to improve habitat, recycle nutrients, and maintain diverse communities.
Land management agencies light "prescribed fires" under controlled conditions for specific management objectives.

guentermanaus/Shutterstock

FIGURE 6.41 Landslides wipe out roads without warning.

Landslides occur in all U.S. states and territories and can be caused by a variety of factors including earthquakes, storms, volcanic eruptions, fireand by human modification of land. Landslides can occur quickly, often with little notice and the best way to prepare is to stay informed about changes in and around your home that could signal that a landslide is likely to occur.

In a landslide, masses of rock, earth or debris move down a slope. Debris and mud flows are rivers of rock, earth, and other debris saturated with water. They develop when water rapidly accumulates in the ground, during heavy rainfall or rapid snowmelt, changing the earth into a flowing river of mud or "slurry." They can flow rapidly, striking with little or no warning at avalanche speeds. They also can travel several miles from their source, growing in size as they pick up trees, boulders, cars and other materials.

Landslide problems can be caused by land mismanagement, particularly in mountain, canyon and coastal regions. In areas burned by forest and brush fires, a lower threshold of precipitation may initiate landslides. Land-use zoning, professional inspections, and proper design can minimize many landslide, mudflow, and debris flow problems.

6.12 Geologic-Related Hazards

Geologic hazards also tend to be dramatic and often even more unpredictable than weather-related hazards. Our Earth is a dynamic planet, with shifting tectonic plates and hot spots where magma flows to the surface. While scientists are working hard on predicting geologic events that pose hazards to human life, there is still a long way to go. In the meantime, we should know the hazards where we live and how to prepare for them.

6.13 Earthquakes

One of the most frightening and destructive phenomena of nature is a severe earthquake and its terrible aftereffects (Figures 6.42 & 6.43). An earthquake is a sudden movement of the Earth, caused by the abrupt release of strain that has accumulated over a long time. For hundreds of millions of years, the forces of plate tectonics have shaped the Earth as the huge plates that form the Earth's surface slowly move over, under, and past each other. Sometimes the movement is gradual. At other times, the plates are locked together, unable to release the accumulating energy. When the accumulated energy grows strong enough, the plates break free. If the earthquake occurs in a populated area, it may cause many deaths and injuries and extensive property damage.

Today we are challenging the assumption that earthquakes must present an uncontrollable and unpredictable hazard to life and property. Scientists have begun to estimate the locations and likelihoods of future damaging earthquakes. Sites of greatest hazard are being identified, and definite progress is being made in designing structures that will withstand the effects of earthquakes.

Image Courtesy of USGS

FIGURE 6.42   The power of an earthquake.

Where Earthquakes Occur

The Earth is formed of several layers that have very different physical and chemical properties. The outer layer, which averages about 70 kilometers in thickness, consists of about a dozen large, irregularly shaped plates that slide over, under and past each other on top of the partly molten inner layer (Figure 6.44). Most earthquakes occur at the boundaries where the plates meet. In fact, the locations of earthquakes and the kinds of ruptures they produce help scientists define the plate boundaries.

There are three types of plate boundaries: spreading zones, transform faults, and subduction zones (Figure 6.45). At spreading zones, molten rock rises, pushing two plates apart and adding new material at their edges. Most spreading zones are found in oceans; for example, the North American and Eurasian plates are spreading apart along the mid-Atlantic ridge. Spreading zones usually have earthquakes at shallow depths (within 30 kilometers of the surface).

Image Courtesy of USGS

FIGURE 6.43 Many buildings in Charleston, South Carolina, were damaged or destroyed by the large earthquake that occured August 31, 1886.

How Earthquakes Happen

An earthquake is the vibration, sometimes violent, of the Earth's surface that follows a release of energy in the Earth's crust. This energy can be generated by a sudden dislocation of segments of the crust, by a volcanic eruption, or event by manmade explosions. Most destructive quakes, however, are caused by dislocations of the crust. The crust may first bend and then, when the stress exceeds the strength of the rocks, break and "snap" to a new position. In the process of breaking, vibrations called "seismic waves" are generated. These waves travel outward from the source of the earthquake along the surface and through the Earth at varying speeds depending on the material through which they move. Some of the vibrations are of high enough frequency to be audible, while others are of very low frequency. These vibrations cause the entire planet to quiver or ring like a bell or tuning fork.

A *fault* is a fracture in the Earth's crust along which two blocks of the crust have slipped with respect to each other. Faults are divided into three main groups, depending on how they move. *Normal faults* occur in response to pulling or tension; the overlying block moves down the dip of the fault plane. *Thrust (reverse) faults* occur

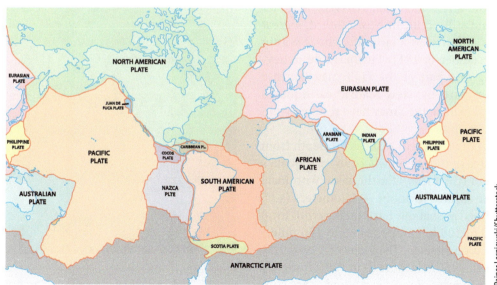

Rainer Lesniewski/Shutterstock

FIGURE 6.44 Plate boundaries.

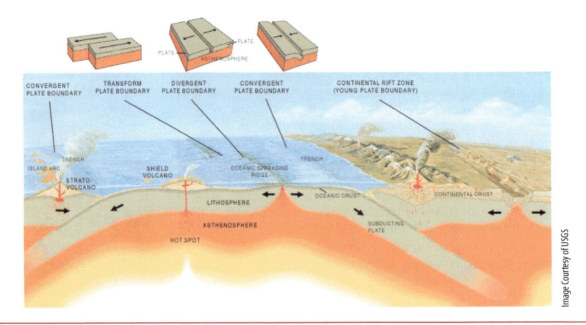

FIGURE 6.45 Types of plate boundaries.

in response to squeezing or compression; the overlying block moves up the dip of the fault plane. *Strike-slip (lateral) faults* occur in response to either type of stress; the blocks move horizontally past one another. Most faulting along spreading zones is normal, along subduction zones is thrust, and along transform faults is strike-slip.

Geologists have found that earthquakes tend to reoccur along faults, which reflect zones of weakness in the Earth's crust. Even if a fault zone has recently experienced an earthquake, however, there is no guarantee that all the stress has been relieved. Another earthquake could still occur. In New Madrid, a great earthquake was followed by a large aftershock within 6 hours on December 6, 1811. Furthermore, relieving stress along one part of the fault may increase stress in another part; the New Madrid earthquakes in January and February 1812 may have resulted from this phenomenon.

The *focal depth* of an earthquake is the depth from the Earth's surface to the region where an earthquake's energy originates (the focus). Earthquakes with focal depths from the surface to about 70 kilometers (43.5 miles) are classified as shallow. Earthquakes with focal depths from 70 to 300 kilometers (43.5 to 186 miles) are classified as intermediate. The focus of deep earthquakes may reach depths of more than 700 kilometers (435 miles). The focuses of most earthquakes are concentrated in the crust and upper mantle. The depth to

FIGURE 6.46 An aerial view of the San Andreas fault in the Carrizo Plain, Central California.

the center of the Earth's core is about 6,370 kilometers (3,960 miles), so event the deepest earthquakes originate in relatively shallow parts of the Earth's interior.

The *epicenter* of an earthquake is the point on the Earth's surface directly above the focus. The location of an earthquake is commonly described by the geographic position of its epicenter and by its focal depth.

Earthquakes beneath the ocean floor sometimes generate immense sea waves or tsunamis (Japan's dread "huge wave"). These waves travel across the ocean at speeds as great as 960 kilometers per hour (597 miles per hour) and may be 15 meters (49 feet) high or higher by the time they reach the shore. During the 1964 Alaskan earthquake, tsunamis engulfing coastal areas caused most of the destruction at Kodiak, Cordova, and Seward and caused severe damage along the west coast of North America, particularly at Crescent City, California. Some waves raced across the ocean to the coasts of Japan.

Liquefaction, which happens when loosely packed, water-logged sediments lose their strength in response to strong shaking, causes major damage during earthquakes. During the 1989 Loma Prieta earthquake, liquefaction of the soils and debris used to fill in a lagoon caused major subsidence, fracturing, and horizontal sliding of the ground surface in the Marina district in San Francisco.

Landslides triggered by earthquakes often cause more destruction than the earthquakes themselves. During the 1964 Alaska quake, shock-induced landslides devastated the Turnagain Heights residential development and many downtown areas in Anchorage. An observer gave a vivid report of the breakup of the unstable earth materials in the Turnagain Heights region: *I got out of my car, ran northward toward my driveway, and then saw that the bluff had broken back approximately 300 feet southward from its original edge. Additional slumping of the bluff caused me to return to my car and back southward approximately 180 feet to the corner of McCollie and Turnagain Parkway. The bluff slowly broke until the corner of Turnagain Parkway and McCollie had slumped northward.*

6.14 Volcanoes

A volcano is a mountain that opens downward to a reservoir of molten rock below the surface of the earth. Unlike most mountains, which are pushed up from below, volcanoes are vents through which molten rock escapes to the earth's surface. When pressure from gases within the molten rock becomes too great, an eruption occurs. Eruptions can be quiet or explosive. There may be lava flows, flattened landscapes, poisonous gases, and flying rock and ash that can sometimes travel hundreds of miles downwind.

Because of their intense heat, lava flows are great fire hazards. Lava flows destroy everything in their path, but most move slowly enough that people can move out of the way.

Fresh volcanic ash, made of pulverized rock, can be abrasive, acidic, gritty, gassy and odorous. While not immediately dangerous to most adults, the acidic gas and ash can cause lung damage to small infants, to older adults and to those suffering from severe respiratory illnesses. Volcanic ash also can damage machinery, including engines and electrical equipment. Ash accumulations mixed with water become heavy and can collapse roofs. Volcanic ash can affect people hundreds of miles away from the cone of a volcano.

Sideways directed volcanic explosions, known as "lateral blasts," can shoot large pieces of rock at very high speeds for several miles. These explosions can kill by impact, burial or heat. They have been known to knock down entire forests.

Volcanic eruptions can be accompanied by other natural hazards, including earthquakes, mudflows and flash floods, rock falls andlandslides, acid rain, fire, and (under special conditions) tsunamis.

Active volcanoes in the U.S. are found mainly in Hawaii, Alaska and the Pacific Northwest. The danger area around a volcano covers approximately a 20-mile radius however some danger may exist 100 miles or more from a volcano.

Volcanic eruptions are one of Earth's most dramatic and violent agents of change. Not only can powerful explosive eruptions drastically alter land and water for tens of kilometers around a volcano, but tiny liquid droplets of sulfuric acid erupted into the stratosphere can change our planet's climate temporarily. Eruptions often force people living near volcanoes to abandon their land and homes, sometimes forever. Those living

Pablo Hidalgo-Fotos593/Shutterstock

FIGURE 6.47 Tungurahua Volcano eruption, Ecuador. This is one of South America's most active volcanoes.

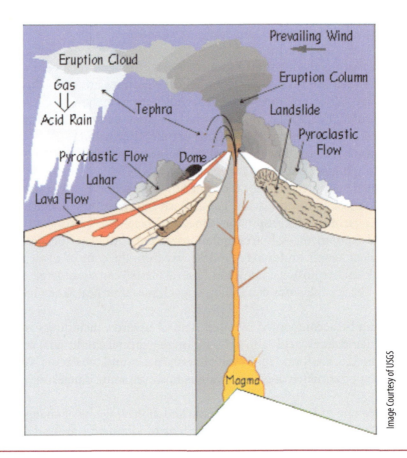

Image Courtesy of USGS

FIGURE 6.48 Volcano Hazards.

farther away are likely to avoid complete destruction, but their cities and towns, crops, industrial plants, transportation systems, and electrical grids can still be damaged by tephra, ash, lahars, and flooding.

Fortunately, volcanoes exhibit precursory unrest that if detected and analyzed in time allows eruptions to be anticipated and communities at risk to be forewarned with reliable information in sufficient time to implement response plans and mitigation measures.

6.15 Tsunamis

Tsunamis (pronounced soo-ná-mees), also known as seismic sea waves (mistakenly called "tidal waves"), are a series of enormous waves created by an underwater disturbance such as an earthquake, landslide, volcanic eruption, or meteorite. A tsunami can move hundreds of miles per hour in the open ocean and smash into land with waves as high as 100 feet or more.

From the area where the tsunami originates, waves travel outward in all directions. Once the wave approaches the shore, it builds in height. The topography of the coastline and the ocean floor will influence the size of the wave. There may be more than one wave and the succeeding one may be larger than the one before. That is why a small tsunami at one beach can be a giant wave a few miles away.

All tsunamis are potentially dangerous, even though they may not damage every coastline they strike. A tsunami can strike anywhere along most of the U.S. coastline. The most destructive tsunamis [in the U.S.] have occurred along the coasts of California, Oregon, Washington, Alaska and Hawaii.

Earthquake-induced movement of the ocean floor most often generates tsunamis. If a major earthquake or landslide occurs close to shore, the first wave in a series could reach the beach in a few minutes, even before a warning is issued. Areas are at greater risk if they are less than 25 feet above sea level and within a mile of the shoreline. Drowning is the most common cause of death associated with a tsunami. Tsunami waves and the receding water are very destructive to structures in the run-up zone. Other hazards include flooding, contamination of drinking water, and fires from gas lines or ruptured tanks.

FIGURE 6.49 Tsunami warning signs are commonly seen along coastal cities, especially in the Pacific countries/states.

WELCOME TO THE BIOSPHERE

CHAPTER 7

Ecology & Ecosystems

7.1 Earth's Biomes

The diversity of life on Earth is vast. Yet ecologists have found that areas on different continents that share similar climate conditions tend to have similar ecosystem structures and functions. As a result, ecologists use the concept of a **biome** *to classify large areas of the planet into a small number of similar units. Biomes include both terrestrial (land-based) and aquatic (in water) communities. Biomes display huge differences in the number or diversity of species present and how these species interact with one another. The following section, which has been excerpted from* Habitable Planet: A Systems Approach to Environmental Science *by the Annenberg Learner Project discusses the different types of biomes and how they are classified.*

> **biome**: a major ecological community type characterized by a similar climate, soil, plants, and animals

The reading points out that scientists have determined that a handful of factors—namely temperature, availability of moisture, abundance of light, and availability of nutrients—are the key influences on the number and variety of organisms in a given ecosystem. Generally speaking, tropical regions with their warm temperatures, abundance of moisture, and relatively constant levels of daylight have the highest number and diversity of organisms. Indeed, tropical, moist forest ecosystems make up the terrestrial biome with the highest productivity and diversity of life. In contrast, polar regions with their frigid temperatures, low moisture conditions, and months of the year with little or no natural light tend to have the lowest levels of productivity and diversity. Scientists study all types of biomes in order to learn about the life cycle and optimal conditions within different types of climates.

By *the Annenberg Learner Project*

Geography has a profound impact on ecosystems because global circulation patterns and climate zones set basic physical conditions for the organisms that inhabit a given area. The most important factors are temperature ranges, moisture availability, light, and nutrient availability, which together determine what types of life are most likely to flourish in specific regions and what environmental challenges they will face.

Earth is divided into distinct climate zones that are created by global circulation patterns. The tropics are the warmest, wettest regions of the globe, while subtropical high-pressure zones create dry zones at about 308 latitude north and south. Temperatures and precipitation are lowest at the poles. These conditions create **biome**—broad geographic zones whose plants and animals are adapted to different climate patterns (Figure 7.1). Since temperature and precipitation vary by latitude, Earth's major terrestrial biomes are broad zones that stretch around the globe. Each biome contains many ecosystems (smaller communities) made up of organisms adapted for life in their specific settings.

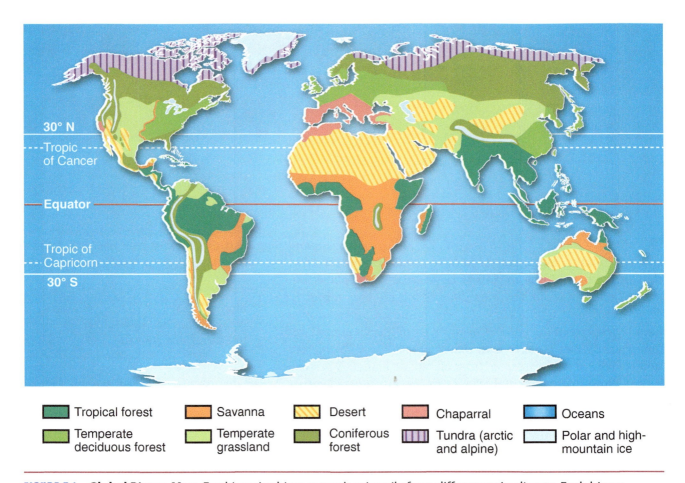

| Tropical forest | Savanna | Desert | Chaparral | Oceans |
| Temperate deciduous forest | Temperate grassland | Coniferous forest | Tundra (arctic and alpine) | Polar and high-mountain ice |

FIGURE 7.1 **Global Biome Map.** Earth's major biomes result primarily from differences in climate. Each biome contains many ecosystems made up of species adapted for life in their specific biome. Illustration by Maury Aaseng.

Land biomes are typically named for their characteristic types of vegetation, which in turn influence what kinds of animals will live there. Soil characteristics also vary from one biome to another, depending on local climate and geology.

Aquatic biomes (marine and freshwater) cover three-quarters of the Earth's surface and include rivers, lakes, coral reefs, estuaries, and open ocean. Oceans account for almost all of this area. Large bodies of water (oceans and lakes) are stratified into layers: surface waters are warmest and contain most of the available light, but depend on mixing to bring up nutrients from deeper levels. The distribution of temperature, light, and nutrients set broad conditions for life in aquatic biomes in much the same way that climate and soils do for land biomes.

Marine and freshwater biomes change daily or seasonally. For example, in the intertidal zone where the oceans and land meet, areas are submerged and exposed as the tide moves in and out. During the winter months lakes and ponds can freeze over, and wetlands that are covered with water in late winter and spring can dry out during the summer months.

There are important differences between marine and freshwater biomes. The oceans occupy large continuous areas, while freshwater habitats vary in size from small ponds to lakes covering thousands of square kilometers. As a result, organisms that live in isolated and temporary freshwater environments must be adapted to a wide range of conditions and able to disperse between habitats when their conditions change or disappear.

Biomes and Biodiversity

Since biomes represent consistent sets of conditions for life, they will support similar kinds of organisms wherever they exist, although the species in the communities in different places may not be taxonomically (the science of classifying animals) related. For example, large areas of Africa, Australia, South America, and India are covered by savannas (grasslands with scattered trees). The various grasses, shrubs, and trees that grow on savannas all are generally adapted to hot climates with distinct rainy and dry seasons and periodic fires, although they may also have characteristics that make them well-suited to specific conditions in the areas where they appear.

Species are not uniformly spread among Earth's biomes. Tropical areas generally have more plant and animal biodiversity (the diversity of animal and plant life in a region) than high latitudes, measured in species richness (the total number of species present). This pattern, known as the latitudinal biodiversity gradient, exists in marine, freshwater, and terrestrial ecosystems in both hemispheres (Figure 7.2).

Why is biodiversity distributed in this way? Ecologists have proposed a number of explanations:

- Higher productivity in the tropics allows for more species;
- The tropics were not severely affected by glaciation and thus have had more time for species to develop and adapt;
- Environments are more stable and predictable in the tropics, with fairly constant temperatures and rainfall levels year-round;
- More predators and pathogens limit competition in the tropics, which allows more species to coexist; and
- Disturbances occur in the tropics at frequencies that promote high successional diversity.

© luoman/iStock/Thinkstock

FIGURE 7.2 Tropical rainforests—the world's most species-rich biome—produce their own moisture. Scientists believe that as these ecosystems are cleared for agriculture and livestock pasture, there is a threshold beyond which they will no longer produce enough moisture to sustain themselves. The result could be conversion of rainforests to drier savannas.

Of these hypotheses, evidence is strongest for the proposition that a stable, predictable environment over time tends to produce larger numbers of species. For example, both tropical ecosystems on land and deep sea marine ecosystems—which are subject to much less physical fluctuation than other marine ecosystems, such as estuaries—have high species diversity. Predators that seek out specific target species may also play a role in maintaining species richness in the tropics.

Text of the unit "Ecosystems" by Paul R. Moorecroft from The Habitable Planet used with permission from Annenberg Learner (http://www.learner.org/).

A Closer Look at Terrestrial Biomes

When ecologists refer to biomes, they are usually referring to terrestrial (land-based) biomes. The following descriptions of Earth's terrestrial biomes are from the NASA Earth Observatory webpage. (www.earthobservatory.nasa.gov/Experiments/Biomes/)

Courtesy of NASA

FIGURE 7.3   A map and summary of Earth's terrestrial biomes.

Coniferous Forest

Protasov AN/Shutterstock

FIGURE 7.4   Coniferous forests grow in the mid-latitudes, or regions with cold winters and warm summers.

Description

Between the tundra to the north and the deciduous forest to the south lies the large area of coniferous forest. One type of coniferous forest, the northern boreal forest, is found in 50° to 60°N latitudes. Another type, temperate coniferous forests, grows in lower latitudes of North America, Europe, and Asia, in the high elevations of mountains.

Coniferous forests consist mostly of conifers, trees that grow needles instead of leaves, and cones instead of flowers. Conifers tend to be evergreen, that is, they bear needles all year long. These adaptations help conifers survive in areas that are very cold or dry. Some of the more common conifers are spruces, pines, and firs.

Precipitation in coniferous forests varies from 300 to 900 mm (12–35 inches) annually, with some temperate coniferous forests receiving up to 2,000 mm (78 inches, or 6.5 feet). The amount of precipitation depends on the forest location. In the northern boreal forests, the winters are long, cold and dry, while the short summers are moderately warm and moist. In the lower latitudes, precipitation is more evenly distributed throughout the year.

In Brief

Temperature

−40°C to 20°C (−40 to 68 degrees F), average summer temperature is 10°C

Precipitation

300 to 900 millimeters of rain per year (12–35 inches)

Vegetation

Coniferous-evergreen trees (trees that produce cones and needles; some needles remain on the trees all year long)

Location

Canada, Europe, Asia, and the United States

Other

Coniferous forest regions have cold, long, snowy winters, and warm, humid summers; well-defined seasons, at least four to six frost-free months

Courtesy of NASA

FIGURE 7.5   Location map (coniferous forests).

Temperate Deciduous Forest

FIGURE 7.6 Trees in temperature deciduous forests lose their leaves every Fall season.

Description

Temperate deciduous forests are located in the mid-latitude areas which means that they are found between the polar regions and the tropics. The deciduous forest regions are exposed to warm and cold air masses, which cause this area to have four seasons. The temperature varies widely from season to season with cold winters and hot, wet summers. The average yearly temperature is about 10°C. The areas in which deciduous forests are located get about 750 to 1,500 mm (30–60 inches) of precipitation spread fairly evenly throughout the year.

FIGURE 7.7 Location map (Temperate Deciduous Forests).

During the fall, trees change color and then lose their leaves. This is in preparation for the winter season. Because it gets so cold, the trees have adapted to the winter by going into a period of dormancy or sleep. They also have thick bark to protect them from the cold weather. Trees flower and grow during the spring and summer growing season.

Many different kinds of trees, shrubs, and herbs grow in deciduous forests. Most of the trees are broadleaf trees such as oak, maple, beech, hickory and chestnut. There are also several different kinds of plants like mountain laurel, azaleas and mosses that live on the shady forest floor where only small amounts of sunlight get through.

In Brief

Temperature

−30°C to 30°C (−20 to 85 degrees F), yearly average is 10°C, hot summers, cold winters

Precipitation

750 to 1,500 mm (30 to 60 inches) of rain per year

Vegetation

Broadleaf trees (oaks, maples, beeches), shrubs, perennial herbs, and mosses

Location

Eastern United States, Canada, Europe, China, and Japan

Other

Temperate deciduous forests are most notable because they go through four seasons. Leaves change color in autumn, fall off in the winter, and grow back in the spring; this adaptation allows plants to survive cold winters.

Desert

FIGURE 7.8   Desert landscapts are characterized by low rainfall and low vegetation densities.

Description

Desert biomes are the driest of all the biomes. In fact, the most important characteristic of a desert is that it receives very little rainfall. Most deserts receive less than 300 mm (about 12 inches, or 1 foot) a year compared to rainforests, which receive over 2,000 mm (6.5 feet). That means that the desert only gets 10 percent of the rain that a rainforest gets! The temperature in the desert can change drastically from day to night because the air is so dry that heat escapes rapidly at night. The daytime temperature averages 38°C (100 degrees F) while in some deserts it can get down to −4°C (25 degrees F) at night. The temperature also varies greatly depending on the location of the desert.

Since desert conditions are so severe, the plants that live there need to have adaptations to compensate for the lack of water. Some plants, such as cacti, store water in their stems and use it very slowly, while others like bushes conserve water by growing few leaves or by having large root systems to gather water or few leaves. Some desert plant species have a short life cycle of a few weeks that lasts only during periods of rain.

In Brief

Temperature

Average of 38°C (100 degrees F, day), average of −3.9°C (25 degrees F, night)

Precipitation

About 250 mm (about 10 inches) of rain per year

Vegetation

Cacti, small bushes, short grasses

Location

Between 15° and 35° latitude (North and South of the equator); examples are Mojave, Sonoran, Chihuahua, and Great Basin (North America); Sahara (Africa); Negev (Middle East); and Gobi (Asia)

Other

Perennials survive for several years by becoming dormant and flourishing when water is available. Annuals are referred to as ephemerals because some can complete an entire life cycle in weeks.

Courtesy of NASA

FIGURE 7.9   Location map (Deserts).

Grassland

Iakov Kalinin/Shutterstock

FIGURE 7.10 Grasslands receive more rainfall than a typical desert, but less than forests. They are often found between deserts and forests in mid-latitudes.

Description

Grasslands are generally open and continuous, fairly flat areas of grass. They are often located between temperate forests at high latitudes and deserts at subtropical latitudes. Grasses vary in size from 2.1 m (7 ft) tall with roots extending down into the soil 1.8 m (6 ft), to the short grasses growing to a height of only 20 to 25 cm (8 to 10 in) tall. These short grasses can have roots that extend 1 m (about 3 ft) deep.

The height of grass correlates with the amount of rainfall it receives. Grasslands receive about 500 to 900 mm (about 20 to 35 inches) of rain per year compared to deserts, which receive less than 300 mm (12 inches) and tropical forests, which receive more than 2,000 mm (6.5 feet). While temperatures are often extreme in some grasslands, the average temperatures are about −20°C to 30°C (−5 to 85 degrees F). Tropical grasslands have dry and wet seasons that remain warm all the time. Temperate grasslands have cold winters and warm summers with some rain.

The grasses die back to their roots annually and the soil and the sod protect the roots and the new buds from the cold of winter or dry conditions. A few trees may be found in this biome along the streams, but not many due to the lack of rainfall.

In Brief

Temperature

Dependent on latitude, yearly range can be between −20°C to 30°C (−5 to 85 degrees F)

Precipitation

About 500 to 900 mm (about 20 to 35 inches) of rain per year

Vegetation

Grasses (prairie clover, salvia, oats, wheat, barley, coneflowers)

Courtesy of NASA

Location

The prairies of the Great Plains of North America, the pampas of South America, the veldt of South Africa, the steppes of Central Eurasia, and surrounding the deserts in Australia

Other

Found on every continent except Antarctica

FIGURE 7.11 Location map (Grasslands).

Rainforest

Pablo Hidalgo-Fotos593/Shutterstock

FIGURE 7.12 Rainforests are warm and moist. Life in the rainforest is incredibly diverse.

Description

There are two types of rainforests, tropical and temperate. Tropical rainforests are found closer to the equator where it is warm. Temperate rainforests are found near the cooler coastal areas further north or south of the equator.

The tropical rainforest is a hot, moist biome where it rains all year long. It is known for its dense canopies of vegetation that form three different layers. The top layer or canopy contains giant trees that grow to heights of 75 m (about 250 ft) or more. This layer of vegetation prevents much of the sunlight from reaching the ground. Thick, woody vines are also found in the canopy. They climb trees in the canopy to reach for sunlight. The middle layer, or understory, is made up of vines, smaller trees, ferns, and palms. A large number of plants from this level are used as common houseplants. Because of the small amount of sunlight and rainfall these plants receive, they adapt easily to home environments. The bottom layer or floor of the rainforest is covered with wet leaves and leaf litter. This material decomposes rapidly in the wet, warm conditions (like a compost pile) sending nutrients back into the soil. Few plants are found on the floor of the forest due to the lack of sunlight. However, the hot, moist atmosphere and all the dead plant material create the perfect conditions in which bacteria and other microorganisms can thrive.

In Brief

Temperature

20°C to 25°C (68 to 75 degrees F), must remain warm and frost-free

Precipitation

2,000 to 10,000 millimeters (78 to 395 inches, or 6.5 to 33 feet) of rain per year

Vegetation

Vines, palm trees, orchids, ferns

FIGURE 7.13 Location map (Rainforests).

Location

Between the Tropic of Cancer and the Tropic of Capricorn

Other

There are two types of rainforests, tropical and temperate. Tropical rainforests are found closer to the equator and temperate rainforests are found farther north near coastal areas. The majority of common houseplants come from the rainforest.

Shrubland

FIGURE 7.14 Shrublands get more rain than deserts, but still not enough to support larger trees.

Description

Shrublands include regions such as chaparral, woodland and savanna. Shrublands are the areas that are located in west coastal regions between 30° and 40° North and South latitude. Some of the places would include southern California, Chile, Mexico, areas surrounding the Mediterranean Sea, and southwest parts of Africa and Australia. These regions are usually found surrounding deserts and grasslands.

Shrublands usually get more rain than deserts and grasslands but less than forested areas. Shrublands typically receive between 200 to 1,000 millimeters (about 8 to 39 inches) of rain a year. This rain is unpredictable, varying from month to month. There is a noticeable dry season and wet season.

The shrublands are made up of shrubs or short trees. Many shrubs thrive on steep, rocky slopes. There is usually not enough rain to support tall trees. Shrublands are usually fairly open so grasses and other short plants grow between the shrubs.

In the areas with little rainfall, plants have adapted to drought-like conditions. Many plants have small, needle-like leaves that help to conserve water. Some have leaves with waxy coatings and leaves that reflect the sunlight. Several plants have developed fire-resistant adaptations to survive the frequent fires that occur during the dry season.

In Brief

Temperature

Hot and dry in the summer, cool and moist in the winter

Courtesy of NASA

FIGURE 7.15 Location map (Shrublands).

Precipitation

200 to 1,000 mm (about 8 to 39 inches) of rain per year

Vegetation

Aromatic herbs (sage, rosemary, thyme, oregano), shrubs, acacia, chamise, grasses

Location

West coastal regions between 30° and 40° North and South latitude

Other

Plants have adapted to fire caused by the frequent lightning that occurs in the hot, dry summers.

Tundra

BMJ/Shutterstock

FIGURE 7.16 Soils of the tundra are frozen much of the year, and they only thaw a few inches deep during warm months. Only small plants with shallow roots can grow in the short growing season of the tundra.

Description

The tundra is the coldest of the biomes. It also receives low amounts of precipitation, making the tundra similar to a desert. Tundra is found in the regions just below the ice caps of the Arctic, extending across North America, to Europe, and Siberia in Asia. Much of Alaska and about half of Canada are in the tundra biome. Tundra is also found at the tops of very high mountains elsewhere in the world. Temperatures are frequently extremely cold, but can get warm in the summers.

Tundra winters are long, dark, and cold, with mean temperatures below 0°C (32 degrees F) for six to 10 months of the year. The temperatures are so cold that there is a layer of permanently frozen ground below the surface, called permafrost. This permafrost is a defining characteristic of the tundra biome. In the tundra summers, the top layer of soil thaws only a few inches down, providing a growing surface for the roots of vegetation.

Precipitation in the tundra totals 150 to 250 mm (about 6 to 10 inches) a year, including melted snow. That's less than most of the world's greatest deserts! Still, the tundra is usually a wet place because the low temperatures cause evaporation of water to be slow. Much of the arctic has rain and fog in the summers, and water gathers in bogs and ponds. Vegetation in the tundra has adapted to the cold and the short growing season. Mosses, sedges, and lichens are common, while few trees grow in the tundra. The trees that do manage to grow stay close to the ground so they are insulated by snow during the cold winters.

In Brief

Temperature

−40°C to 18°C (−40 to 65 degrees F)

Precipitation

150 to 250 mm (about 6 to 10 inches) of rain per year

Vegetation

Almost no trees due to short growing season and permafrost; lichens, mosses, grasses, sedges, shrubs

Location

Regions south of the ice caps of the Arctic and extending across North America, Europe, and Siberia (high mountain tops)

Other

Tundra comes from the Finnish word tunturia, meaning "treeless plain"; it is the coldest of the biomes

Courtesy of NASA

FIGURE 7.17 Location map (Tundra).

7.2 Energy Flows Through Ecosystems

photosynthesis: The biological process that captures light energy and uses it to convert carbon dioxide and water into glucose (sugars).

trophic levels: sequential stages in a food chain where organisms that are the same number of steps away from the original energy source are grouped together.

With a few exceptions, Earth's ecosystems are powered by solar energy. Primary producers such as plants and algae use sunlight in a process known as **photosynthesis** *to convert carbon dioxide and water into glucose (sugars). Glucose represents a form of stored energy that is used by plants for their own growth and maintenance. Other organisms can then consume this plant material and use it as a source of energy. Animals, in turn, can eat the organisms that ate the plants in order to acquire energy. Ecologists use the concept of* **trophic levels** *to study how energy moves through ecosystems. The following selection adapted from Habitable Planet: A Systems Approach to Environmental Science, by the Annenberg Learner Project explains how energy flows through ecosystems and discusses the impact on the environment.*

Trophic levels can be best visualized as a series of steps, with the base made up of large amounts of primary producers such as plants and algae. These primary producers have the unique ability to transform solar energy from the sun into stored energy in the form of sugars through the process of photosynthesis. Animals that feed on primary producers are known as primary consumers. An example of a primary consumer is a rabbit that eats grass and then utilizes much of the energy stored in the grass for its own growth and bodily functions. In order to sustain the rabbit there must be a huge amount of available grass for it to eat. This is why the trophic level comprised of primary producers is the largest. However, the animals that eat rabbits and other primary consumers are fewer in number than there are rabbits, so their step is smaller than the one below it that represents plants and algae.

Ecologists study dynamics between and among trophic levels as well as the concept of primary productivity to figure out how much energy is available to support the organisms within a particular ecosystem. For example, net primary productivity is the amount of energy available as plant matter for primary consumers, or the amount left over after plants use some of the energy from photosynthesis for themselves. The last section made clear that tropical, moist forests are the most productive of terrestrial biomes. That's the same as saying that tropical forests have the highest net primary productivity (NPP). Since it is the NPP of an ecosystem that supports all life at higher trophic levels, the high NPP in tropical forests explains the abundance and diversity of life in these ecosystems.

By *the Annenberg Learner Project*

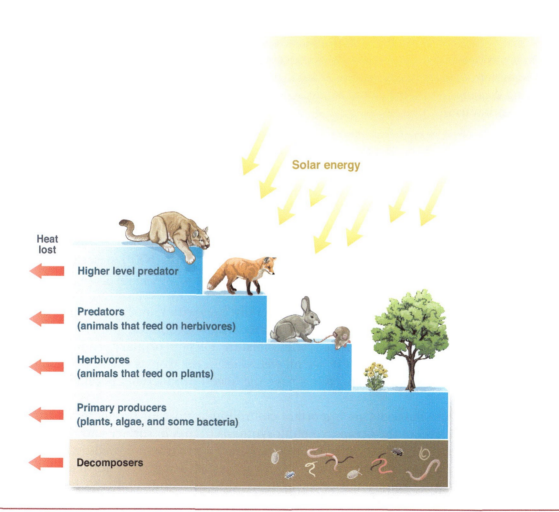

FIGURE 7.18 Trophic Levels. Energy enters an ecosystem through an external source (the sun) and flows through the progressive trophic levels of a food chain. On average, about 10% of the net energy produced at one trophic level is passed on to the next level; the rest is lost as heat energy. Illustration by Maury Aaseng.

FIGURE 7.19 The sun's energy drives the process of photosynthesis, which ultimately supports the remarkable diversity of species that inhabit Earth's ecosystems.

*E*cosystems maintain themselves by cycling energy and nutrients obtained from external sources. At the first trophic level, primary producers (plants, algae, and some bacteria) use solar energy to produce organic plant material through photosynthesis. Herbivores—animals that feed solely on plants—make up the second trophic level. Predators that eat herbivores comprise the third trophic level; if larger predators are present, they represent still higher trophic levels. Organisms that feed at several trophic levels (for example, grizzly bears that eat berries and salmon) are classified at the highest of the trophic levels at which they feed. Decomposers, which include bacteria, fungi, molds, worms, and insects, break down wastes and dead organisms and return nutrients to the soil.

On average about 10 percent of net energy production at one trophic level is passed on to the next level. Processes that reduce the energy transferred between trophic levels include respiration, growth and reproduction, defecation, and nonpredatory death (organisms that die but are not eaten by consumers). The nutritional quality of material that is consumed also influences how efficiently energy is transferred, because consumers can convert high-quality food sources into new living tissue more efficiently than low-quality food sources.

The low rate of energy transfer between trophic levels makes decomposers generally more important than producers in terms of energy flow. Decomposers process large amounts of organic material and return nutrients to the ecosystem in inorganic form, which are then taken up again by primary producers. Energy is not recycled during decomposition, but rather is released, mostly as heat (this is what makes compost piles and fresh garden mulch warm).

gross primary productivity: the total amount of new organic matter produced by photosynthesis

net primary productivity: the amount of energy that remains in an ecosystem after cellular respiration has occurred

Gross and Net Primary Productivity in Ecosystems

An ecosystem's **gross primary productivity** (GPP) is the total amount of organic matter that it produces through photosynthesis. **Net primary productivity** (NPP) describes the amount of energy that remains available for plant growth after subtracting the fraction that plants use for respiration. Productivity in land ecosystems generally rises with temperature up to about 30°C (or 85 degrees F), after which it declines, and is positively correlated [related] with moisture. On land primary productivity thus is highest in warm, wet zones in the tropics where tropical forest biomes are located. In contrast, desert scrub ecosystems have the lowest productivity because their climates are extremely hot and dry.

In the oceans, light and nutrients are important controlling factors for productivity. Light penetrates only into the uppermost level of the oceans, so photosynthesis occurs in surface and near-surface waters. Marine primary productivity is high near coastlines and other areas where upwelling brings nutrients to the surface, promoting plankton blooms. Runoff from land is also a source of nutrients in estuaries and along the continental shelves. Among aquatic ecosystems, algal beds and coral reefs have the highest net primary production, while the lowest rates occur in the open due to a lack of nutrients in the illuminated surface layers.

How many trophic levels can an ecosystem support? The answer depends on several factors, including the amount of energy entering the ecosystem, energy loss between trophic levels, and the form, structure, and

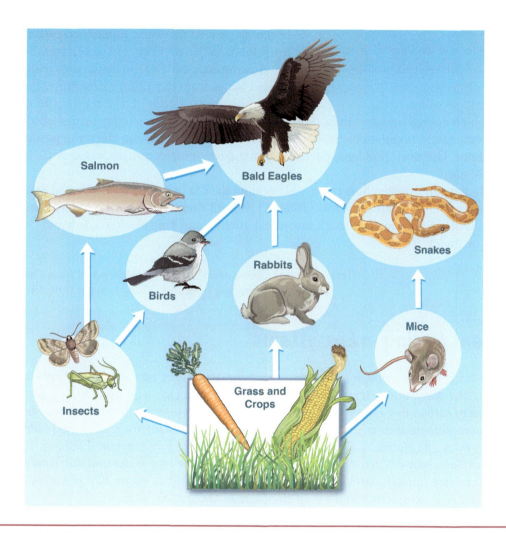

FIGURE 7.20 **Food Web.** This food web demonstrates how contaminants introduced at lower trophic levels can bioaccumulate and affect species at higher trophic levels, as was the case with eagles and raptors in the 1940s. Illustration by Maury Aaseng.

physiology [functioning] of organisms at each level. At higher trophic levels, predators generally are physically larger and are able to utilize a fraction of the energy that was produced at the level beneath them, so they have to forage over increasingly large areas to meet their caloric needs.

Because of these energy losses, most terrestrial ecosystems have no more than five trophic levels, and marine ecosystems generally have no more than seven.

Food Webs and Bioaccumulation

The simplest way to describe the flux of energy through ecosystems is as a food chain in which energy passes from one trophic level to the next, without factoring in more complex relationships between individual species. Some very simple ecosystems may consist of a food chain with only a few trophic levels. For example, the ecosystem of the remote wind-swept Taylor Valley in Antarctica consists mainly of bacteria and algae that are eaten by nematode worms. More commonly, however, producers and consumers are connected in intricate food webs with some consumers feeding at several trophic levels.

An important consequence of the loss of energy between trophic levels is that contaminants collect in animal tissues—a process called **bioaccumulation**. As contaminants bioaccumulate up the food web, organisms at

> **bioaccumulation**: the buildup of a substance, such as a toxin, in an organism's body

higher trophic levels can be threatened even if the pollutant is introduced to the environment in very small quantities.

The insecticide DDT, which was widely used in the United States from the 1940s through the 1960s, is a famous case of bioaccumulation. DDT built up in eagles and other raptors to levels high enough to affect their reproduction, causing the birds to lay thin-shelled eggs that broke in their nests. Fortunately, populations have rebounded over several decades since the pesticide was banned in the United States. However, problems persist in some developing countries where toxic bioaccumulating pesticides are still used.

Bioaccumulation can threaten humans as well as animals. For example, in the United States many federal and state agencies currently warn consumers to avoid or limit their consumption of large predatory fish that contain high levels of mercury, such as shark, swordfish, tilefish, and king mackerel, to avoid risking neurological damage and birth defects.

Text of the unit "Ecosystems" by Paul R. Moorecroft from The Habitable Planet used with permission from Annenberg Learner (http://www.learner.org/).

7.3 Nutrient Cycling in Ecosystems

Whereas energy tends to flow through ecosystems—entering as sunlight and leaving as heat—nutrients such as carbon, nitrogen, and phosphorous tend to cycle in ecosystems. Ecologists who study this cycle have learned that the same molecule of carbon that is used by a tree outside your window for photosynthesis may have been exhaled by a human or animal thousands of years ago. Ecologists also closely study the hydrologic (water) cycle, and the following excerpt from The Habitable Planet: A Systems Approach to Environmental Science, by the Annenberg Learner Project explains why ecologists must have an appreciation for how water and nutrients cycle. Indeed, the study of nutrient cycling through various ecosystems has made ecologists aware that pollution or contaminants released in one part of an ecosystem can show up elsewhere in undesirable ways.

The interesting fact about nutrient and water cycles is that we are talking about the same material cycling over time. In other words, due to conservation of matter—matter can be transformed and combined in different ways but cannot be created, nor destroyed—nutrient and water cycles are working with a fixed amount of material. Water can be transformed to ice or mist; it can end up in the ocean only to come down later as rain and soak into the ground to become groundwater; but it is always water and there is only so much of it. Likewise, a carbon molecule could be absorbed from the atmosphere by a plant during photosynthesis, transferred to a rabbit that eats the leaves of the plant, transferred to a fox that eats the rabbit, and then returned to the atmosphere when the fox exhales carbon dioxide.

This principle of conservation of matter is sometimes described by ecologists as "there is no away." When we burn fossil fuels that contain mostly carbon (such as coal or oil) we are moving that carbon from one place, where it had been buried for millions of years, to another. When we mine phosphate deposits to make fertilizer and some of that fertilizer runs off into streams we are moving that phosphorous from one place to another, but it does not go away. These and other human actions that move matter from one place to another can often have unintended consequences, as we will see in upcoming chapters.

By the Annenberg Learner Project

Along with energy, water and several other chemical elements cycle through ecosystems and influence the rates at which organisms grow and reproduce. About 10 major nutrients and six trace nutrients are essential to all animals and plants, while others play important roles for selected species. The most important biogeochemical cycles [movement of matter, such as nitrogen, between living and non-living components of an ecosystem] affecting ecosystem health are the water, carbon, nitrogen, and phosphorus cycles.

As noted earlier, most of the Earth's area that is covered by water is ocean. In terms of volume, the oceans dominate further still: nearly all of Earth's water inventory is contained in the oceans (about 97 percent) or in ice caps and glaciers (about 2 percent), with the rest divided among groundwater, lakes, rivers, streams, soils, and the atmosphere. In addition, water moves very quickly through land ecosystems. These two factors mean that water's residence time in land ecosystems is generally short, on average one or two months as soil moisture, weeks or months in shallow groundwater, or up to six months as snow cover.

But land ecosystems process a lot of water: almost two-thirds of the water that falls on land as precipitation annually is transpired [conversion of water to water vapor through plant tissue] back into the atmosphere by plants, with the rest flowing into rivers and then to the oceans. Because cycling of water is central to the functioning of land ecosystems, changes that affect the hydrologic cycle are likely to have significant impacts on land ecosystems.

Both land and ocean ecosystems are important sinks for carbon, which is taken up by plants and algae during photosynthesis and fixed as plant tissue.

Carbon cycles relatively quickly through land and surface-ocean ecosystems, but may remain locked up in the deep oceans or in sediments for thousands of years. The average residence time that a molecule of carbon spends in a terrestrial ecosystem is about 17.5 years, although this varies widely depending on the type of ecosystem: carbon can be held in old-growth forests for hundreds of years, but its residence time in heavily grazed ecosystems where plants and soils are repeatedly turned over may be as short as a few months.

Shutterstock/Iko

FIGURE 7.21 In nature, all waste is food. As dead trees fall to the ground and decompose, they release nutrients that become the next generation of trees, plants, and even animals. Nitrate may flow into a stream, where it is assimilated by a plant that is consumed by a macroinvertebrate, which is food for a fish. Perhaps the fish is eaten by an osprey, whose N-rich droppings fall to the ground and fertilize a tree. The cycle continues.

Nitrogen and Phosphorous Cycles

Nitrogen and phosphorus are two of the most essential mineral nutrients for all types of ecosystems and often limit growth if they are not available in sufficient quantities. (This is why the basic ingredients in plant fertilizer are nitrogen, phosphorus, and potassium, commonly abbreviated as NPK.)

Because atmospheric nitrogen (N_2) is inert [does not react chemically] and cannot be used directly by most organisms, microorganisms that convert it into usable forms of nitrogen play central roles in the nitrogen cycle. So-called nitrogen-fixing bacteria take inert nitrogen (N_2) from the atmosphere and convert it to ammonia (NH_4) nitrate (NO_3) and another nitrogen compounds, which in turn are taken up by plants. Some of these bacteria live in mutualistic relationships [an interaction between two species that benefits both] on the roots of plants, mainly legumes (peas and beans), and provide nitrogen directly to the plants; farmers often plant these crops to restore nitrogen to depleted soils. At the back end of the cycle, decomposers break down dead organisms and wastes, converting organic materials to inorganic nutrients (Figures 7.21 & 7.22). Other bacteria carry out denitrification, breaking down nitrate to gain oxygen and returning gaseous nitrogen to the atmosphere.

Human activities, including fossil fuel combustion, cultivation of nitrogen-fixing crops, and rising use of nitrogen fertilizer, are altering the natural nitrogen cycle. Together these activities add roughly as much nitrogen to terrestrial ecosystems each year as the amount fixed by natural processes; in other words, anthropogenic [human-caused] inputs are doubling annual nitrogen fixation in land ecosystems. The main effect of

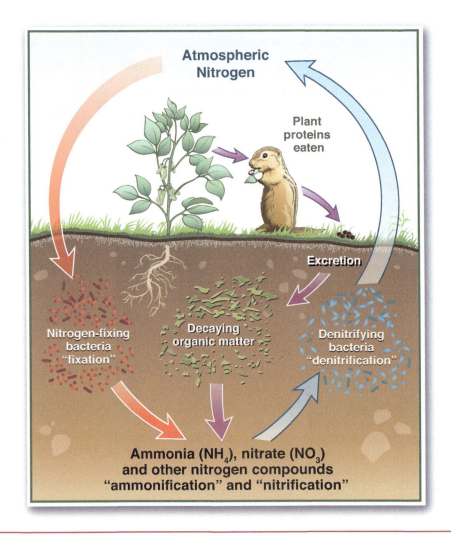

FIGURE 7.22 **The Nitrogen Cycle.** Nitrogen circulates from the environment to living organisms and back to the environment. This cycle involves nitrogen-fixing bacteria that convert nitrogen into forms usable by living organisms, and denitrifying bacteria, which break down nitrogen compounds and return gaseous nitrogen to the atmosphere. Illustration by Maury Aaseng.

this extra nitrogen is over-fertilization of aquatic ecosystems. Excess nitrogen promotes algal blooms, which then deplete oxygen from the water when the algae die and decompose. Additionally, airborne nitrogen emissions from fossil fuel combustion promote the formation of ground-level ozone, particulate emissions, and acid rain.

Phosphorus, the other major plant nutrient, does not have a gaseous phase like carbon or nitrogen. As a result it cycles more slowly through the biosphere. Most phosphorus in soils occurs in forms that organisms cannot use directly, such as calcium and iron phosphate. Usable forms (mainly orthophosphate, or PO_4) are produced mainly by decomposition [disintegration] of organic material, with a small contribution from weathering [breaking down] of rocks.

Excessive phosphorus can also contribute to over-fertilization and eutrophication [excessive growth of algae] of rivers and lakes. Human activities that increase phosphorus concentrations in natural ecosystems include fertilizer use, discharges from wastewater treatment plants, and use of phosphate detergents.

Text of the unit "Ecosystems" by Paul R. Moorecroft from The Habitable Planet used with permission from Annenberg Learner (http://www.learner.org/).

7.4 Population Biology

The excerpted selection below from The Habitable Planet: A Systems Approach to Environmental Science *by the Annenberg Learner Project explains that different organisms grow and reproduce in very different ways. Some organisms are characterized by very short life spans but high rates of reproduction, whereas others have long life spans and low rates of reproduction. The difference depends on environmental conditions in a particular area and the organism's life history strategy—such as how fast it develops, age of sexual maturity, and number of offspring. Ecologists study differences in life history strategies to better determine how to manage a species. For example, an insect pest with high rates of reproduction will require one approach to management whereas an endangered mammal with low rates of reproduction will require a different approach.*

Since ecosystems vary greatly from each other, organisms must be able to adapt to the conditions they face in order to survive. Ecologists refer to organisms that reproduce quickly as r-selected, and note that these kinds of species are found in areas that are relatively unstable, such as flood plains. K-selected species, in contrast, reproduce more slowly and are found in more stable ecosystems such as old-growth forests. Ecologists and resource managers can use their understanding of an organism's population biology, such as the degree to which it is r-selected or K-selected, to try and manage a given population. This approach can be utilized for commercial purposes, such as in the management of wild fish populations.

By *the Annenberg Learner Project*

Anup Shah/Digital Vision/Thinkstock

FIGURE 7.23 As wildlife populations grow toward their carrying capacity, competition for limited resources such as food and space increases. Populations that exceed their carrying capacity experience increased death rates, reduced birth rates, and sometimes sudden, catastrophic collapses.

*E*very organism in an ecosystem divides its energy among three competing goals: growing, surviving, and reproducing. Ecologists refer to an organism's allocation of energy among these three ends throughout its lifetime as its life history strategy. There are tradeoffs between these functions: for example, an organism that spends much of its energy on reproduction early in life will have lower growth and survival rates, and thus a lower reproductive level later in life. An optimal life history strategy maximizes the organism's contribution to population growth.

Understanding how the environment shapes organisms' life histories is a major question in ecology. Compare the conditions for survival in an unstable area, such as a flood plain near a river that frequently overflows its banks, to those in a stable environment, such as a remote old-growth forest. On the flood plain, there is a higher chance of being killed early in life, so the organisms that mature and reproduce earlier will be most likely to survive and add to population growth. Producing many offspring increases the chance that some will survive. Conversely, organisms in the forest will mature later and have lower early reproductive rates. This allows them to put more energy into growth and competition for resources.

carrying capacity: the maximum population of a particular species that an environment can support over a given period of time

Ecologists refer to organisms at the first of these two extremes (those adapted to unstable environments) as r-selected (able to reproduce quickly). These organisms live in settings where population levels are well below the maximum number that the environment can support—the **carrying capacity**—so their numbers are growing exponentially at the maximum rate at which that population can increase if resources are not limited (often abbreviated as r).

The other extreme, organisms adapted to stable environments, are termed K-selected because they live in environments in which the number of individuals is at or near the environment's carrying capacity (often abbreviated as K).

Organisms that are r-selected tend to be small, short-lived, and opportunistic, and to grow through irregular boom-and-bust population cycles. They include many insects, annual plants, bacteria, and larger species such as frogs and rats. Species considered pests typically are r-selected organisms that are capable of rapid growth when environmental conditions are favorable. In contrast, K-selected species are typically larger, grow more slowly, have fewer offspring and spend more time parenting them. Examples include large mammals, birds, and long-lived plants such as redwood trees. K-selected species are more prone to extinction than r-selected species because they mature later in life and have fewer offspring with longer gestation times.

Many organisms fall between these two extremes and have some characteristics of both types. As we will see below, ecosystems tend to be dominated by r-selected species in their early stages with the balance gradually shifting toward K-selected species.

In a growing population, survival and reproduction rates will not stay constant over time. Eventually resource limitations will reduce one or both of these variables. Populations grow fastest when they are near zero and the species is uncrowded. A simple mathematical model of population growth implies that the maximum population growth rate occurs when the population size (N) is at one-half of the environment's carrying capacity, K (i.e., at N = K/2).

In theory, if a population is harvested at exactly its natural rate of growth, the population will not change in size, and the harvest (yield) can be sustained at that level. In practice, however, it can be very hard to estimate population sizes and growth rates in the wild accurately enough to achieve this maximum sustainable yield.

Text of the unit "Ecosystems" by Paul R. Moorecroft from The Habitable Planet
used with permission from Annenberg Learner (http://www.learner.org/).

7.5 Ecosystem Functions

An organism's ability to survive and thrive depends on the availability of resources, including light, water, and nutrients. Survival also depends on how that organism interacts with other organisms and what sort of **niche** *it occupies within that ecosystem. Different species compete for limited resources, but generally over time a species will evolve to occupy a particular niche in an ecosystem. The excerpted selection below from* The Habitable Planet: A Systems Approach to Environmental Science *by the Annenberg Learner Project explains how disturbances to ecosystems, often from human activity, can disrupt the availability of resources and alter relationships between species, often with disastrous consequences.*

niche: the way of life or role of a species in an ecosystem

A key concept in understanding how ecosystems work is that of the **limiting factor**. *Consider that a plant needs sunlight, water, carbon dioxide, and certain essential nutrients to sustain it. Because a plant needs all of these factors in some combination, increasing one factor, such as carbon dioxide, may not result in increased plant growth. A more familiar analogy might be making pancakes, which requires a certain amount of flour, milk, and eggs. Doubling the amount of flour available will not result in any more pancakes unless you also increase the amount of milk and eggs.*

limiting factor: a variable that limits the reproduction, growth, and/or survival of organisms

An impact at one point in an ecosystem can have a ripple effect throughout an entire ecosystem. For example, phosphorous is often a limiting factor in many aquatic ecosystems. When excess phosphorous enters an aquatic ecosystem, such as through fertilizer runoff, it can cause an explosive growth in algae and eventually result in a sharp drop in oxygen levels in that system. This drop in oxygen can kill or drive off fish species, and this can have ripple effects throughout the entire food chain. Likewise, ripple effects can begin at higher trophic levels through the removal of a top predator in a food chain. The removal of that predator can

keystone species: a species upon which many other species depend and whose disappearance initiates significant changes in an ecosystem

lead to a population explosion at lower trophic levels, known as a trophic cascade. For this reason top predators are often considered **keystone species** *because their removal can trigger impacts throughout an ecosystem.*

Different species in an ecosystem occupy ecological niches or positions within that ecosystem. Put another way, various species pursue specific survival strategies, make use of particular resources, occupy different regions, and engage with other species in prescribed ways in order to meet their survival needs. Because many species will end up competing for the same resources, their realized niche (positions they actually occupy) will be smaller than their fundamental niche (the full range of positions they could occupy in the absence of competition). Likewise, species that are specialists that depend on a very narrow range of food sources and conditions will generally be smaller in number than those species that are generalists that can take advantage of a wider range of food sources. Specialist species are therefore more prone to ecosystem disturbances, and most endangered species tend to be specialists.

By *the Annenberg Learner Project*

A key question for ecologists studying growth and productivity in ecosystems is which factors limit ecosystem activity. Availability of resources, such as light, water, and nutrients, is a key control on growth and reproduction. Some nutrients are used in specific ratios. For example, the ratio of nitrogen to phosphorus in the organic tissues of algae is about 16 to 1, so if the available nitrogen concentration is greater than 16 times the phosphorus concentration, then phosphorus will be the factor that limits growth; if it is less, then nitrogen will be limiting. To understand how a specific ecosystem functions, it thus is important to identify what factors limit ecosystem activity.

Resources influence ecosystem activity differently depending on whether they are essential, substitutable, or complementary. Essential resources limit growth independently of other levels: if the minimum quantity needed for growth is not available, then growth does not occur. In contrast, if two resources are substitutable, then population growth is limited by an appropriately weighted sum of the two resources in the environment. For example, glucose and fructose are substitutable food sources for many types of bacteria. Resources may also be complementary, which means that a small amount of one resource can substitute for a relatively large amount of another, or can be complementary over a specific range of conditions.

Resource availability serves as a so-called "bottom-up" control on an ecosystem: the supply of energy and nutrients influences ecosystem activities at higher trophic levels by affecting the amount of energy that moves up the food chain. In some cases, ecosystems may be more strongly influenced by so-called "top-down" controls—namely, the abundance of organisms at high trophic levels in the ecosystem. Both types of effects can be at work in an ecosystem at the same time, but how far bottom-up effects extend in the food web, and the extent to which the effects of trophic interactions at the top of the food web are felt through lower levels, vary over space and time and with the structure of the ecosystem.

Trophic Cascades and Keystone Species

Many ecological studies seek to measure whether bottom-up or top-down controls are more important in specific ecosystems because the answers can influence conservation and environmental protection strategies. For example, a study by Benjamin S. Halpern [a marine ecologist at the Center for Ocean Solutions, University of California—Santa Barbara] and others of food web controls in kelp forest ecosystems off the coast of Southern California found that variations in predator abundance explained a significant proportion of variations in the abundance of algae and the organisms at higher trophic levels that fed on algae and plankton. In contrast, they found no significant relationship between primary production by algae and species abundance at higher trophic levels. The most influential predators included spiny lobster, Kellet's whelk, rockfish, and sea perch. Based on

© Nathan Hobbs/iStock/Thinkstock

FIGURE 7.24 Reintroducing wolves to Yellowstone Park has had a positive effect on the ecosystem.

these findings, the authors concluded that "[e]fforts to control activities that affect higher trophic levels (such as fishing) will have far larger impacts on community dynamics than efforts to control, for example, nutrient input, except when these inputs are so great as to create anoxic (dead) zones."

> **trophic cascades**: the cascading effect that a change in the size of one population at the top of a food web has on the populations below it

Drastic changes at the top of the food web can trigger **trophic cascades**, or domino effects that are felt through many lower trophic levels. The likelihood of a trophic cascade depends on the number of trophic levels in the ecosystem and the extent to which predators reduce the abundance of a trophic level to below their resource-limited carrying capacity. Some species are so important to an entire ecosystem that they are referred to as keystone species, connoting that they occupy an ecological niche that influences many other species. Removing or seriously impacting a keystone species produces major impacts throughout the ecosystem.

Many scientists believe that the reintroduction of wolves into Yellowstone National Park in 1995, after they had been eradicated from the park for decades through hunting, has caused a trophic cascade with results that are generally positive for the ecosystem. Wolves have sharply reduced the population of elk, allowing willows to grow back in many riparian areas [the banks of rivers or streams] where the elk had grazed the willows heavily. Healthier willows are attracting birds and small mammals in large numbers (Figure 7.24).

"Species, like riparian songbirds, insects, and in particular, rodents, have come back into these preferred habitat types, and other species are starting to respond," says biologist Robert Crabtree of the Yellowstone Ecological Research Center. "For example, fox and coyotes are moving into these areas because there's more prey for them. There's been an erupting trophic cascade in some of these lush riparian habitat sites."

Ecological Niches

Within ecosystems, different species interact in different ways. These interactions can have positive, negative, or neutral impacts on the species involved.

Each species in an ecosystem occupies a niche, which comprises the sum total of its relationships with the biotic [living] and abiotic [non-living] elements of its environment—more simply, what it needs to survive. In a 1957 address, zoologist George Evelyn Hutchinson framed the view that most ecologists use today when he defined the niche as the intersection of all of the **ranges of tolerance** under which an organism can live. This approach makes ecological niches easier to quantify and analyze because they can be described as specific ranges of variables like temperature, latitude, and altitude. For example, the African Fish Eagle occupies a very similar ecological niche to the American Bald Eagle. In practice it is hard to measure all of the variables that a species needs to survive, so descriptions of an organism's niche tend to focus on the most important limiting factors.

The full range of habitat types in which a species can exist and reproduce without any competition from other species is called its **fundamental niche**. The presence of other species means that few species live in such conditions. A species' realized niche can be thought of as its niche in practice—the range of habitat types from

> **fundamental niche**: the full range of habitat types and resources a species can possibly occupy and use without competition from other species

> **realized niche**: the range of habitat types and resources from which a species is not excluded from by competition from other species

which it is not excluded by competing species. **Realized niches** are usually smaller than fundamental niches, since competitive interactions exclude species from at least some conditions under which they would otherwise grow. Species may occupy different realized niches in various locations if some constraint, such as a certain predator, is present in one area but not in another.

In a classic set of laboratory experiments, Russian biologist G. F. Gause showed the difference between fundamental and realized niches. Gause compared how two strains of *Paramecium* grew when they were cultured separately in the same type of medium to their growth rates when cultured together. When cultured separately both strains reproduced rapidly, which indicated that they were adapted to living and reproducing under the same conditions. But when they were cultured together, one strain out-competed and eventually eliminated the other. From this work Gause developed a fundamental concept in community ecology: the competitive exclusion principle, which states that if two competitors try to occupy the same realized niche, one species will eliminate the other.

Specialists and Generalists

Many key questions about how species function in ecosystems can be answered by looking at their niches. Species with narrow niches tend to be **specialists**, relying on comparatively few food sources. As a result, they are highly sensitive to changes in key environmental conditions, such as water temperature in aquatic ecosystems.

> **specialists**: species with a narrow ecological niche who can only eat a few types of food or live in limited types of habitats

For example, pandas, which only eat bamboo, have a highly specialized diet. Many endangered species are threatened because they live or forage in particular habitats that have been lost or converted to other uses. One well-known case, the northern spotted owl lives in cavities of trees in old-growth forests (forests with trees that are more than 200 years old and have not been cut, pruned, or managed), but these forests have been heavily logged, reducing the owl's habitat.

In contrast, species with broad niches are **generalists** that can adapt to wider ranges of environmental conditions within their own lifetimes (i.e., not through evolution over generations, but rather through changes in their behavior or physiologic functioning) and survive on diverse types of prey. Coyotes once were found only on the Great Plains

> **generalists**: species with a broad ecological niche who can eat a variety of food and live in many types of environments

and in the western United States, but have spread through the eastern states in part because of their flexible lifestyle. They can kill and eat large, medium, or small prey, from deer to house cats, as well as other foods such as invertebrates [an animal without a backbone] and fruit, and can live in a range of habitats, from forests to open landscapes, farmland, and suburban neighborhoods.

Overlap between the niches of two species (more precisely, overlap between their resource use curves) causes the species to compete if resources are limited. One might expect to see species constantly dying off as a result, but in many cases competing species can coexist without either being eliminated. This happens through niche partitioning (also referred to as resource partitioning), in which two species divide a limiting resource such as light, food supply, or habitat.

Text of the unit "Ecosystems" by Paul R. Moorecroft from The Habitable Planet used with permission from Annenberg Learner (http://www.learner.org/).

7.6 Evolution and Natural Selection in Ecosystems

Just as the species within an ecosystem change and evolve, ecosystems themselves go through natural changes in a process known as **succession**. *Successional changes alter conditions for various species over time, favoring some at the expense of others. For example, a tract of forest blown over by a hurricane or burned to the ground in a wildfire will usually return to a forested state given enough time. Initially, r-selected species, those that reproduce quickly and colonize disturbed areas, will dominate the site and take advantage of the lack of competition from other species. However, over time K-selected species will move in and the disturbed ecosystem will move through successional stages to a more diverse and complex system.*

> **succession**: the sequence of changes in an ecosystem over time

It might be tempting to look at an ecosystem, such as a forest, as relatively static and unchanging over time. However, ecosystems are constantly changing. The theory of evolution, first developed by Charles Darwin and other 19th-century scientists, suggested that competition between species for scarce resources will favor some species over others. Likewise, natural selection within species is a process where certain individuals in a given population will be better suited for survival and will pass on their genes and the traits to future generations. Competition between species and natural selection within species results in an ongoing process of constant change in order to ensure the survival of a given species. For species that form a predator–prey relationship—such as bats and insects—ecologists have observed a process known as co-evolution. Co-evolution has been compared to an ecological arms race where the prey species might, over time, develop a defensive mechanism to thwart the predator that then is overcome by an evolved predator. The adapted section below from The Habitable Planet: A Systems Approach to Environmental Science by the Annenberg Learner Project explains how constant species interactions with each other and their environment in a given ecosystem result in a process of natural selection and evolution.

By *the Annenberg Learner Project*

As species interact, their relationships with competitors, predators, and prey contribute to natural selection and thus influence their evolution over many generations. To illustrate this concept, consider how evolution has influenced the factors that affect the foraging efficiency of predators. This includes the predator's search time (how long it takes to find prey), its handling time (how hard it has to work to catch and kill it), and its prey profitability (the ratio of energy gained to energy spent handling prey). Characteristics that help predators to find, catch, and kill prey will enhance their chances of surviving and reproducing. Similarly, prey will profit from attributes that help avoid detection and make organisms harder to handle or less biologically profitable to eat.

These common goals drive natural selection for a wide range of traits and behaviors, including:

- **Mimicry** by either predators or prey. A predator such as a praying mantis that blends in with surrounding plants is better able to surprise its target (Figure 7.25). However, many prey species also engage in mimicry, developing markings similar to those of unpalatable species so that predators avoid them. For example, harmless viceroy butterflies have similar coloration to monarch butterflies, which store toxins in their tissues, so predators avoid viceroy butterflies.

- Optimal foraging strategies enable predators to obtain a maximum amount of net energy per unit of time spent foraging. Predators are more likely to survive and reproduce if they restrict their diets to **prey** that provide the most energy per unit of handling time and focus on areas that are rich with prey or that are close together.

- Avoidance/escape features help prey elude predators. These attributes may be behavioral patterns, such as animal herding or fish schooling to make individual organisms harder to pick out. Markings can confuse and disorient predators: for example, the automeris moth has false eye spots on its hind wings that misdirect predators.

- Features that increase handling time help to discourage predators. Spines serve this function for many plants and animals, and shells make crustaceans and mollusks

mimicry: the ability of certain species to blend in with their environment or to develop a resemblance to other species for the purpose of hunting prey or for protection from a predator

prey: organism that is captured and serves as a source of food for another organism (predator)

FIGURE 7.25 The ability to blend in with, or mimic, its environment aids the praying mantis in capturing its prey.

harder to eat. Behaviors can also make prey harder to handle: squid and octopus emit clouds of ink that distract and confuse attackers, while hedgehogs and porcupines increase the effectiveness of their protective spines by rolling up in a ball to conceal their vulnerable underbellies.

■ Some plants and animals emit noxious chemical substances to make themselves less profitable as prey. These protective substances may be bad-tasting, antimicrobial, or toxic. Many species that use noxious substances as protection have evolved bright coloration that signals their identity to would-be predators—for example, the black and yellow coloration of bees, wasps, and yellowjackets. The substances may be generalist defenses that protect against a range of threats, or specialist compounds developed to ward off one major predator. Sometimes specialized predators are able to overcome these noxious substances: for example, ragwort contains toxins that can poison horses and cattle grazing on it, but it is the exclusive food of cinnabar moth caterpillars. Ragwort toxin is stored in the caterpillars' bodies and eventually protects them as moths from being eaten by birds.

Co-evolution and Competition

Natural selection based on features that make predators and prey more likely to survive can generate predator–prey "arms races," with improvements in prey defenses triggering counter-improvements in predator attack tools and vice versa over many generations. Many cases of predator–prey arms races have been identified. One widely known case is bats' use of echolocation [the use of echoes to determine the location of something] to find insects. Tiger moths respond by emitting high-frequency clicks to "jam" bats' signals, but some bat species have overcome these measures through new techniques such as flying erratically to confuse moths or sending echolocation chirps at frequencies that moths cannot detect. This type of pattern involving two species that interact in important ways and evolve in a series of reciprocal genetic steps is called **coevolution** and represents an important factor in adaptation and the evolution of new biological species.

> **coevolution**: the evolution of two or more species that occurs as a result of their interactions with each other over a period of time

Other types of relationship, such as competition, also affect evolution and the characteristics of individual species. For example, if a species has an opportunity to move into a vacant niche, the shift may facilitate evolutionary changes over succeeding generations because the species plays a different ecological role in the new niche. By the early 20th century, large predators such as wolves and puma had been largely eliminated from the eastern United States. This has allowed coyotes, who compete with wolves where they are found together, to spread throughout urban, suburban, and rural habitats in the eastern states, including surprising locations such as Cape Cod in Massachusetts and Central Park in New York City. Research suggests that northeastern coyotes are slightly larger than their counterparts in western states, although it is not yet clear whether this is because the northeastern animals are hybridizing [cross-breeding] with wolves and domestic dogs or because they have adapted genetically to preying on larger species such as white-tailed deer.

Natural Ecosystem Change

Just as relationships between individual species are dynamic, so too is the overall makeup of ecosystems. The process by which one natural community changes into another over a time scale of years to centuries is called succession. Common succession patterns include plant colonization of sand dunes and the regrowth of forests on abandoned farmland. While the general process is widely recognized, ecologists have offered differing views of what drives succession and how to define its end point. By analyzing the natural succession process, scientists seek to measure how stable ecosystems are at different stages in their trajectory of development, and how they respond to disturbances in their physical environment or changes in the frequency at which they are disturbed.

In the early 20th century, plant biologist Frederic Clements described two types of succession: primary (referring to colonization of a newly exposed landform, such as sand dunes or lava flows after a volcanic eruption) and secondary (describing the return of an area to its natural vegetation following a disturbance such

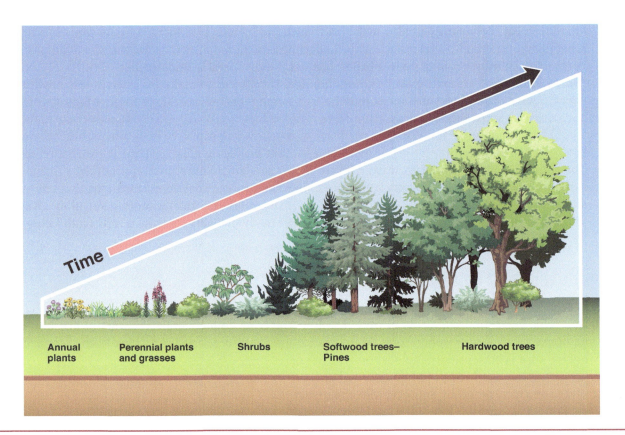

Annual
plants

Perennial plants
and grasses

Shrubs

Softwood trees–
Pines

Hardwood trees

FIGURE 7.26 Through the process of succession certain organisms that initially colonize an area are replaced over time by other organisms. Illustration by Maury Aaseng.

as fire, treefall, or forest harvesting). British ecologist Arthur Tansley distinguished between autogenic succession—change driven by the inhabitants of an ecosystem, such as forests regrowing on abandoned agricultural fields—and allogenic succession, or change driven by new external geophysical conditions such as rising average temperatures resulting from global climate change.

As discussed above, ecologists often group species depending on whether they are better adapted for survival at low or high population densities (r-selected versus K-selected). Succession represents a natural transition from r- to K-selected species. Ecosystems that have recently experienced traumatic extinction events such as floods or fires are favorable environments for r-selected species because these organisms, which are generalists and grow rapidly, can increase their populations in the absence of competition immediately after the event. Over time, however, they will be out-competed by K-selected species, which often derive a competitive advantage from the habitat modification that takes place during early stages of primary succession.

For example, when an abandoned agricultural field transitions back to forest, sun-tolerant weeds and herbs appear first, followed by dense shrubs like hawthorn and blackberry. After about a decade, birches and other small fast-growing trees move in, sprouting wherever the wind blows their lightweight seeds. In 30 to 40 years,

FIGURE 7.27 Evolution and natural selection are the drivers of species diversity—one of nature's most precious gifts, and a critical component of ecosystem stability. Scientists often find that more species-rich ecosystems are more resilient following disturbance.

Shutterstock/Ethan Daniels

slower-spreading trees like ash, red maple, and oak take root, followed by shade-tolerant trees such as beech and hemlock.

A common observation is that as ecosystems mature through successional stages, they tend to become more diverse and complex. The number of organisms and species increases and niches become narrower as competition for resources increases. Primary production rates and nutrient cycling may slow as energy moves through a longer sequence of trophic levels.

Many natural disturbances have interrupted the process of ecosystem succession throughout Earth's history, including natural climate fluctuations, the expansion and retreat of glaciers, and local factors such as fires and storms. An understanding of succession is central for conserving and restoring ecosystems because it identifies conditions that managers must create to bring an ecosystem back into its natural state. The Tallgrass Prairie National Preserve in Kansas, created in 1996 to protect 11,000 acres of prairie habitat, is an example of a conservation project that seeks to approximate natural ecosystem succession. A herd of grazing buffalo tramples on tree seedlings and digs up the ground, creating bare patches where new plants can grow, just as millions of buffalo maintained the grassland prairies that covered North America before European settlement.

Text of the unit "Ecosystems" by Paul R. Moorecroft from The Habitable Planet used with permission from Annenberg Learner (http://www.learner.org/).

CHAPTER 8

Biodiversity

8.1 Case History—The Sixth Great Extinction

Human domination of the Earth's ecosystems inevitably means we have put pressure on other species. Whether from deforestation or loss of open space due to sprawl and urban development, human actions have a profound effect on many organisms. Ecologists worry that human impacts are driving natural extinction rates—which are as low as one species per one million each year—to 100 or 1,000 times this rate. As a result, there is concern that we could be facing a Sixth Great Extinction event. In this article, Janet Larsen, Director of Research at The Earth Policy Institute, reviews the evidence for such a concern and discusses what solutions are being investigated to avoid it.

Scientists believe that since life began on Earth there have been five great extinction events—periods where a significant percentage (70–95 percent) of species present on the planet were wiped out. The first, known as the Ordovician-Silurian extinction event occurred roughly 440 million years ago. The most recent, known as the Cretaceous-Tertiary extinction event occurred 65 million years ago. It takes millions of years to bounce back from extinction events and reach comparable levels of species diversity. But under relatively stable conditions, evolutionary processes (see section 1.6) create new species faster than others go extinct—and so species diversity will increase over time. Since the last great extinction, the number of species on Earth has grown into the tens of millions. Of these, we know the most about numbers of mammals and birds, but far less about the status of fish, reptiles, amphibians, plants, and invertebrates (organisms without a backbone, such as insects). Today, as extinction rates increase and far surpass the rate at which evolution develops new species, we could be losing hundreds if not thousands of species before we have had a chance to fully understand and study their place in an ecosystem.

Unlike the first five great extinction events, which were caused by natural forces like mass volcanic eruptions and meteor strikes, the current crisis is a direct result of human actions. The most significant cause of species extinction today is habitat destruction of the kind already reviewed earlier in this chapter. Other human actions that are speeding up extinctions are pollution, over-harvesting and hunting, and the impacts of human-caused climate change.

By Janet Larsen

Almost 440 million years ago, some 85 percent of marine animal species were wiped out in the Earth's first known mass **extinction**. Roughly 367 million years ago, once again many species of fish and 70 percent of marine invertebrates perished in a major extinction event. Then about 245 million years ago, up to 95 percent of all animals—nearly the entire animal kingdom—were lost in what is thought to be the worst extinction in history.

> **extinction:** complete disappearance of a species from the Earth, which occurs when a species cannot adapt and successfully reproduce under new environmental conditions

Some 208 million years ago, another mass extinction took a toll primarily on sea creatures, but also some land animals. And 65 million years ago, three quarters of all species—including the dinosaurs—were eliminated.

Among the possible causes of these mass extinctions are volcanic eruptions, meteorites colliding with the earth, and a changing climate. After each extinction, it took upwards of 10 million years for biological richness to recover. Yet once a species is gone, it is gone forever.

Minden Pictures/SuperStock

FIGURE 8.1 Sumatran rhinos are now highly endangered due to large-scale destruction of their Indonesian rainforest habitat. They are just one of nearly 200 threatened mammal species that are critically endangered—the highest threat level.

The consensus among biologists is that we now are moving toward another mass extinction that could rival the past big five. This potential sixth great extinction is unique in that it is caused largely by the activities of a single species. It is the first mass extinction that humans will witness firsthand—and not just as innocent bystanders.

While scientists are not sure how many species inhabit the planet today, their estimates top 10 million. Yet each year thousands of species, ranging from the smallest microorganisms to larger mammals, are lost for good. Some disappear even before we know of their existence.

The average extinction rate is now some 1,000 to 10,000 times faster than the rate that prevailed over the past 60 million years. Throughout most of geological history, new species evolved faster than existing species disappeared, thus continuously increasing the planet's biological diversity. Now evolution is falling behind.

Only a small fraction of the world's plant species has been studied in detail, but as many as half are threatened with extinction. South and Central America, Central and West Africa, and Southeast Asia—all home to diverse tropical forests—are losing plants most rapidly.

Today nearly 5,500 animal species are known to be threatened with extinction. The IUCN [International Union for Conservation of Nature]—World Conservation Union's 2003 Red List survey of the world's flora and fauna shows that almost one in every four mammal species and one in eight bird species are threatened with extinction within the next several decades. (For access to IUCN's Red List of Threatened Species database, see www.redlist.org).

endangered: seriously at risk of extinction

Of 1,130 threatened mammal species, 16 percent are critically **endangered**—the highest threat level. This means that 184 mammal species have suffered extreme and rapid reduction in population or habitat and may not survive this decade. Their remaining numbers range from under a few hundred to, at most, a few thousand individuals. For birds, 182 of the 1,194 threatened species are critically endangered.

Although the status of most of the world's mammals and birds is fairly well documented, we know relatively little about the rest of the world's fauna. Only 5 percent of fish, 6 percent of reptiles, and 7 percent of amphibians have been evaluated. Of those studied, at least 750 fish species, 290 reptiles, and 150 amphibians are at risk. Worrisome signs—like the mysterious disappearance of entire amphibian populations and fishers' nets that come up empty more frequently—reveal that there may be more species in trouble. Of invertebrates, including insects, mollusks, and crustaceans, we know the least. But what is known is far from reassuring.

The Human Threat to Nature

At the advent of agriculture some 11,000 years ago, the world was home to 6 million people. Since then our ranks have grown a thousandfold. Yet the increase in our numbers has come at the expense of many other species.

The greatest threat to the world's living creatures is the degradation and destruction of habitat, affecting 9 out of 10 threatened species. Humans have transformed nearly half of the planet's ice-free land areas, with serious effects on the rest of nature. We have made agricultural fields out of prairies and forests. We have dammed rivers and drained wetlands. We have paved over soil to build cities and roads.

Each year the earth's forest cover shrinks by 16 million hectares (40 million acres), with most of the loss occurring in tropical forests, where levels of biodiversity are high. Ecologically rich wetlands have been cut in half over the past century. Other freshwater and terrestrial ecosystems have been degraded by pollution. Deserts have expanded to overtake previously vegetated areas, accelerated in some cases by overgrazing of domesticated animals.

A recent study of 173 species of mammals from around the world showed that their collective geographical ranges have been halved over the past several decades, signifying a loss of breeding and foraging area. Overall, between 2 and 10 percent of mammal populations (groups of a single species in a specific geographical location) are thought to have disappeared along with their habitat.

Direct human exploitation of organisms, such as through hunting and harvesting, threatens more than a third of the listed birds and mammals. Other threats to biodiversity include exotic species, often transported by humans, which can outcompete and displace native species.

A recent survey of some 1,100 animal and plant species found that climate change could wipe out between 15 and 37 percent of them by 2050. Yet the actual losses may be greater because of the complexity of natural systems. The extinction of key species could have cascading effects throughout the food web. As John Donne wrote, "no man is an island." The same is true for the other species we share this planet with: the loss of any single species from the web of life can affect many others.

Healthy ecosystems support us with many services—most fundamentally by supplying the air we breathe and filtering the water we drink. They provide us with food, medicine, and shelter. When ecosystems lose biological richness, they also lose resilience, becoming more susceptible to the effects of climate change, invasions of alien species, and other disturbances.

What's Next?

Consciously avoiding habitat destruction and mitigating the effects of land use change, reducing the direct exploitation of plants and wildlife, and slowing climate change can help us stop weakening the very life-support systems we depend on. While this may be the first time in history that a single species can precipitate a mass extinction event, it is also the first time in history that a single species can act to prevent it.

Adapted with permission from: Larsen, J. (2004). The Sixth Great Extinction: A Status Report. Earth Policy Institute, Plan B Updates. http://www.earth-policy.org/index.php?/plan_b_updates/2004/update35.

8.2 Tropical Deforestation

Tropical forests make up less than 10% of the Earth's land surface, yet these ecosystems are home to roughly half of the planet's biodiversity. The rapid loss of tropical forests in Latin America, Asia, and Africa is therefore of great concern to ecologists and environmental scientists. In the following article, Rebecca Lindsey, a technical writer for the National Aeronautics and Space Administration (NASA), examines the causes and impact of deforestation in tropical regions, as well as possible approaches to slow or reverse this process.

As we learned in tropical, moist forests have the highest rates of biodiversity and primary productivity of any of the world's terrestrial ecosystems. This high level of diversity provides direct benefits to humans in the form of various useful products (nuts, fruit, latex, medicines). This specialized diversity also represents a store of genetic information that could be useful to medical advances for current and future generations. Tropical forests have been likened to the Library of Congress in that they contain a vast store of information. And like this massive library, we may never know even a fraction of the information that is contained within it.

The causes of deforestation are complex and vary from region to region, but it is generally clearing forests, or cutting down trees, to open land for crops and livestock that has the biggest impact. Efforts to conserve tropical forests have focused on finding markets for products that can be harvested from standing forests. Such a sustainable use approach is based on the idea that if managed properly, forests may yield greater economic and social value over time instead of simply cutting them down. Indeed, tropical forests provide many critical ecosystem services— services provided by ecosystems that are necessary to sustain life. Perhaps the most important of these is the role that forests play in the carbon cycle and in maintaining climate.

By *Rebecca Lindsey*

Shutterstock/Frontpage

FIGURE 8.2 This is the edge of a clear-cut in the Brazilian Amazon. Tropical rainforests are the most species-rich ecosystems in the world. As they disappear, the planet and humanity alike lose nature's riches, some of which are known but others have yet to be discovered.

> **microhabitats**: a very small, specialized habitat within a larger ecosystem

> **volcanism**: any of the various processes and phenomena associated with the discharge of molten rock or hot water and steam from the Earth's surface

Stretching out from the equator on all Earth's land surfaces is a wide belt of forests of amazing diversity and productivity. Tropical forests include dense rainforests, where rainfall is abundant year-round; seasonally moist forests, where rainfall is abundant, but seasonal; and drier, more open woodlands. Tropical forests of all varieties are disappearing rapidly as humans clear the natural landscape to make room for farms and pastures, to harvest timber for construction and fuel, and to build roads and urban areas Figure 8.2. Although deforestation meets some human needs, it also has profound, sometimes devastating, consequences, including social conflict, extinction of plants and animals, and climate change—challenges that aren't just local, but global.

Impacts of Deforestation: Biodiversity Impacts

Although tropical forests cover only about 7 percent of the Earth's dry land, they probably harbor about half of all species on Earth. Many species are so specialized to **microhabitats** within the forest that they can only be found in small areas. Their specialization makes them vulnerable to extinction. In addition to the species lost when an area is totally deforested, the plants and animals in the fragments of forest that remain also become increasingly vulnerable, sometimes even committed, to extinction. The edges of the fragments dry out and are buffeted by hot winds; mature rainforest trees often die standing at the margins. Cascading changes in the types of trees, plants, and insects that can survive in the fragments rapidly reduces biodiversity in the forest that remains. People may disagree about whether the extinction of other species through human action is an ethical issue, but there is little doubt about the practical problems that extinction poses.

First, global markets consume rainforest products that depend on sustainable harvesting: latex, cork, fruit, nuts, timber, fibers, spices, natural oils and resins, and medicines. In addition, the genetic diversity of tropical forests is basically the deepest end of the planetary gene pool (Figure 8.3). Hidden in the genes of plants, animals, fungi, and bacteria that have not even been discovered yet may be cures for cancer and other diseases or the key to improving the yield and nutritional quality of foods—which the U.N. [United Nations] Food and Agriculture Organization says will be crucial for feeding the nearly ten billion people the Earth will likely need to support in coming decades. Finally, genetic diversity in the planetary gene pool is crucial for the resilience of all life on Earth to rare but catastrophic environmental events, such as meteor impacts or massive, sustained **volcanism**.

FIGURE 8.3 **Biodiversity Hotspots.** Ecologists have identified certain areas of the Earth as biodiversity "hotspots." These areas are important centers of biodiversity that contain a large number of native species that are not found anywhere else. Identifying and saving these areas is critical to protecting our planet's biodiversity. Illustration by Maury Aaseng.

Soil Impacts

With all the lushness and productivity that exist in tropical forests, it can be surprising to learn that tropical soils are actually very thin and poor in nutrients. The underlying "parent" rock weathers rapidly in the tropics' high temperatures and heavy rains, and over time, most of the minerals have washed from the soil. Nearly all the nutrient content of a tropical forest is in the living plants and the decomposing litter on the forest floor.

When an area is completely deforested for farming, the farmer typically burns the trees and vegetation to create a fertilizing layer of ash. After this **slash-and-burn** deforestation, the nutrient reservoir is lost, flooding and erosion rates are high, and soils often become unable to support crops in just a few years. If the area is then turned into cattle pasture, the ground may become compacted as well, slowing down or preventing forest recovery.

> **slash and burn:** a form of agriculture in which an area of forest is cleared by cutting and burning and is then planted for a few seasons until the soil is depleted of nutrients

Social Impacts

Tropical forests are home to millions of native (indigenous) people who make their livings through subsistence agriculture, hunting and gathering, or through low-impact harvesting of forest products like rubber or nuts. Deforestation in indigenous territories by loggers, colonizers, and refugees has sometimes triggered violent conflict. Forest preservation can be socially divisive, as well. National and international governments and aid agencies struggle with questions about what level of human presence, if any, is compatible with conservation goals in tropical forests, how to balance the needs of indigenous peoples with expanding rural populations and national economic development, and whether establishing large, pristine, uninhabited protected areas—even if that means removing current residents—should be the highest priority of conservation efforts in tropical forests.

Climate Impacts: Rainfall and Temperature

Up to thirty percent of the rain that falls in tropical forests is water that the rainforest has recycled into the atmosphere. Water evaporates from the soil and vegetation, condenses into clouds, and falls again as rain in a perpetual self-watering cycle. In addition to maintaining tropical rainfall, the evaporation cools the Earth's surface. In many computer models of future climate, replacing tropical forests with a landscape of pasture and crops creates a drier, hotter climate in the tropics. Some models also predict that tropical deforestation will disrupt rainfall pattern far outside the tropics, including China, northern Mexico, and the south-central United States.

The Carbon Cycle and Global Warming

In the Amazon alone, scientists estimate that the trees contain more carbon than 10 years worth of human-produced greenhouse gases. When people clear the forests, usually with fire, carbon stored in the wood returns to the atmosphere, enhancing the greenhouse effect and global warming (Figure 8.4). Once the forest is cleared for crop or grazing land, the soils can become a large source of carbon emissions, depending on how farmers and ranchers manage the land. In places such as Indonesia, the soils of swampy lowland forests are rich in partially decayed organic matter, known as peat. During extended droughts, such as during **El Niño** events, the forests and the peat become flammable, especially if they have been degraded by logging or accidental fire. When they burn, they release huge volumes of carbon dioxide and other greenhouse gases.

© Wijnand Loven/iStock/Thinkstock

FIGURE 8.4 By clearing forests with fire, carbon stored in the wood returns to the atmosphere and the soil's nutrient reservoir is lost.

El Niño: a warming of the surface water of the eastern and central Pacific Ocean, occurring every 4 to 12 years and causing unusual global weather patterns; said to occur when the trade winds that usually push warm surface water westward weaken, allowing the warm water to pool as far eastward as the western coast of South America

Causes of Deforestation: Direct Causes

People have been deforesting the Earth for thousands of years, primarily to clear land for crops or livestock. Although tropical forests are largely confined to developing countries, they aren't just meeting local or national needs; economic globalization means that the needs and wants of the global population are bearing down on them as well. Direct causes of deforestation are agricultural expansion, wood extraction (e.g., logging or wood harvest for domestic fuel or charcoal), and infrastructure expansion such as road building and urbanization. Rarely is there a single direct cause for deforestation. Most often, multiple processes work simultaneously or sequentially to cause deforestation.

The single biggest direct cause of tropical deforestation is conversion to cropland and pasture, mostly for subsistence, which is growing crops or raising livestock to meet daily needs. The conversion to agricultural land usually results from multiple direct factors. For example, countries build roads into remote areas to improve overland transportation of goods. The road development itself causes a limited amount of deforestation. But roads also provide entry to previously inaccessible—and often unclaimed—land. Logging, both legal and illegal, often follows road expansion (and in some cases is the reason for the road expansion). When loggers have harvested an area's valuable timber, they move on. The roads and the logged areas become a magnet for settlers—farmers and ranchers who slash and burn the remaining forest for cropland or cattle pasture, completing the deforestation chain that began with road building. In other cases, forests that have been degraded by logging become fire-prone and are eventually deforested by repeated accidental fires from adjacent farms or pastures.

Although subsistence activities have dominated agriculture-driven deforestation in the tropics to date, large-scale commercial activities are playing an increasingly significant role. In the Amazon, industrial-scale cattle ranching and soybean production for world markets are increasingly important causes of deforestation, and in Indonesia, the conversion of tropical forest to commercial palm tree plantations to produce **bio-fuels** for export is a major cause of deforestation on Borneo and Sumatra.

> **bio-fuels**: gas or liquid fuel made from plant material

Underlying Causes

Although poverty is often cited as *the* underlying cause of tropical deforestation, analyses of multiple scientific studies indicate that that explanation is an oversimplification. Poverty does drive people to migrate to forest frontiers, where they engage in slash and burn forest clearing for subsistence. But rarely does one factor alone bear the sole responsibility for tropical deforestation.

State policies to encourage economic development, such as road and railway expansion projects, have caused significant, unintentional deforestation in the Amazon and Central America. Agricultural subsidies and tax breaks, as well as timber concessions, have encouraged forest clearing as well. Global economic factors such as a country's foreign debt, expanding global markets for rainforest timber and pulpwood, or low domestic costs of land, labor, and fuel can encourage deforestation over more sustainable land use.

Rates of Tropical Deforestation

The scope and impact of deforestation can be viewed in different ways. One is in absolute numbers: total area of forest cleared over a certain period. By that metric, all three major tropical forest areas, including South America, Africa, and Southeast Asia, are represented near the top of the list. Brazil led the world in terms of total deforested area between 1990 and 2005. The country lost 42,330,000 hectares (163,436 square miles) of forest, roughly the size of California. Rounding out the top five tropical countries with the greatest total area of deforestation were Indonesia, Sudan, Myanmar, and the Democratic Republic of Congo.

Another way to look at deforestation is in terms of the percent of a country's forest that was cleared over time. By this metric, the island nation of Comoros (north of Madagascar) fared the worst, clearing nearly 60 percent of its forests between 1990 and 2005. Landlocked Burundi in central Africa was second, clearing 47 percent of its forests. The other top five countries that cleared large percentages of their forests were Togo, in West Africa (44 percent); Honduras (37 percent); and Mauritania (36 percent). Thirteen other tropical countries or island territories cleared 20 percent or more of their forests between 1990–2005.

Sustaining Tropical Forests

Strategies for preserving tropical forests can operate on local to international scales. On a local scale, governments and non-governmental organizations are working with forest communities to encourage low-impact agricultural activities, such as shade farming, as well as the sustainable harvesting of non-wood forest products such as rubber, cork, produce, or medicinal plants. Parks and protected areas that draw tourists—ecotourism— can provide employment and educational opportunities for local people as well as creating or stimulating related service-sector economies.

On the national scale, tropical countries must integrate existing research on human impacts on tropical ecosystems into national land use and economic development plans. For tropical forests to survive, governments must develop realistic scenarios for future deforestation that take into account what scientists already know

Shutterstock/Kjersi Joergensen

FIGURE 8.5 Species such as the Orangutan (*Pongo pygmaeus*)—pictured in Borneo, Indonesia—are highly endangered due to destruction of their tropical rainforest habitats. When rainforests are cut down and converted to oil palm plantations, some workers brutally kill orangutans found consuming oil palm fruits.

about the causes and consequences of deforestation, including the unintended deforestation that results from road-building, accidental fire, selective logging, and economic development incentives such as timber concessions and agricultural subsidies.

Several scientists are encouraging the conservation community to re-consider the belief that vast, pristine parks and protected areas are the holy grail of forest conservation. In 2005, for example, scientists using satellite and ground-based data in the Amazon demonstrated that far less "unfettered" deforestation occurred in recent decades within territories occupied and managed by indigenous people than occurred in parks and other protected areas. The year before, scientists studying Indonesia's tropical forests documented a 56 percent decline in tropical lowland forests in protected areas of Borneo between 1985 and 2001 (Figure 8.5). They concluded that the deforestation in the protected areas resulted from a combination of illegal logging and devastating fires that raged through logging-damaged forests during the 1997–1998 El Niño-triggered drought. While some might argue that these losses could be prevented in the future through better enforcement of environmental laws, it may also be true that inhabited forest reserves are a more realistic strategy for preserving the majority of biodiversity in larger areas than parks alone can accomplish.

Finally, on the national and international scale, an increasing value in the global marketplace for products that are certified as sustainably produced or harvested—timber, beef, coffee, soy—may provide incentives for landowners to adopt more forest-friendly practices, and for regional and national governments to create and enforce forest-preservation policies. Direct payments to tropical countries for the ecosystem services that intact tropical forest provide, particularly for carbon storage to offset greenhouse gas emissions, are likely to become an important international mechanism for sustaining tropical forests as more countries begin to seriously tackle the problem of global warming.

Adapted from Lindsey, R. 2007. Tropical Deforestation. NASA Earth Observatory Feature Article.
http://earthobservatory.nasa.gov/Features/Deforestation/

CHAPTER 9

Human Population Dynamics

9.1 Population Growth and Demographics

The following excerpt from Chapter 6 of AP Environmental Science: History and Global Distribution, Encyclopedia of Earth *explains that for much of history the human race has lived a precarious existence. Disease, natural disasters, and famines kept human population low and, on occasion, close to extinction. However, in just the past 200 years, human population numbers have grown from roughly one billion to almost seven billion. This population explosion is due mainly to better nutrition, advances in health care, and improved sanitation practices. These advances have doubled human life expectancy and have sharply decreased death rates. The study of how human populations change over time is known as demographics, and demographers examine trends in birth rates, death rates, and other factors to predict future population trends.*

Some of the same concepts of population biology or population dynamics that applied to other species in section 1.4 also apply to humans. When conditions are right, human populations can experience exponential growth, doubling from 1 to 2 units, 2 to 4, 4 to 8, 8 to 16, 16 to 32, and so on. For most of human history, however, such conditions did not apply. In fact, it has just been in the last few hundred years that conditions have favored exponential growth of the human population on this scale.

Demographers typically measure population changes by examining births and deaths per 1,000 people in a given time period. If a random group of 1,000 people from a population were locked away in a gymnasium or shipped to a remote island for a year, we could determine the growth rate and final population of that group by looking at births and deaths over that time. For example, if 20 children were born over the course of that year (some of the women picked at random may have already been pregnant) and 10 individuals in the group were to die, the so-called natural rate of population change would be (20 − 10)/1,000 × 100 = 1%. In reality, populations are not locked away in gymnasiums or stranded on remote islands, so demographers also have to consider factors such as immigration and emigration in determining population change.

The key variables in determining changes to a population thus turn out to be birth rates (births per 1,000 people), death rates (deaths per 1,000 people), and fertility rates (the average number of children a woman will have over her lifetime). Throughout most of human history, fertility and birth rates were typically quite high, with fertility rates of 6–8 children per woman and birth rates of 40 per 1,000 annually. Because death rates were also quite high, also around 40 per 1,000, human population numbers did not grow very much or at all. The first variable to change significantly within the past few hundred years was death rates. This was due to advances in science, medicine, sanitation, and nutrition—all of which helped people live longer lives. As death rates declined, fertility rates and birth rates stayed the same, and human populations began to grow.

Eventually, however, fertility rates started to decline as did the birth rate. When and if the birth rate declines to the point where it equals the death rate, population will begin to move toward stabilization. This process of declining death rates, followed by declining birth rate—with population growth occurring in the interim—is known as the **demographic transition**. *Most industrialized countries like the United States have already completed a demographic transition; however, many developing countries are still in the midst of one. Thus, it is in these regions that most of the world's population growth continues to occur.*

> **demographic transition**: the process through which a country moves from relatively high birth and death rates to relatively low birth and death rates

By *Peter Saundry,* Topic Editor

A population is a group of individuals living together in a given area at a given time. Changes in populations are termed **population dynamics.** The current human population is made up of all of the people who currently share the earth. The first humans walked the planet millions of years ago. Since that time, the number of humans living on the planet and where they live has constantly changed over time. Every birth and death is a part of human population dynamics. Each time a person moves from one location to another, the spatial arrangement of the population is changed, and this, too, is an element of population dynamics. While humans are unique in many ways as a species, they are subject to many of the same limiting forces and unexpected events of all populations of organisms.

> **population dynamics**: patterns or processes that affect growth or change within a population

> **human population dynamics**: patterns or processes that affect growth or change with the human population

> **exponential growth**: the accelerating population growth that occurs when optimal conditions allow a constant rate of growth over a period of time

> **J curve**: a curve on a graph resembling the letter "J" that can represent prolonged exponential growth

Population Growth

Human populations are not stagnant. They naturally change in size, density and predominance of age groups in response to environmental factors such as resources availability and disease, as well as social and cultural factors. The increases and decreases in human population size make up what is known as **human population dynamics**. If resources are not limited, then populations experience exponential growth. A plot of **exponential growth** over time resembles a **J curve**. Absolute numbers are relatively small at first along the base of the J curve, but the population rapidly skyrockets when the critical time near the stem of the J curve is reached.

For most of the history of modern humans (*Homo sapiens*), people were hunter-gatherers. Food, especially meat from large mammals, was usually plentiful. However, populations were small because the nomadic life did not favor large family sizes. During those times, the human population was probably not more than a few million worldwide. It was still in the base of the J growth curve.

With the end of the last Ice Age, roughly 10,000 years ago, the climates worldwide changed, and many large mammals that had been the mainstay of human diet became extinct. This forced a change in diet and lifestyle, from one of the nomadic hunter-gatherer to that of a more stationary agricultural society.

Humans began cultivating food and started eating more plants and less meat. Having larger families was possible with the more stationary lifestyle. In fact, having a large family increasingly became an asset, as extra hands were needed for maintaining crops and homes. As agriculture became the mainstay of human life, the population increased (Figure 9.1).

As the population increased, people began living in villages, then in towns and finally in cities. This led to problems associated with overcrowded conditions, such as the build up of wastes, poverty and disease. Large families were no longer advantageous. Infanticide [intentionally killing an infant] was common during medieval times in Europe, and communicable diseases also limited the human population numbers. Easily spread

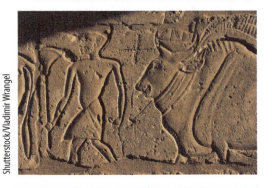

Shutterstock/Vladimir Wrangel

FIGURE 9.1 As populations transitioned from nomadic hunter-gatherer societies, to larger and more stationary groups, agriculture developed to help meet food demands.

in crowded, rat-infested urban areas, Black Death, the first major outbreak of the Bubonic Plague (1347–1351) drastically reduced the populations in Europe and Asia, possibly by as much as 50 percent.

Starting in the 17th Century, advances in science, medicine, agriculture and industry allowed rapid growth of human population and infanticide again became a common practice.

The next big influence on the human population occurred with the start of the Industrial Revolution in the late 18th century. With the advent of factories, children became valuable labor resources, thereby contributing to survival, and family sizes increased. The resulting population boom was further aided by improvements in agricultural technology that led to increased food production (Figure 9.2). Medical advancements increased control over disease and lengthened the average lifespan. By the early 19th century, the human population worldwide reached one billion. It was now in the stem of the J curve graph. As the world approached the 20th century, the human population was growing at an exponential rate.

FIGURE 9.2 With the rise of fossil fuels in the early 20th century, food production systems were able to support unprecedented human population growth.

During the 20th century, another important event in human population dynamics occurred. The birth rates in the highly developed countries decreased dramatically. Factors contributing to this decrease included: a rise in the standard of living, the availability of practical birth control methods and the establishment of child education and labor laws. These factors made large families economically impractical.

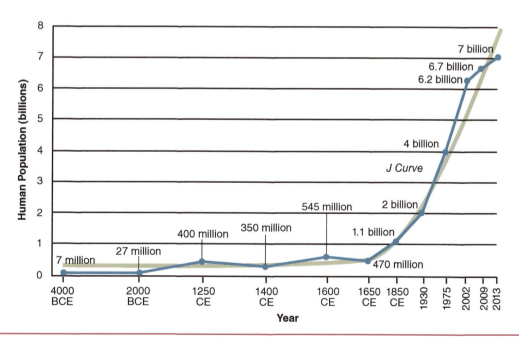

FIGURE 9.3 **Human Population Growth through History.** The human population grew slowly at first, declined during the spread of the Black Death, and accelerated beginning in the 17th century. Exponential population growth is represented by a J curve.

demography: the study of statistics—births, deaths, income, or the incidence of disease—that illustrate the changing structure of human populations

immigration: movement of people into a region, thus increasing the region's population

emigration: movement of people out of a region, thus decreasing the region's population

natural population change: the increase or decrease of a population size based on birth rates and death rates

human fertility: the average number of children a woman will have over her lifetime

Population Demographics

Human **demography** (population change) is usually described in terms of the births and deaths per 1000 people. When births of an area exceed deaths, population increases. When the births of an area are fewer than deaths, the population decreases. The annual rate at which the size of a population changes is:

Natural Population Change Rate (%) =
[(Births – Deaths) / 1000] × 100

During the year 2000, the birth rate for the world was 22 and the death rate was 9. Thus, the world's population grew at a rate of 1.3 percent. The annual rate of population change for a particular city or region is also affected by **immigration** (movement of people into a region) and **emigration** (movement out of a region).

Population Change Rate (%) =
(Birth rate + Immigration rate) – (Death rate + Emigration rate)

Highly industrialized nations, like the United States, Canada, Japan, and Germany, generally have low birth and death rates. Annual rates of **natural population change** vary from –0.1% to 0.5%. In some industrial nations (e.g. Germany and Russia) death rates exceed birth rates so the net population decreases over time. Newly industrialized countries (e.g. South Korea, Mexico and China) have moderate birth rates and low death rates. The low death rates result from better sanitation, better heath care and stable food production that accompany industrialization.

The annual rates of natural population change are about 1 percent to 2 percent in these countries. Countries with limited industrial development (e.g. Pakistan and Ethiopia) tend to have high birth rates and moderate to low death rates. These nations are growing rapidly with annual rates of natural population change exceeding 2 percent.

Several factors influence **human fertility**. Important factors influencing birth and fertility rates in human populations are: affluence, average marriage age, availability of birth control, family labor needs, cultural beliefs, religious beliefs, and the cost of raising and educating children.

The rapid growth of the world's population over the past 100 years mainly results from a decline in death rates. Reasons for the drop in death rates include: better nutrition, fewer infant deaths, increased average life span and improvements in medical technology.

Demographic Transition

As countries become developed and industrialized, they experience a movement from high population growth to low population growth. Both death and birth rates decline.

These countries usually move from rapid population growth, to slow growth, to zero growth and finally to a reduction in population. This shift in growth rate with development is called the "demographic transition." Four distinct stages occur during the transition: pre-industrial, transitional, industrial and post-industrial.

During the **pre-industrial stage**, harsh living conditions result in a high birth rate and a high death rate. The population grows very slowly, if at all. The **transitional stage** begins shortly after industrialization. During this phase, the death rate drops because of increased food production and better sanitation and

pre-industrial stage: first stage of demographic transition characterized by high birth and death rates, wherein the population grows slowly, if at all

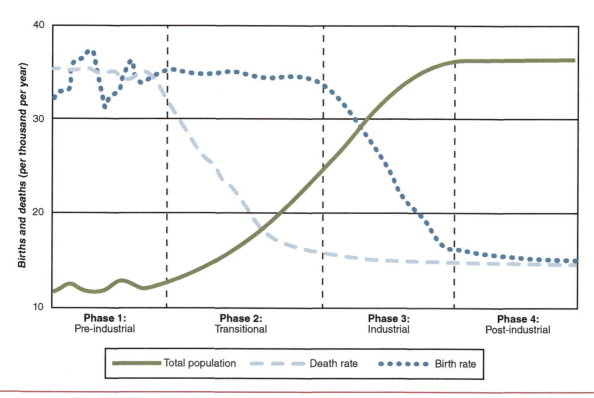

FIGURE 9.4 Demographic Transition. The four stages of demographic transition show the change in population growth that a country experiences as it becomes developed and industrialized.

health conditions, but, the birth rate remains high. Therefore, the population grows rapidly.

During the **industrial stage,** industrialization is well established in the country. The birth rate drops and eventually approaches the death rate. Couples in cities realize that children are expensive to raise and that having large families restrict their job opportunities. The **post-industrial stage** occurs when the birth rate declines even further to equal the death rate, thus population growth reaches zero. The birth rate may eventually fall below the death rate, resulting in negative population growth.

The United States and most European countries have experienced this gradual transition over the past 150 years. The transition moves much faster for today's developing countries. This is because improvements in preventive health and medical care in recent decades have dramatically reduced mortality—especially infant mortality—and increased life expectancy.

In a growing number of countries, couples are having fewer children than the two they need to "replace" themselves. However, even if the level of **"replacement fertility"** were reached today, populations would continue to grow for several decades because of the large numbers of people now entering their reproductive years.

As a result of reduced fertility and mortality, there will be a gradual demographic shift in all countries over the next few decades towards an older population. In developed countries,

> **transitional stage**: second stage of demographic transition during which the birth rate remains high but the death rate decreases, resulting in rapid population growth

> **industrial stage**: third stage of demographic transition during which the birth rate drops, thereby causing population growth to slow

> **post-industrial stage**: fourth stage of demographic transition during which the birth rate continues to fall till it is equal to the death rate, thereby causing no or negative population growth

> **replacement fertility**: the average number of children a couple must produce in order to "replace" themselves; this number is greater than two because some children die before reaching reproductive age

the proportion of people over age 65 has increased from 8 to 14 percent since 1950, and is expected to reach 25 percent by 2050. Within the next 35 years, those over age 65 will represent 30 percent or more of the populations in Japan and Germany. In some countries, the number of residents over age 85 will more than double.

Adapted from AP Environmental Science Chapter 6 – History and Global Distribution. In: Encyclopedia of Earth. Eds. Cutler J. Cleveland (Washington, D.C.: Environmental Information Coalition, National Council for Science and the Environment). [First published in the Encyclopedia of Earth, November 25, 2008; Last revised date November 25, 2008; Retrieved February 26, 2011. http://www.eoearth.org/article/AP_Environmental_Science_Chapter_6-_History_and_Global_Distribution

9.2 Current Population Situation

The following section is a summary from the United Nations Population Division Prospectus, 2012. (http:// esa.un.org/unpd/wpp/Documentation/pdf/WPP2012_Press_Release.pdf).

In 2013, the world population was 7.2 billion people. Projections by the United Nations Population Division indicate that in 2025 the population will be 8.1 billion people and then 9.6 billion people by 2050. Overall growth is projected to be highest in countries with high fertility (mainly in Africa) and in countries with large populations (India, Indonesia, Pakistan, the Philippines and the United States). The highest rate of growth is projected to take place in the 49 least developed countries in the world, while the population of developed regions will remain fairly static.

Population projections are higher than they have been in past reports. Fertility has increased in some countries in recent years, especially in some parts of Africa and Asia. Additionally, people are living longer in general, which increases population (a good thing!). Indeed, life expectancy increased over the course of the 20th century from 49 years to 69 years worldwide, and life expectancy is expected to reach 76 years by 2050.

In the coming century, population distribution across the globe is expected to be very different. India is expected to pass China in population by 2028, the population of Nigeria will be greater than that of the United States, and Europe's population will decline. Of course, all of these predictions depend upon current conditions and trends staying the same as they are today. If fertility rates increase or decline in some regions, this will impact the models greatly.

9.3 Population Policy and Future Population Trends

By the 1960s, human population growth was so rapid that fears of a "population bomb" and global famine became widespread. Of particular concern were high fertility rates and population growth in poorer countries of the Third World where women, on average, were often having seven or eight children each. Efforts were made in population policy around the world to reduce fertility rates through a variety of means, some more controversial than others. Today, most of these countries have moved through a demographic transition and have seen fertility rates decline to only two or three children per woman. According to the National Geographic article below, "Population 7 Billion," by Robert Kunzig, demographers were projecting that the human population will stabilize at roughly nine billion by later this century.

The study of how human population has grown from the millions to billions presents something of a paradox. At a fundamental level, the exponential growth in human numbers over the past two hundred years was triggered by humans living longer, healthier lives. Because fertility rates and birth rates at first stayed high while death rates dropped, human numbers grew. Longer lives and lower death rates are of course a good thing. However, the population growth that began with these changes is still adding close to 80 million new people to the planet each year (for comparison, New York City's population is approximately 8.4 million), even as fertility rates continue to drop around the world. The only way to stabilize population is for average fertility rates to reach approximately 2 children per woman. This figure is known as the replacement fertility rate because it represents the number of children needed to replace a set of parents.

Policy and program efforts to bring fertility rates down and "speed up" the demographic transition have taken a variety of different forms. Some governments have used coercive means to reduce fertility or provided economic incentives such as better housing and jobs for couples with fewer children. Others have taken a more holistic view based on research that has shown that women with better access to education, health care, and economic opportunities will tend to have fewer children even without coercion or explicit incentives. These differences in approach are examined in this article through examples from two regions of India. In addition, although the number of human beings living on Earth affects the well-being of the planet, how those people use resources also matters. This issue of resource consumption rates will be examined in this section as well as in section 9.4.

By *R.J. Kunzig*

Historians now estimate that in the 1670s there were only half a billion or so humans on Earth. After rising very slowly for millennia, the number was just starting to take off. A century and a half later the world's population had doubled to more than a billion. A century after that, around 1930, it had doubled again to two billion. The acceleration since then has been astounding. Before the 20th century, no human had lived through a doubling of the human population, but there are people alive today who have seen it triple.

And the explosion, though it is slowing, is far from over. Not only are people living longer, but so many women across the world are now in their childbearing years—1.8 billion—that the global population will keep growing for another few decades at least, even though each woman is having fewer children than she would have had a generation ago. By 2050 the total number could reach 10.5 billion, or it could stop at eight billion—the difference is about one child per woman. UN demographers consider the middle road their best estimate: They now project that the population may reach nine billion before 2050—in 2045. The eventual tally will depend on the choices individual couples make.

With the population still growing by about 80 million each year, it's hard not to be alarmed. Right now on Earth, water tables are falling, soil is eroding, glaciers are melting, and fish stocks are vanishing. Close to a billion people go hungry each day. Decades from now, there will likely be two billion more mouths to feed, mostly in poor countries. There will be billions more people wanting and deserving to boost themselves out of poverty. If they follow the path blazed by wealthy countries—clearing forests, burning coal and oil, freely scattering fertilizers and pesticides—they too will be stepping hard on the planet's natural resources. How exactly is this going to work?

Early Warnings

In 1798 Thomas Malthus, an English priest and economist, enunciated his general law of population: that it necessarily grows faster than the food supply, until war, disease, and famine arrive to reduce the number of people. As it turned out, the last plagues great enough to put a dent in global population had already happened when Malthus wrote. World population hasn't fallen, historians think, since the Black Death of the 14th century.

In the two centuries after Malthus declared that population couldn't continue to soar, that's exactly what it did. The process started in what we now call the developed countries, which were then still developing. The spread of New World crops like corn and the potato, along with the discovery of chemical fertilizers, helped banish starvation in Europe. Growing cities remained cesspools of disease at first, but from the mid-19th century on, sewers began to channel human waste away from drinking water, which was then filtered and chlorinated; that dramatically reduced the spread of cholera and typhus.

Moreover in 1798, the same year that Malthus published his dyspeptic [troubling] tract, his compatriot Edward Jenner described a vaccine for smallpox—the first and most important in a series of vaccines and antibiotics that, along with better nutrition and sanitation, would double life expectancy in the industrializing countries, from 35 years to 77 today. It would take a cranky person to see that trend as gloomy: "The development of medical science was the straw that broke the camel's back," wrote Stanford population biologist Paul Ehrlich in 1968.

Ehrlich's book, *The Population Bomb,* made him the most famous of modern Malthusians. In the 1970s, Ehrlich predicted, "hundreds of millions of people are going to starve to death," and it was too late to do anything about it. "The cancer of population growth must be cut out," Ehrlich wrote, "by compulsion if voluntary methods fail." The very future of the United States was at risk. In spite or perhaps because of such language, the book was a best seller, as Malthus's had been. And this time too the bomb proved a dud. The green revolution—a combination of high-yield seeds, irrigation, pesticides, and fertilizers that enabled grain production to double—was already under way. Today many people are undernourished, but mass starvation is rare.

Changes in Birth and Death Rates

Ehrlich was right, though, that population would surge as medical science spared many lives. After World War II the developing countries got a sudden transfusion of preventive care, with the help of institutions like the World Health Organization and UNICEF. Penicillin, the smallpox vaccine, DDT (which, though later controversial, saved millions from dying of malaria)—all arrived at once. In India life expectancy went from 38 years in 1952 to 64 today; in China, from 41 to 73. Millions of people in developing countries who would have died in childhood survived to have children themselves. That's why the population explosion spread around the planet: because a great many people were saved from dying.

And because, for a time, women kept giving birth at a high rate. In 18th-century Europe or early 20th-century Asia, when the average woman had six children, she was doing what it took to replace herself and her mate, because most of those children never reached adulthood. When child mortality declines, couples eventually have fewer children—but that transition usually takes a generation at the very least. Today in developed countries, an average of 2.1 births per woman would maintain a steady population; in the developing world, "replacement fertility" is somewhat higher. In the time it takes for the birthrate to settle into that new balance with the death rate, population explodes.

Demographers call this evolution the demographic transition. All countries go through it in their own time. It's a hallmark of human progress: In a country that has completed the transition, people have wrested from nature at least some control over death and birth. The global population explosion is an inevitable side effect, a huge one that some people are not sure our civilization can survive. But the growth rate was actually at its peak just as Ehrlich was sounding his alarm. By the early 1970s, fertility rates around the world had begun dropping faster than anyone had anticipated. Since then, the population growth rate has fallen by more than 40 percent.

The Fertility Transition

In industrialized countries it took generations for fertility to fall to the replacement level or below. As that same transition takes place in the rest of the world, what has astonished demographers is how much faster it is happening there. Though its population continues to grow, China, home to a fifth of the world's people, is already below replacement fertility and has been for nearly 20 years, thanks in part to the coercive one-child policy

FIGURE 9.5 Japan's population rose by only 0.2 percent between 2005 and 2010, marking its slowest growth on record. As a country develops from a pre-industrial to an industrialized economic system, it typically experiences a transition from high birth and death rates to low birth and death rates.

Shutterstock/aslysun

implemented in 1979; Chinese women, who were bearing an average of six children each as recently as 1965, are now having around 1.5. In Iran, with the support of the Islamic regime, fertility has fallen more than 70 percent since the early '80s. In Catholic and democratic Brazil, women have reduced their fertility rate by half over the same quarter century. "We still don't understand why fertility has gone down so fast in so many societies, so many cultures and religions. It's just mind-boggling," says Hania Zlotnik, director of the UN Population Division.

"At this moment, much as I want to say there's still a problem of high fertility rates, it's only about 16 percent of the world population, mostly in Africa," says Zlotnik. South of the Sahara, fertility is still five children per woman; in Niger it is seven. But then, 17 of the countries in the region still have life expectancies of 50 or less; they have just begun the demographic transition. In most of the world, however, family size has shrunk dramatically. The UN projects that the world will reach replacement fertility by 2030. "The population as a whole is on a path toward nonexplosion—which is good news," Zlotnik says.

The bad news is that 2030 is two decades away and that the largest generation of adolescents in history will then be entering their childbearing years. Even if each of those women has only two children, population will coast upward under its own momentum for another quarter century.

The Case of India

The Indian government tried once before to push vasectomies, in the 1970s, when anxiety about the population bomb was at its height. Prime Minister Indira Gandhi and her son Sanjay used state-of-emergency powers to force a dramatic increase in sterilizations. From 1976 to 1977 the number of operations tripled, to more than eight million. Over six million of those were vasectomies. Family planning workers were pressured to meet quotas; in a few states, sterilization became a condition for receiving new housing or other government benefits. In some cases the police simply rounded up poor people and hauled them to sterilization camps.

The excesses gave the whole concept of family planning a bad name. "Successive governments refused to touch the subject," says Shailaja Chandra, former head of the National Population Stabilisation Fund (NPSF). Yet fertility in India has dropped anyway, though not as fast as in China, where it was nose-diving even before the draconian [unusually severe] one-child policy took effect. The national average in India is now 2.6 children per woman, less than half what it was when Ehrlich visited [in 1966]. The southern half of the country and a few states in the northern half are already at replacement fertility or below.

In Kerala, on the southwest coast, investments in health and education helped fertility fall to 1.7. The key, demographers there say, is the female literacy rate: At around 90 percent, it's easily the highest in India. Girls who go to school start having children later than ones who don't. They are more open to contraception and more likely to understand their options.

So far this approach held up as a model internationally, has not caught on in the poor states of northern India—in the "Hindi belt" that stretches across the country just south of Delhi. Nearly half of India's population growth is occurring in Rajasthan, Madhya Pradesh, Bihar, and Uttar Pradesh, where fertility rates still hover between three and four children per woman. More than half the women in the Hindi belt are illiterate, and many marry well before reaching the legal age of 18. They gain social status by bearing children—and usually don't stop until they have at least one son.

The Andhra Pradesh Model

As an alternative to the Kerala model, some point to the southern state of Andhra Pradesh, where sterilization "camps"—temporary operating rooms often set up in schools—were introduced during the '70s and where sterilization rates have remained high as improved hospitals have replaced the camps. In a single decade beginning in the early 1990s, the fertility rate fell from around three to less than two. Unlike in Kerala, half of all women in Andhra Pradesh remain illiterate.

Amarjit Singh, the current executive director of the NPSF, calculates that if the four biggest states of the Hindi belt had followed the Andhra Pradesh model, they would have avoided 40 million births—and considerable suffering. "Because 40 million were born, 2.5 million children died," Singh says. He thinks if all India were to adopt high-quality programs to encourage sterilizations, in hospitals rather than camps, it could have 1.4 billion people in 2050 instead of 1.6 billion.

Critics of the Andhra Pradesh model, such as the Population Foundation's Nanda, say Indians need better health care, particularly in rural areas. They are against numerical targets that pressure government workers to sterilize people or cash incentives that distort a couple's choice of family size. "It's a private decision," Nanda says.

In Indian cities today, many couples are making the same choice as their counterparts in Europe or America. Sonalde Desai, a senior fellow at New Delhi's National Council of Applied Economic Research, introduced me to five working women in Delhi who were spending most of their salaries on private-school fees and after-school tutors; each had one or two children and was not planning to have more. In a nationwide survey of 41,554 households, Desai's team identified a small but growing vanguard of urban one-child families. "We were totally blown away at the emphasis parents were placing on their children," she says. "It suddenly makes you understand—that is why fertility is going down." Indian children on average are much better educated than their parents.

That's less true in the countryside. With Desai's team I went to Palanpur, a village in Uttar Pradesh—a Hindi-belt state with as many people as Brazil. Walking into the village we passed a cell phone tower but also rivulets of raw sewage running along the lanes of small brick houses. Under a mango tree, the keeper of the grove said he saw no reason to educate his three daughters. Under a neem tree in the center of the village, I asked a dozen farmers what would improve their lives most. "If we could get a little money, that would be wonderful," one joked.

The Demographer's Perspective

The annual meeting of the Population Association of America (PAA) is one of the premier gatherings of the world's demographers. Last April [2010] the global population explosion was not on the agenda. "The problem has become a bit passé," Hervé Le Bras [a French demographer] says. Demographers are generally confident that by the second half of this century we will be ending one unique era in history—the population explosion— and entering another, in which population will level out or even fall.

But will there be too many of us? At the PAA meeting, in the Dallas Hyatt Regency, I learned that the current population of the planet could fit into the state of Texas, if Texas were settled as densely as New York City. If in 2045 there are nine billion people living on the six habitable continents, the world population density will be a little more than half that of France today. France is not usually considered a hellish place. Will the world be hellish then?

Some parts of it may well be; some parts of it are hellish today. There are now 21 cities with populations larger than ten million, and by 2050 there will be many more. Delhi adds hundreds of thousands of migrants each year, and those people arrive to find that "no plans have been made for water, sewage, or habitation," says Shailaja Chandra. Dhaka in Bangladesh and Kinshasa in the Democratic Republic of the Congo are 40 times larger today than they were in 1950. Their slums are filled with desperately poor people who have fled worse poverty in the countryside.

Many people are justifiably worried that Malthus will finally be proved right on a global scale—that the planet won't be able to feed nine billion people. Lester Brown, founder of Worldwatch Institute and now head of the Earth Policy Institute in Washington, believes food shortages could cause a collapse of global civilization. Human beings are living off natural capital, Brown argues, eroding soil and depleting groundwater faster than they can be replenished. All of that will soon be cramping food production. Brown's Plan B to save civilization would put the whole world on a wartime footing, like the U.S. after Pearl Harbor, to stabilize climate and repair the ecological damage. "Filling the family planning gap may be the most urgent item on the global agenda," he writes, so if we don't hold the world's population to eight billion by reducing fertility, the death rate may increase instead.

Eight billion corresponds to the UN's lowest projection for 2050. In that optimistic scenario, Bangladesh has a fertility rate of 1.35 in 2050, but it still has 25 million more people than it does today. Rwanda's fertility rate also falls below the replacement level, but its population still rises to well over twice what it was before the genocide [1994 mass killing of Tutsis and moderate Hutus]. If that's the optimistic scenario, one might argue, the future is indeed bleak.

Hope for the Future?

But one can also draw a different conclusion—that fixating on population numbers is not the best way to confront the future. People packed into slums need help, but the problem that needs solving is poverty and lack of infrastructure, not overpopulation. Giving every woman access to family planning services is a good idea—"the one strategy that can make the biggest difference to women's lives," Chandra calls it. But the most aggressive population control program imaginable will not save Bangladesh from sea level rise, Rwanda from another genocide, or all of us from our enormous environmental problems.

FIGURE 9.6 Overcrowding most significantly affects those living in poverty and areas where overcrowding and lack of infrastructure collide. Here, volunteers hand out free meals in overcrowded Ethiopia.

Global warming is a good example. Carbon emissions from fossil fuels are growing fastest in China, thanks to its prolonged economic boom, but fertility there is already below replacement; not much more can be done to control population. Where population is growing fastest, in sub-Saharan Africa, emissions per person are only a few percent of what they are in the U.S.—so population control would have little effect on climate. Brian O'Neill of the National Center for Atmospheric Research has calculated that if the population were to reach 7.4 billion in 2050 instead of 8.9 billion, it would reduce emissions by 15 percent. "Those who say the whole problem is population are wrong," Joel Cohen [Professor of Populations and head of the Laboratory of Populations at the Rockefeller University and Columbia University] says. "It's not even the dominant factor." To stop global warming we'll have to switch from fossil fuels to alternative energy—regardless of how big the population gets.

The number of people does matter, of course. But how people consume resources matters a lot more. Some of us leave much bigger footprints than others. The central challenge for the future of people and the planet is how to raise more of us out of poverty—the slum dwellers in Delhi, the subsistence farmers in Rwanda—while reducing the impact each of us has on the planet.

The World Bank has predicted that by 2030 more than a billion people in the developing world will belong to the "global middle class," up from just 400 million in 2005. That's a good thing. But it will be a hard thing for the planet if those people are eating meat and driving gasoline-powered cars at the same rate as Americans now do. It's too late to keep the new middle class of 2030 from being born; it's not too late to change how they and the rest of us will produce and consume food and energy. "Eating less meat seems more reasonable to me than saying, 'Have fewer children!'" Le Bras says.

How many people can the Earth support? Cohen spent years reviewing all the research, from Leeuwenhoek on. "I wrote the book thinking I would answer the question," he says. "I found out it's unanswerable in the present state of knowledge." What he found instead was an enormous range of "political numbers, intended to persuade people" one way or the other.

Conclusion

For centuries population pessimists have hurled apocalyptic warnings at the congenital optimists, who believe in their bones that humanity will find ways to cope and even improve its lot. History, on the whole, has so far favored the optimists, but history is no certain guide to the future. Neither is science. It cannot predict the outcome of *People* v. *Planet,* because all the facts of the case—how many of us there will be and how we will live—depend on choices we have yet to make and ideas we have yet to have. We may, for example, says Cohen, "see to it that all children are nourished well enough to learn in school and are educated well enough to solve the problems they will face as adults." That would change the future significantly.

The debate was present at the creation of population alarmism, in the person of Rev. Thomas Malthus himself. Toward the end of the book in which he formulated the iron law by which unchecked population growth leads to famine, he declared that law a good thing: It gets us off our duffs. It leads us to conquer the world. Man, Malthus wrote, and he must have meant woman too, is "inert, sluggish, and averse from labour, unless compelled by necessity." But necessity, he added, gives hope:

"The exertions that men find it necessary to make, in order to support themselves or families, frequently awaken faculties that might otherwise have lain for ever dormant, and it has been commonly remarked that new and extraordinary situations generally create minds adequate to grapple with the difficulties in which they are involved."

Seven billion of us now, nine billion in 2045. Let's hope that Malthus was right about our ingenuity.

Adapted from, Kunzig, R. J. 2011. Population 7 Billion. National Geographic, January. http://ngm.nationalgeographic.com/2011/01/seven-billion/kunzig-text.

9.4 Population Growth, Material Consumption, and the IPAT Equation

As mentioned in the previous essay, as a general rule, a larger human population means a greater environmental impact on our planet. Indeed, more people create an increased demand for food, fuel, water, and minerals, all of which cause some environmental impact during their production. However, demographers and environmental scientists realize that the impact a population has on the environment is determined not just by the number of people involved but by the kinds of technologies used and their relative levels of affluence and material consumption. This has led to the development of a formula known as IPAT to study the environmental impact of the different variables, as explained in this article by Paul R. and Anne H. Ehrlich.

The IPAT equation expresses the idea that the environmental impact (I) of a given population will be determined by the interactions of the size of that population (P), the average affluence or consumption rate of individuals in that population (A), and the kinds of technologies that population makes use of (T). In recent decades most ecologists and environmental scientists have fixated on the P factor, population, as human numbers have grown from less than two billion in 1900 to almost seven billion by 2010. However, with population growth rates beginning to slow and global population projected to peak at nine or ten billion later this century, more attention is being paid to the A and T factors – how much we consume and what technologies we use to enable that consumption.

The IPAT concept is sometimes expressed through the idea of an ecological footprint—a measure of how much land and water is required to support the consumption of an individual. You might live in a small apartment in the city, but your lifestyle requires forestland for the production of paper and farmland for the production of food (among other things), and so your actual footprint will be larger than the space you occupy. While concerns over growing populations have focused on the poorer developing countries of the world, IPAT and footprint analysis have shown that high consumption rates in more wealthy developed countries also result in significant environmental impacts.

By *P.R. and A.H. Ehrlich.*

Over some 60 million years, *Homo sapiens* has evolved into the dominant animal on the planet, acquiring binocular vision, upright posture, large brains, and—most importantly—language with syntax and that complex store of non-genetic information we call culture. However, in the last several centuries we've increasingly been using our relatively newly acquired power, especially our culturally evolved technologies, to deplete the natural capital of Earth—in particular its deep, rich agricultural soils, its groundwater stored during ice ages, and its biodiversity—as if there were no tomorrow.

The point, all too often ignored, is that this trend is being driven in large part by a combination of population growth and increasing per capita consumption, and it cannot be long continued without risking a collapse of our now-global civilization. Too many people—and especially too many politicians and business executives—are under the delusion that such a disastrous end to the modern human enterprise can be avoided by technological fixes that will allow the population and the economy to grow forever. But if we fail to bring population growth and over-consumption under control the number of people on Earth is expected to grow from more than 7 billion today to 9 billion by the second half of the 21st century—then we will inhabit a planet where life becomes increasingly untenable because of two looming crises: global heating, and the degradation of the natural systems on which we all depend.

I = PAT

Our species' negative impact on our own life-support systems can be approximated by the equation **I = PAT**. In that equation, the size of the population (P) is multiplied by the average affluence or consumption per individual (A), and that in turn is multiplied by some measure of the technology (T) that services and drives the consumption. Thus commuting in automobiles powered by subsidized fossil fuels on proliferating freeways creates a much greater T factor than commuting on bikes using simple paths or working at home on a computer network. The product of P, A, and T is Impact (I), a rough estimate of how much humanity is degrading the ecosystem services it depends upon.

> **I = PAT**: an equation that shows the relationship between environmental impacts and the forces that cause them, where I represents the impact, P the size of the population, A the average affluence or consumption per individual, and T a measure of technology that drives the consumption

The equation is not rocket science. Two billion people, all else being equal, put more greenhouse gases into the atmosphere than one billion people. Two billion rich people disrupt the climate more than two billion poor people. Three hundred million Americans consume more petroleum than 1.3 billion Chinese. And driving an SUV is using a far more environmentally malign transportation technology than riding mass transit.

The technological dimensions of our predicament—such as the need for alternatives to fossil fuel energy—are frequently discussed if too little acted upon. Judging from media reports and the statements of politicians, environmental problems, to the degree they are recognized, can be solved by minor changes in technologies and recycling (T). Switching to ultra-light, fuel-efficient cars will obviously give some short-term advantage, but as population and consumption grow, they will pour still more carbon dioxide (and vaporized rubber) into the atmosphere and require more natural areas to be buried under concrete. More recycling will help, but many of our society's potentially most dangerous effluents [pollutants that flow out into the environment] (such as hormone-mimicking chemicals) cannot practically be recycled. There is no technological change we can make that will permit growth in either human numbers or material affluence to continue to expand. In the face of this, the neglect of the intertwined issues of population and consumption is stunning.

Population

Many past human societies have collapsed under the weight of overpopulation and environmental neglect, but today the civilization in peril is global. The population factor in what appears to be a looming catastrophe is even greater than most people suppose. Each person added today to the population on average causes more damage to humanity's critical life-support systems than did the previous addition—everything else being equal. The reason is simple: *Homo sapiens* became the dominant animal by being smart. Farmers didn't settle first on poor soils where water was scarce, but rather in rich river valleys. That's where most cities developed, where rich soils are now being paved over for roads and suburbs, and where water supplies are being polluted or overexploited.

FIGURE 9.7 Will bustling cities such as Shanghai, approaching a population of 23 million in 2011, continue to have the resources required to sustain themselves long into the future?

Shutterstock/TonyV3112

As a result, to support additional people it is necessary to move to ever poorer lands, drill wells deeper, or tap increasingly remote sources to obtain water—and then spend more energy to transport that water ever greater distances to farm fields, homes, and factories. Our distant ancestors could pick up nearly pure copper on Earth's surface when they started to use metals; now people must use vast amounts of energy to mine and refine gigantic amounts of copper ore of ever poorer quality, some in concentrations of less than one percent. The same can be said for other important metals. And petroleum can no longer be found easily on or near the surface, but must be gleaned from wells drilled a mile or more deep, often in inaccessible localities, such as under continental shelves beneath the sea. All of the paving, drilling, fertilizer manufacturing, pumping, smelting, and transporting needed to provide for the consumption of burgeoning numbers of people produces greenhouse gases and thus tightens the connection between population and climate disruption.

So why is the topic of overpopulation so generally ignored? There are some obvious reasons. Attempts by governments to limit their nation's population growth are anathema to those on the right who believe the only role for governments in the bedroom is to force women to take unwanted babies to term. Those on the left fear, with some legitimacy, that population control could turn racist or discriminatory in other ways—for example, attempting to reduce the numbers of minorities or the poor. Many fear the specter of more of "them" compared to "us," and all of us fear loss of liberty and economic decline (since population growth is often claimed necessary for economic health). And [some religions] promote [large families], though in much of the world their efforts are largely futile (Catholic countries in Europe tend to be low-birthrate leaders, for example).

FIGURE 9.8 Each child will have material needs, but children born in the United States will place a much greater demand on the Earth's natural resources than those born in developing countries.

Affluence and Technology

Silence on the overconsumption (Affluence) factor in the I = PAT equation is more readily explained. Consumption is still viewed as an unalloyed [pure] good by many economists, along with business leaders and politicians, who tend to see jacking up consumption as a cure-all for economic ills. Too much unemployment? Encourage people to buy an SUV or a new refrigerator. Perpetual growth is the creed of the cancer cell, but third-rate economists can't think of anything else. Some leading economists *are* starting to tackle the issue of overconsumption, but the problem and its cures are tough to analyze.

And, of course, there are the vexing problems of consumption of people in poor countries. On one hand, a billion or more people have problems of *underconsumption*. Unless their basic needs are met, they are unlikely to be able to make important contributions to attaining sustainability. On the other hand, there is also the issue of the "new consumers" in developing economies such as China and India, where the wealth of a sizable minority is permitting them to acquire the consumption habits (e.g., eating a lot of meat and driving automobiles) of the rich nations. Consumption regulation is a lot more complex than population regulation, and it is much more difficult to find humane and equitable solutions to the problem.

The dominant animal is wasting its brilliance and its wonderful achievements; civilization's fate is being determined by decision makers who determinedly look the other way in favor of immediate comfort and profit. Thousands of scientists recently participated in a Millennium Ecosystem Assessment that outlined our current environmental dilemma, but the report's dire message made very little impact. Absent attention to that message, the fates of Easter Island, the Classic Maya civilization, and Nineveh—all of which collapsed following environmental degradation—await us all.

We believe it is possible to avoid that global denouement [outcome]. Such mobilization means developing some consensus on goals—perhaps through a global dialogue in which people discuss the human predicament and decide whether they would like to see a maximum number of people living at a minimum standard of living, or perhaps a much lower population size that gives individuals a broad choice of lifestyles. We have suggested a forum for such a dialogue, modeled partly on the Intergovernmental Panel on Climate Change [intergovernmental body that studies and reports scientific findings on global climate change], but with more "bottom up" participation. It is clear that only widespread changes in norms can give humanity a chance of attaining a sustainable and reasonably conflict-free society.

How to achieve such change—involving everything from demographic policies and transformation of planet-wide energy, industrial, and agricultural systems, to North-South and interfaith relationships and military postures—is a gigantic challenge to everyone. Politicians, industrialists, ecologists, social scientists, everyday citizens, and the media must join this debate. Whether it is possible remains to be seen; societies have managed to make major transitions in the recent past, as the civil rights revolution in the United States and the collapse of communism in the Soviet Union clearly demonstrate.

We'll continue to hope and work for a cultural transformation in how we treat each other and the natural systems we depend upon. We can create a peaceful and sustainable global civilization, but it will require realistic thinking about the problems we face and a new mobilization of political will.

Adapted from Ehrlich, P.R. and A.H. Ehrlich. 2008. Too Many People, Too Much Consumption. Yale Environment 360. Copyright © 2008 Ehrlich, P.R. and A.H. Ehrlich. Available online at http://e360.yale.edu/feature/too_many_people_too_much_consumption/2041/

Summary

Though it may be difficult to imagine, the human race likely came close to extinction on several different occasions during our millions of years of existence on this planet. Only in the last few hundred years, a mere blip on the record of human history, have we created the conditions that have allowed the human population to grow at an exponential rate. Our evolutionary success has changed the world fundamentally with a population growth of perhaps a few million (at the start of the agricultural revolution 12,000 years ago) to almost ten billion this century. We are now the dominant species on the planet in all respects. And the ecological impacts of our dominance, as evidenced by our staggering consumption and technological progress, can be felt in every corner of the world. Such changes in population and their impact require careful study so that scientists can understand and propose ideas for resource management.

In order to keep track of population trends over time, researchers developed the field of demography. Demographers focus on a few key variables—including birth rates, death rates, and fertility rates—to learn and predict how a population changes through time. Demographers have identified the main cause of the exponential growth in human population of the past two hundred years as an imbalance between death rates and birth rates. Throughout most of human history, both birth and death rates remained high, but roughly even, and so the population stayed the same as well. Advances in science, medicine, sanitation, and nutrition over the past two hundred years have led to longer life expectancies and a decline in death rates. Because birth rates did not immediately decline, as death rates went down, total human population increased. Today, birth rates in many countries have dropped to levels close to death rates, and their populations have begun to stabilize. However, other countries are still in the midst of a demographic transition toward low overall birth and death rates.

Reducing birth rates is the focus of most population policy, and there is much debate over what approaches to use. Indeed, debates between and among political, social, and religious spheres rage over whether stronger, more coercive measures are needed to reduce birth rates. Also, a growing number of environmental scientists now recognize that in addition to absolute numbers, affluence and the consumption patterns of individuals in a given population also have an impact on the environment. The links between growing human numbers, rising levels of affluence and consumption, and impacts on the environment will be examined more closely in the following chapters.

CHAPTER 10

Feeding the World

10.1 The Global Food Crisis—Feeding Nine Billion

By the 1960s, rising populations and stagnant world grain production combined to create the specter of massive famine. In response, scientists and development organizations launched what came to be known as the green revolution. This revolution involved the development of new varieties of wheat, rice, and other grains that doubled yields and allowed farmers in tropical regions to grow two crops a year instead of just one. The results were staggering; famine was largely avoided in certain regions of the world and **green revolution** *grain varieties came to dominate farming in many regions of the world.*

However, in order to grow green revolution varieties, farmers were required to use much larger quantities of irrigated water, fertilizers for plant growth, and pesticides and herbicides to control insect pests and weeds. These varieties also did better when they were planted in large blocks of a single variety, known as **monocultures***. This, in turn, necessitated even more irrigated water, fertilizers, pesticides, and herbicides to be used. Today, crop yields from green revolution varieties have peaked and are no longer responding the way they once did to increased applications of fertilizer and other inputs. Furthermore, over-pumping of groundwater for irrigation and over-use of synthetic fertilizers, pesticides, and herbicides are taking an increasing environmental toll.*

In the following article, Joel K. Bourne, Jr., of National Geographic Magazine reviews the history of the first Green Revolution and explains why it might be time for another one. With crop yields stagnant and the population still growing (though at a slower rate than 50 years ago)—and increasing affluence in countries like China and India spurring increased food consumption—Bourne argues that we could be on the verge of a global food crisis, especially for the world's poorest. The question is whether the next revolution in agriculture will focus on high-tech approaches such as **genetic engineering** *or on new ways of farming in a less environmentally destructive manner sometimes referred to as* **agroecology** *or sustainable agriculture, or both.*

green revolution: term for the introduction of scientifically bred or selected varieties of grain that, with adequate inputs of fertilizer and water, can greatly increase crop yields

monocultures: cultivation of a single crop, usually on a large area of land

genetic engineering: the deliberate modification of the characteristics of an organism by manipulating its genetic material

agroecology: an ecological approach to agriculture that views agricultural areas as ecosystems and is concerned with the ecological impact of agricultural practices

By Joel K. Bourne, Jr.

It is the simplest, most natural of acts, akin to breathing and walking upright. We sit down at the dinner table, pick up a fork, and take a juicy bite, oblivious to the double helping of global ramifications on our plate. Our beef comes from Iowa, fed by Nebraska corn. Our grapes come from Chile, our bananas from Honduras, our olive oil from Sicily, our apple juice—not from Washington State but all the way from

China. Modern society has relieved us of the burden of growing, harvesting, even preparing our daily bread, in exchange for the burden of simply paying for it. Only when prices rise do we take notice. And the consequences of our inattention are profound.

Last year the skyrocketing cost of food was a wake-up call for the planet. Between 2005 and the summer of 2008, the price of wheat and corn tripled, and the price of rice climbed fivefold, spurring food riots in nearly two dozen countries and pushing 75 million more people into poverty. But unlike previous shocks driven by short-term food shortages, this price spike came in a year when the world's farmers reaped a record grain crop. This time, the high prices were a symptom of a larger problem tugging at the strands of our worldwide food web, one that's not going away anytime soon. Simply put: For most of the past decade, the world has been consuming more food than it has been producing. After years of drawing down stockpiles, in 2007 the world saw global carryover stocks fall to 61 days of global consumption, the second lowest on record.

"Agricultural productivity growth is only one to two percent a year," warned Joachim von Braun, director general of the International Food Policy Research Institute in Washington, D.C., at the height of the crisis. "This is too low to meet population growth and increased demand."

> **agflation**: an increase in the price of food that occurs as a result of increased demand from human consumption

High prices are the ultimate signal that demand is outstripping supply, that there is simply not enough food to go around. Such **agflation** hits the poorest billion people on the planet the hardest, since they typically spend 50 to 70 percent of their income on food. Even though prices have fallen with the imploding world economy, they are still near record highs, and the underlying problems of low stockpiles, rising population, and flattening yield growth remain. Climate change—with its hotter growing seasons and increasing water scarcity—is projected to reduce future harvests in much of the world, raising the specter of what some scientists are now calling a perpetual food crisis.

The High Cost of Meat

It's no coincidence that as countries like China and India prosper and their people move up the food ladder, demand for grain has increased. For as tasty as that sweet-and-sour pork may be, eating meat is an incredibly inefficient way to feed oneself. It takes up to five times more grain to get the equivalent amount of calories

> **biofuels**: gas or liquid fuel made from plant material

from eating pork as from simply eating grain itself—ten times if we're talking about grain-fattened U.S. beef. As more grain has been diverted to livestock and to the production of **biofuels** for cars, annual worldwide consumption of grain has risen from 815 million metric tons in 1960 to 2.16 billion in 2008. Since 2005, the mad rush to biofuels alone has pushed grain-consumption growth from about 20 million tons annually to 50 million tons, according to Lester Brown of the Earth Policy Institute.

Even China, the second largest corn-growing nation on the planet, can't grow enough grain to feed all its pigs. Most of the shortfall is made up with imported soybeans from the U.S. or Brazil, one of the few countries with the potential to expand its cropland—often by plowing up rain forest. Increasing demand for food, feed, and biofuels has been a major driver of deforestation in the tropics. Between 1980 and 2000 more than half of new cropland acreage in the tropics was carved out of intact rain forests; Brazil alone increased its soybean acreage in Amazonia 10 percent a year from 1990 to 2005.

Some of those Brazilian soybeans may end up in the troughs of Guangzhou Lizhi Farms, the largest CAFO [concentrated animal feeding operation] in Guangdong Province. Tucked into a green valley just off a four-lane highway that's still being built,

Shutterstock/Ronnie Chua

FIGURE 10.1 Higher demand for meat means a higher need for grain to feed livestock. The soybeans used to feed the livestock shown in this market in China were imported from Brazil because China does not produce enough grain to feed all of its livestock.

some 60 white hog houses are scattered around large ponds, part of the waste-treatment system for 100,000 hogs. The city of Guangzhou is also building a brand-new meatpacking plant that will slaughter 5,000 head a day. By the time China has 1.5 billion people, sometime in the next 20 years, some experts predict they'll need another 200 million hogs just to keep up. And that's just China. World meat consumption is expected to double by 2050. That means we're going to need a whole lot more grain.

The First Green Revolution

This isn't the first time the world has stood at the brink of a food crisis—it's only the most recent iteration. At 83, Gurcharan Singh Kalkat has lived long enough to remember one of the worst famines of the 20th century. In 1943 as many as four million people died in the "**Malthusian correction**" known as the Bengal Famine. For the following two decades, India had to import millions of tons of grain to feed its people.

> **Malthusian correction**: Theory put forth by the Reverend Thomas Robert Malthus (1766–1834) that population growth is eventually curtailed by famine, disease, or other factors

Then came the green revolution. In the mid-1960s, as India was struggling to feed its people during yet another crippling drought, an American plant breeder named Norman Borlaug was working with Indian researchers to bring his high-yielding wheat varieties to Punjab. The new seeds were a godsend, says Kalkat, who was deputy director of agriculture for Punjab at the time. By 1970, farmers had nearly tripled their production with the same amount of work. "We had a big problem with what to do with the surplus," says Kalkat. "We closed schools one month early to store the wheat crop in the buildings."

Borlaug was born in Iowa and saw his mission as spreading the high-yield farming methods that had turned the American Midwest into the world's breadbasket to impoverished places throughout the world. His new dwarf wheat varieties, with their short, stocky stems supporting full, fat seed heads, were a startling breakthrough. They could produce grain like no other wheat ever seen—as long as there was plenty of water and synthetic fertilizer and little competition from weeds or insects. To that end, the Indian government subsidized canals, fertilizer, and the drilling of tube wells for irrigation and gave farmers free electricity to pump the water. The new wheat varieties quickly spread throughout Asia, changing the traditional farming practices of millions of farmers, and were soon followed by new strains of "miracle" rice. The new crops matured faster and enabled farmers to grow two crops a year instead of one. Today a double crop of wheat, rice, or cotton is the norm in Punjab, which, with neighboring Haryana, recently supplied more than 90 percent of the wheat needed by grain-deficient states in India.

The green revolution Borlaug started had nothing to do with the eco-friendly green label in vogue today. With its use of synthetic fertilizers and pesticides to nurture vast fields of the same crop, a practice known as monoculture, this new method of industrial farming was the antithesis of today's organic trend. Rather, William S. Gaud, then administrator of the U.S. Agency for International Development, coined the phrase in 1968 to describe an alternative to Russia's red revolution, in which workers, soldiers, and hungry peasants had rebelled violently against the tsarist government. The more pacifying green revolution was such a staggering success that Borlaug won the Nobel Peace Prize in 1970.

Today, though, the miracle of the green revolution is over in Punjab: Yield growth has essentially flattened since the mid-1990s. Overirrigation has led to steep drops in the water table, now tapped by 1.3 million tube wells, while thousands of hectares of productive land have been lost to salinization [soils becoming salty] and waterlogged soils. Forty years of intensive irrigation, fertilization, and pesticides have not been kind to the loamy gray fields of Punjab. Nor, in some cases, to the people themselves.

FIGURE 10.2 The green revolution, although it eased hunger, created a number of other problems, including the overuse of fertilizers, pesticides, and irrigation, which lead to removal of nutrients from the soil.

Shutterstock/Rawpixel

"The green revolution has brought us only downfall," says Jarnail Singh, a retired schoolteacher in Jajjal village. "It ruined our soil, our environment, our water table. Used to be we had fairs in villages where people would come together and have fun. Now we gather in medical centers. The government has sacrificed the people of Punjab for grain."

Others, of course, see it differently. Rattan Lal, a noted soil scientist at Ohio State who graduated from Punjab Agricultural University in 1963, believes it was the abuse—not the use—of green revolution technologies that caused most of the problems. That includes the overuse of fertilizers, pesticides, and irrigation and the removal of all crop residues from the fields, essentially strip-mining soil nutrients. "I realize the problems of water quality and water withdrawal," says Lal. "But it saved hundreds of millions of people. We paid a price in water, but the choice was to let people die."

In terms of production, the benefits of the green revolution are hard to deny. India hasn't experienced famine since Borlaug brought his seeds to town, while world grain production has more than doubled. Some scientists credit increased rice yields alone with the existence of 700 million more people on the planet.

The Next Green Revolution

Many crop scientists and farmers believe the solution to our current food crisis lies in a second green revolution, based largely on our newfound knowledge of the gene. Plant breeders now know the sequence of nearly all of the 50,000 or so genes in corn and soybean plants and are using that knowledge in ways that were unimaginable only four or five years ago, says Robert Fraley, chief technology officer for the agricultural giant Monsanto. Fraley is convinced that genetic modification, which allows breeders to bolster crops with beneficial traits from other species, will lead to new varieties with higher yields, reduced fertilizer needs, and drought tolerance—the holy grail for the past decade. He believes biotech will make it possible to double yields of Monsanto's core crops of corn, cotton, and soybeans by 2030. "We're now poised to see probably the greatest period of fundamental scientific advance in the history of agriculture."

But is a reprise of the green revolution—with the traditional package of synthetic fertilizers, pesticides, and irrigation, supercharged by genetically engineered seeds—really the answer to the world's food crisis? Last year a massive study called the "International Assessment of Agricultural Knowledge, Science and Technology for Development" concluded that the immense production increases brought about by science and technology in the past 30 years have failed to improve food access for many of the world's poor. The six-year study, initiated by the World Bank and the UN's Food and Agriculture Organization and involving some 400 agricultural experts from around the globe, called for a paradigm shift in agriculture toward more sustainable and ecologically friendly practices that would benefit the world's 900 million small farmers, not just agribusiness.

The green revolution's legacy of tainted soil and depleted aquifers is one reason to look for new strategies. So is what author and University of California, Berkeley, professor Michael Pollan calls the Achilles heel of current green revolution methods: a dependence on fossil fuels. Natural gas, for example, is a raw material for nitrogen fertilizers. "The only way you can have one farmer feed 140 Americans is with monocultures. And monocultures need lots of fossil-fuel-based fertilizers and lots of fossil-fuel-based pesticides," Pollan says. "That only works in an era of cheap fossil fuels, and that era is coming to an end. Moving anyone to a dependence on fossil fuels seems the height of irresponsibility."

So far, genetic breakthroughs that would free green revolution crops from their heavy dependence on irrigation and fertilizer have proved elusive. Engineering plants that can fix their own nitrogen or are resistant to drought "has proven a lot harder than they thought," says Pollan.

A Change in Focus—Agroecology

And so a shift has already begun to small, underfunded projects scattered across Africa and Asia. Some call it **agroecology**, others sustainable agriculture, but the underlying idea is revolutionary: that we must stop focusing on simply maximizing grain yields at any cost and consider the environmental and social impacts of

agroecology: an ecological approach to agriculture that views agricultural areas as ecosystems and is concerned with the ecological impact of agricultural practices

food production. Vandana Shiva is a nuclear physicist turned agroecologist who is India's harshest critic of the green revolution. "I call it monocultures of the mind," she says. "They just look at yields of wheat and rice, but overall the food basket is going down. There were 250 kinds of crops in Punjab before the green revolution." Shiva argues that small-scale, biologically diverse farms can produce more food with fewer petroleum-based inputs. Her research has shown that using compost instead of natural-gas-derived fertilizer increases organic matter in the soil, sequestering carbon and holding moisture—two key advantages for farmers facing climate change. "If you are talking about solving the food crisis, these are the methods you need," adds Shiva.

In northern Malawi one project is getting many of the same results as the Millennium Villages project, at a fraction of the cost. There are no hybrid corn seeds, free fertilizers, or new roads here in the village of Ekwendeni. Instead the Soils, Food and Healthy Communities (SFHC) project distributes legume seeds, recipes, and technical advice for growing nutritious crops like peanuts, pigeon peas, and soybeans, which enrich the soil by fixing nitrogen while also enriching children's diets. The program began in 2000 at Ekwendeni Hospital, where the staff was seeing high rates of malnutrition. Research suggested the culprit was the corn monoculture that had left small farmers with poor yields due to depleted soils and the high price of fertilizer.

Which is why the project's research coordinator, Rachel Bezner Kerr, is alarmed that big-money foundations are pushing for a new green revolution in Africa. "I find it deeply disturbing," she says. "It's getting farmers to rely on expensive inputs produced from afar that are making money for big companies rather than on agroecological methods for using local resources and skills. I don't think that's the solution."

The Challenge Ahead

Regardless of which model prevails—agriculture as a diverse ecological art, as a high-tech industry, or some combination of the two—the challenge of putting enough food in nine billion mouths by 2050 is daunting. Two billion people already live in the driest parts of the globe, and climate change is projected to slash yields in these regions even further. No matter how great their yield potential, plants still need water to grow. And in the not too distant future, every year could be a drought year for much of the globe.

New climate studies show that extreme heat waves, such as the one that withered crops and killed thousands in western Europe in 2003, are very likely to become common in the tropics and subtropics by century's end. Himalayan glaciers that now provide water for hundreds of millions of people, livestock, and farmland in China and India are melting faster and could vanish completely by 2035. In the worst-case scenario, yields for some grains could decline by 10 to 15 percent in South Asia by 2030. Projections for southern Africa are even more dire. In a region already racked by water scarcity and food insecurity, the all-important corn harvest could drop by 30 percent—47 percent in the worst-case scenario. All the while the population clock keeps ticking, with a net of 2.5 more mouths to feed born every second. That amounts to 4,500 more mouths in the time it takes you to read this article.

Adapted from Bourne, J. K., Jr. 2009. The Global Food Crisis: The End of Plenty. National Geographic.
Available online at: http://ngm.nationalgeographic.com/print/2009/06/cheap-food/bourne-text

10.2 Genetically Engineering Foods

The opening article for this chapter asked what the next agricultural revolution would look like—would it be one focused on new forms of genetic engineering or one based on the concept of agroecology or sustainable agriculture. In the following article, Tadlock Cowan of the Congressional Research Service reviews the major environmental, health, and regulatory issues associated with the development and use of genetic engineering or agricultural biotechnology.

Unlike the first green revolution, which was achieved mainly through traditional plant-breeding approaches, genetic modification of crops represents a fundamentally new technology. Traditional plant breeding sought to cross-breed or combine traits from the same plant types to produce a new and better variety. For example, a rice plant that produced a lot of grain but blew over easily in the wind could be cross-bred with another rice plant that produced less grain but had a stronger stem and could withstand the wind. The resulting rice plant, after repeated breeding, would feature both desirable traits—high grain production and stoutness—in a single seed. In contrast, genetic modification works by removing genetic material from one organism and inserting it into the DNA of another, often in "novel" ways or in combinations that would never occur in nature (for example, inserting fish genes into a tomato plant).

As with traditional plant breeding, genetic engineering seeks to develop plants that feature certain desirable traits. These could include developing plants that feature "input" traits such as resistance to pests or resistance to fungus and disease or plants that can withstand frost or drought conditions. This could also include developing "output" traits such as plants that have much higher nutritional content than traditional varieties.

The use of genetically engineered crops has grown rapidly in countries like the United States, especially for soybeans, corn, and cotton where GM crops make up between 70–90% of total production. This rapid growth has raised concerns about the environmental, health, and economic impacts of widespread use of genetically engineered crops—all issues reviewed in this reading.

By *Tadlock Cowan*

Farmers have always modified plants and animals to improve growth rates and yields, create varieties resistant to pests and diseases, and infuse special nutritional or handling characteristics. Such modifications have been achieved by crossbreeding plants and animals with desirable traits, through hybridization [mixing different species], and other methods. Now, using recombinant DNA techniques [combining DNA fragments from different organisms], scientists also genetically modify plants and animals by selecting individual genes that carry desirable traits (e.g., resistance to a pest or disease) from one organism, and inserting them into another, sometimes very different, organism, that can be raised for food, fiber, pharmaceutical, or industrial uses.

Karl Ereky, a Hungarian engineer, coined the term "biotechnology" in 1919 to refer to the science and the methods that permit products to be produced from raw materials with the aid of living organisms. According to the Convention of Biological Diversity, biotechnology is "any technological application that uses biological systems, living organisms, or derivatives thereof, to make or modify products or processes for specific use". According to the FAO's [Food and Agriculture Organization] statement on biotechnology, "interpreted in a narrow sense, [biotechnology] covers a range of different technologies such as gene manipulation and gene transfer, DNA typing and cloning of plants and animals."

Since genetically engineered (GE, sometimes called genetically modified organism or GMO) crop varieties first became commercially available in the mid-1990s, U.S. soybean, cotton, and corn farmers have rapidly adopted them in order to lower production costs and increase crop yields. Proponents point to the emergence of "second generation" GE commodities that could shift the focus of biotechnology from the "input" side (creating traits that benefit crop production, such as pest resistance) to the "output" side (creating traits that benefit consumers, such as lower fat oils). These second generation products could offer enhanced nutritional and processing qualities and also industrial and pharmaceutical uses. Future products are expected to be livestock- as well as crop-based. Critics, meanwhile, complain that biotechnology companies generally have not yet delivered the consumer benefits they have been promising for years.

Current Applications

Crops

In 2008, GE crops were planted on an estimated 308.8 million acres worldwide, a year-over-year increase of 26.4 million acres. The total number of countries growing such crops reached 25 in 2008. Most of the acreage was highly concentrated among four crops—soybeans, corn, cotton, and canola—and six countries.

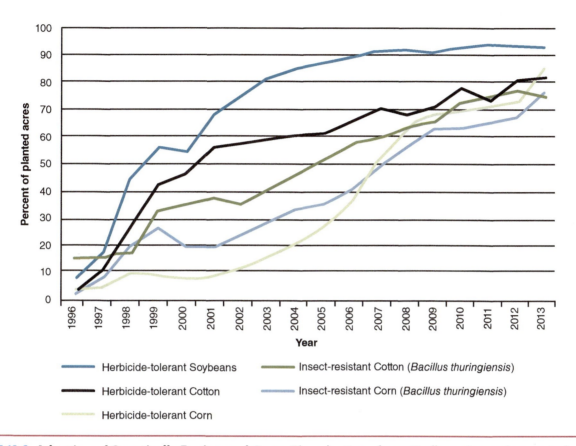

Year

- Herbicide-tolerant Soybeans
- Herbicide-tolerant Cotton
- Herbicide-tolerant Corn
- Insect-resistant Cotton (*Bacillus thuringiensis*)
- Insect-resistant Corn (*Bacillus thuringiensis*)

FIGURE 10.3 **Adoption of Genetically Engineered Crops.** The adoption of genetically engineered crops, including herbicide-tolerant (HT) crops and insect-resistant crops engineered with *Bacillus thuringiensis*, has increased greatly. Illustration by Maury Aaseng.

The United States has approximately 50% of global acreage (154.4 million acres), and Argentina had 16.8% (51.9 million acres). Brazil (12.6%, 39.0 million acres), Canada (6.1%, 18.8 million acres), India (6.1%, 18.8 million acres), and China (3.0%, 9.4 million acres) have the largest shares of the remaining planted acres. In the United States, over 60 GE plant varieties were approved by APHIS [United States Department of Agriculture—Animal and Plant Health Inspection Service] for commercial use through early 2005. By 2009, 70 plant varieties had been deregulated by APHIS. Ninety-one percent of all U.S. soybean, 88% of all upland cotton, and 85% of all corn acres were planted with GE seed varieties in 2009, according to USDA's National Agricultural Statistics Service. Virtually all current commercial applications benefit the production side of agriculture, with weed and insect control by far the most widespread uses of GE crops in the United States and abroad.

FIGURE 10.4 Genetically modified foods can have a number of advantages, including resistance to pests. Genetically modified brinjal is resistant to the shoot borer insect.

Herbicide-tolerant (HT) crops are engineered to tolerate herbicides [a substance used to kill unwanted plants] that would otherwise kill them along with the targeted weeds. These include HT soybeans, HT upland cotton, and to a lesser extent, HT corn. Many of these are referred to as "Roundup Ready" because they are engineered to resist Monsanto's glyphosate herbicide, marketed under the brand name "Roundup." More recently, Monsanto has announced various "stacked trait" varieties—varieties that combine resistance not only to glyphosate/Roundup but also to the herbicides dicamba and glufosinate.

Insect-resistant crops effectively have the pesticide inserted into the plants themselves to control insect pests for the life of the crop. These varieties are often referred to as having a plant-incorporated protectant (PIP). Many of these crops have been genetically engineered with *Bt* (*Bacillus thuringiensis,* a soil bacterium), which produces a naturally occurring pesticide. These insect-resistant varieties are most prevalent in upland cotton to control tobacco budworm, bollworm, and pink bollworm; and in corn to control earworm and several types of corn borers. Monsanto is also developing "stacked trait" varieties of soybeans and sugar cane that are resistant to insects as well as glyphosate/Roundup.

Other crops approved for commercialization have included varieties of flax, papaya, potatoes, radicchio, rapeseed, rice, squash, sugar beets, tobacco, and tomatoes. However, these are either not commercialized or not widely planted. For example, the biotechnology firm Calgene's FlavrSavr tomato, first marketed to consumers from 1995 to 1997, was withdrawn after Calgene determined that the varieties being grown were not of consistently high quality.

In contrast to abandoning certain approved GE products, a variety of white GE corn has recently begun to be used in tortilla making after initial resistance by food processors. Herbicide resistant GE sugar beets were only planted in large acreage in the 2008 crop year. While commercially available since 2000, Western beet growers did not plant them because sugar-using food companies (e.g., Hershey, Mars) and beet sugar industry groups (e.g., American Crystal Sugar) balked at the idea of GE beets, thinking that consumers would be opposed. That opposition—real or potential—has apparently subsided to the point that processors have cleared their growers to plant the GE variety. Nonetheless, the Center for Food Safety filed suit in January 2008 challenging APHIS's deregulation of GE sugar beets arguing that wind-pollinated GE sugar beets will inevitably cross-pollinate with related crops being grown in proximity, contaminating conventional sugar beets and organic chard and table beet crops.

Nonetheless, USDA reported that between 1987 and early 2005, APHIS had approved more than 10,700 applications to conduct field tests of various GE crop varieties (out of 11,600 received from companies and other researchers), which the USDA characterized as "a useful indicator of R&D efforts on crop biotechnology." Nearly 5,000 applications were approved for corn alone, followed by soybeans, potatoes, cotton, tomatoes, and wheat. More than 6,700 applications were for HT and insect resistant varieties; the others were to test product quality, virus or fungal resistance, or agronomic (e.g., drought resistance) properties. By October 2008, APHIS had approved more than 13,000 field trials of GE plants, most of which continued to be crop plants bearing genes conferring resistance to certain insects or tolerance to certain herbicides.

U.S. Food Products Containing GE Crops

An estimated 70% of all processed U.S. foods likely contain some GE material. That is largely because two such plants (corn and soybeans, where farmers have widely adopted GE varieties) are used in many different processed foods. U.S. biotechnology rules do not require segregation and labeling of GE crops and foods, as long as they are substantially equivalent to those produced by more conventional methods.

Soy-based ingredients include oil, flour, lecithin, and protein extracts. Corn-based ingredients include corn meal and corn syrups, used in many processed products. Canola oil (mostly imported from Canada, where GE-canola is grown) and cottonseed oil are used in cooking oils, salad dressings, snack foods, and other supermarket items. No GE-produced animals are yet approved for human consumption, although cheeses may contain chymosin, and dairy products may have been produced from milk containing GE-BST [a growth hormone].

Analysts say some farmers are wary of planting GE crop varieties because their customers may be worried about their safety, although as the case of sugar beets noted above suggests, public opposition to GE products in processed food may be declining. Biotechnology supporters contend that safety concerns are unfounded because scientific reviews have found approved GE crop varieties to be safe, and that foreign governments are simply using such concerns to maintain barriers to imports.

Future GE Applications

"Input" Traits

For farmers, new insect-resistant and herbicide-tolerant GE varieties are under development or have been developed for other crops besides corn, cotton, and soybeans. These include wheat and rice, alfalfa, peanuts, sunflowers, forestry products, sugarcane, apples, bananas, lettuce, strawberries, and eventually other fruits and vegetables. Other traits being developed through genetic engineering include drought and frost tolerance, enhanced photosynthesis, and more efficient use of nitrogen. Tomatoes that can be grown in salty soils, and recreational turf grasses that are herbicide tolerant, pest resistant, and/or more heat and drought tolerant, also are under development. In animal agriculture, pigs have been engineered for increased sow milk output to produce faster-growing piglets. Cloned cattle also have been developed to resist mastitis. APHIS approved field trials in June 2007 for a transgenic sunflower with a carp growth hormone inserted. The GE sunflower would be used in aquaculture feed for farm raised shrimp. Currently awaiting government approval for food use are GE salmon that require as little as half the usual time to grow to market size. Other such fish could follow later.

"Output" Traits

For processors and consumers, research on a range of GE products is continuing: oilseeds low in saturated and transfats; tomatoes with anti-cancer agents; grains with optimal levels of amino acids; rice with elevated iron levels; and rice with beta-carotene, a precursor of Vitamin A ("golden" rice). Other future products could include "low-calorie" sugar beets; strawberries and corn with higher sugar content to improve flavor; colored cotton; improved cotton fiber; delayed-ripening melons, bananas, strawberries, raspberries, and other produce (delayed-ripening tomatoes already are approved); and naturally decaffeinated coffee. Critics, however, point out that, although biotechnology advocates have been forecasting the adoption of various "output" traits for some time, few have actually reached the marketplace.

Other plants being developed could become "factories" for pharmaceutical compounds. The compounds would be extracted and purified for human and animal health uses (among concerns are whether they could "contaminate" food crops. Some varieties of plants under development could also produce "bioindustrials," including plastics and polyurethane. Future transgenic livestock also might yield pharmaceuticals and/or human organ and tissue replacements. To date, none of these innovations have been commercialized, although some have begun field trials.

GMOs in the Developing World

In Asia, particularly China and India, governments view GMOs as a way to produce more food for burgeoning populations, despite some in-country opposition and support for labeling GE products. China has been researching GE corn, cotton, wheat, soy, tomatoes, and peppers since 1986. It has, however, been reluctant to approve commercial varieties of GE, which have been under development there. If so, it would be the first time a GE plant was used widely as a staple food, and may influence the decisions of other Asian countries with regard to accepting GE foods.

In the debate over the potential contribution of biotechnology to food security in developing countries, critics argue that the benefits of biotechnology in such countries have not been established and that the technology poses unacceptable risks. They also suggest that intellectual property rights (IPR) protection gives multinational companies control over developing country farmers. Proponents say that the development of GE technology appears to hold great promise, with the potential to complement other, more traditional research methods, as the new driving force for sustained agricultural productivity in the 21st century. They maintain that IPR difficulties have been exaggerated.

According to a recent report published by the International Service for the Acquisition of Agribiotech Applications, 12 developing nations planted GE crops in 2007. Of the total 114.3 million hectares of GE crops cultivated worldwide, 43% of the global GE crop area is in developing countries.

Differences on this issue were featured in 2002, when the United Nations (UN) World Food Program (WFP) announced an appeal for food aid to meet the needs of some 14 million foodshort people in six southern African countries: Lesotho, Malawi, Mozambique, Swaziland, Zambia, and Zimbabwe. However, a debate over the presence of genetically modified corn in U.S. food aid shipments made the provision of food aid more difficult and costly. Some of the countries expressed reluctance to accept unmilled GE corn on account of perceived environmental and commercial risks associated with potential introduction of GE seeds into southern African agriculture. Zambia refused all shipments of food aid with GE corn out of health concerns as well. In March 2004, Angola said it too would ban imports of GE food aid, including thousands of tons of U.S. corn, despite a need to feed approximately 2 million Angolans.

The United States has blamed EU policies for southern African countries' views on food aid containing GE products. The United States maintains that genetically modified crops are safe to eat and that there is little likelihood of GE corn entering the food supply of African countries for several reasons, including the fact that current bioengineered varieties of corn are not well adapted to African growing conditions. South Africa is the only African country to commercialize biotech crops.

The Food and Agriculture Organization (FAO) of the United Nations has also offered a qualified endorsement of agricultural biotechnology, stating that it "can benefit the poor when appropriate innovations are developed and when poor farmers in poor countries have access to them. Thus far, these conditions are only being met in a handful of developing countries." Biotechnology research and development should complement other agricultural improvements that give priority to the problems of the poor, FAO said, adding: "Regulatory procedures should be strengthened and rationalized to ensure that the environment and public health are protected and that the process is transparent, predictable and science-based." Other groups have been more pointed in criticizing GE crops, arguing that they can have hidden costs that are inadequately examined by biotechnology advocates.

Environmental Concerns

Two main issues continue to drive the science and public debate on the environmental impacts of GMOs. One issue is the transfer of the introduced genes to wild plants and non-GM crops (i.e., gene flow from GE plants). The second issue concerns the indirect effects of the GE crops themselves on the local environment. Aside from the contamination issue to producers who do not want to plant GE crops, there is the concern that introduced genes can lead to herbicide and pesticide resistance [a process where a weed or insect pest develops resistance to chemical sprays] in non-target species [plants that were not targeted for resistance]. There is mounting evidence that some weeds have begun to show resistance to glyphosate/Roundup. Several varieties of rye grass, Palmer amaranth (pigweed), common waterhemp, and giant ragweed are showing signs of resistance to glyphosate, in addition to the common ragweed.

Biotechnology advocates claim that GE crops offer environmental advantages over conventionally produced organisms. They note that the technology is more precise than traditional methods like crossbreeding. The latter methods transfer unwanted and unanticipated characteristics along with the desired new traits from one organism to another. Biotechnology also has made it possible to apply fewer and less toxic chemical herbicides and insecticides and to reduce soil tillage (thereby decreasing erosion and improving soil fertility), supporters of the technology assert.

FIGURE 10.5 Young soybean plants thrive in the residue of a wheat crop, which protects the soil from erosion and helps retain moisture for the new crop. During a 1999 drought, the Rodale Institute's organic no-till soybean field produced nearly twice as much crop as an adjacent conventional field. Rodale's experts say that instead of solving drought with genetic engineering, it is safer to manage the soil to maximize its water holding capacity.

Shutterstock/Fotokostic

Critics counter that genetic engineering is not like traditional breeding. It creates crop and animal varieties that would not otherwise occur in nature, posing unpredictable risks to the environment (and to human health), they point out. Because they are living organisms, GE crops are difficult to control, greatly increasing the potential for escaping into the environment, crossbreeding with and overtaking wild species, and generally disrupting the natural ecosystem, critics believe. For example, GE, herbicide-tolerant seeds or pollen could inadvertently create "superweeds" that outcompete cultivated or wild plants, critics argue.

Adapted from: Cowan, T. 2010. Agricultural Biotechnology: Background and Recent Issues. Congressional Research Service, Washington, D.C. Available online at: http://www.cnie.org/NLE/CRSreports/10Oct/RL32809.pdf

10.3 The Local Foods Movement

Our modern food system not only provides us with an abundance of food at relatively low prices, it also allows us to eat different foods at almost any time of the year, even if they are out of season. Earlier generations of Americans expected to only have fresh strawberries in June or July and fresh apples in September and October. Today, however, fresh raspberries, grapes, peaches, beans, and mangoes are available in supermarkets throughout the year.

*This trend, combined with increased consumption of processed foods and the concentration of meat production, has given rise to a concept known as "**food miles**." Food miles is a measure of how far our food travels on average from where it was produced to where it is consumed. Since transport of food requires the use of fossil fuels, an increase in food miles is likely to increase the overall environmental impact of that product. Recent studies have found that most supermarket produce has traveled an average of 1,500 miles, and one study estimated that it requires 435 calories of fossil fuel energy to transport a 5-calorie strawberry from California to New York.*

> **food miles**: measure of how far food travels on average from where it is produced to where it is consumed

Awareness of the environmental impacts of food miles combined with growing concern over food safety has led more and more Americans to grow their own food or seek out local producers. In this article, Lester Brown of the Worldwatch Institute summarizes these trends and argues that they could be the early signs of a more fundamental shift in the way food is grown, marketed, and consumed in this country. Brown argues that a shift to purchasing more local foods can significantly decrease food miles and reduce other environmental impacts of conventional agriculture. For example, more localized livestock production can address some of the problems caused by concentrated animal feeding operations (CAFOs) and encourage a return to integrated crop-livestock operations that characterized almost all agricultural systems until very recently.

By L. Brown

*I*n the United States, there has been a surge of interest in eating fresh local foods, corresponding with mounting concerns about the climate effects of consuming food from distant places and about the obesity and other health problems associated with junk food diets. This is reflected in the rise in urban gardening, school gardening, and farmers' markets.

With the fast-growing local foods movement, diets are becoming more locally shaped and more seasonal. In a typical supermarket in an industrial country today it is often difficult to tell what season it is because the store tries to make everything available on a year-round basis. As oil prices rise, this will become less common. In essence, a reduction in the use of oil to transport food over long distances—whether by plane, truck, or ship—will also localize the food economy.

This trend toward localization is reflected in the recent rise in the number of farms in the United States, which may be the reversal of a century-long trend of farm consolidation. Between the agricultural census of 2002 and that of 2007, the number of farms in the United States increased by 4 percent to roughly 2.2 million.

The new farms were mostly small, many of them operated by women, whose numbers in farming jumped from 238,000 in 2002 to 306,000 in 2007, a rise of nearly 30 percent.

Many of the new farms cater to local markets. Some produce fresh fruits and vegetables exclusively for farmers' markets or for their own roadside stands. Others produce specialized products, such as the goat farms that produce milk, cheese, and meat or the farms that grow flowers or wood for fireplaces. Others specialize in organic food. The number of organic farms in the United States jumped from 12,000 in 2002 to 18,200 in 2007, increasing by half in five years.

Gardening

Gardening was given a big boost in the spring of 2009 when U.S. First Lady Michelle Obama worked with children from a local school to dig up a piece of lawn by the White House to start a veg-

FIGURE 10.6 More people are purchasing local foods grown by the region's farmers. Many find locally grown food to be fresher than foods shipped thousands of miles, and enjoy purchasing directly from farmers.

etable garden. There was a precedent. Eleanor Roosevelt planted a White House victory garden during World War II. Her initiative encouraged millions of victory gardens that eventually grew 40 percent of the nation's fresh produce.

Although it was much easier to expand home gardening during World War II, when the United States was largely a rural society, there is still a huge gardening potential—given that the grass lawns surrounding U.S. residences collectively cover some 18 million acres. Converting even a small share of this to fresh vegetables and fruit trees could make an important contribution to improving nutrition.

Many cities and small towns in the United States and England are creating community gardens that can be used by those who would otherwise not have access to land for gardening. Providing space for community gardens is seen by many local governments as an essential service, like providing playgrounds for children or tennis courts and other sport facilities.

Local Markets

Many market outlets are opening up for local produce. Perhaps the best known of these are the farmers' markets where local farmers bring their produce for sale. In the United States, the number of these markets increased from 1,755 in 1994 to more than 4,700 in mid-2009, nearly tripling over 15 years. Farmers' markets reestablish personal ties between producers and consumers that do not exist in the impersonal confines of the supermarket. Many farmers' markets also now take food stamps, giving low-income consumers access to fresh produce that they might not otherwise be able to afford. With so many trends now boosting interest in these markets, their numbers may grow even faster in the future.

Schools

In school gardens, children learn how food is produced, a skill often lacking in urban settings, and they may get their first taste of freshly picked peas or vine-ripened tomatoes. School gardens also provide fresh produce for school lunches. California, a leader in this area, has 6,000 school gardens.

Many schools and universities are now making a point of buying local food because it is fresher, tastier, and more nutritious, and it fits into new campus greening programs. Some universities compost kitchen and cafeteria food waste and make the compost available to the farmers who supply them with fresh produce.

Supermarkets are increasingly contracting with local farmers during the season when locally grown produce is available. Upscale restaurants emphasize locally grown food on their menus. In some cases, year-round food markets are evolving that market just locally produced foods, including not only fruit and vegetables but also meat, milk, cheese, eggs, and other farm products.

The Benefits of Local

Food from more distant locations boosts carbon emissions while losing flavor and nutrition. A survey of food consumed in Iowa showed conventional produce traveled on average 1,500 miles, not including food imported from other countries. In contrast, locally grown produce traveled on average 56 miles—a huge difference in fuel investment. And a study in Ontario, Canada, found that 58 imported foods traveled an average of 2,800 miles. Simply put, consumers are worried about food security in a long-distance food economy. This trend has led to a new term: **locavore**, complementing the better known terms herbivore, carnivore, and omnivore.

locavore: one who primarily eats food that is grown or produced within the local community or region

As agriculture localizes, livestock production will likely start to shift away from mega-sized cattle, hog, and poultry feeding operations. The shift from factory farm production of milk, meat, and eggs by returning to mixed crop-livestock operations facilitates nutrient recycling as local farmers return livestock manure to the land. The combination of high prices of natural gas, which is used to make nitrogen fertilizer, and of phosphate, as reserves are depleted, suggests a much greater future emphasis on nutrient recycling—an area where small farmers producing for local markets have a distinct advantage over massive feeding operations.

In combination with moving down the food chain to eat fewer livestock products, reducing the food miles in our diets can dramatically reduce energy use in the food economy. And as world food insecurity mounts, more and more people will be looking to produce some of their own food in backyards, in front yards, on rooftops, in community gardens, and elsewhere, further contributing to the localization of agriculture.

Adapted from Chapter 9, "Feeding Eight Billion People Well," in Lester R. Brown, Plan B 4.0: Mobilizing to Save Civilization. Copyright © Earth Institute 2009, available online at http://www.earth-policy.org/index.php?/book_bytes/2009/pb4ch09_ss5#"

ENERGY & THE ENVIRONMENT

CHAPTER 11

Introduction to Energy

11.1 What Is Energy?

Energy makes change possible. It moves cars along the road and boats through the water. It bakes a cake in the oven, keeps ice frozen in the freezer, and lights our homes.

Scientists define energy as the ability to do work. Modern civilization is possible because we have learned how to change energy from one form to another and then use it to do work for us.

Energy comes in different forms:

- Heat (thermal)
- Light (radiant)
- Motion (kinetic)
- Electrical
- Chemical
- Nuclear energy
- Gravitational

We use energy for everything we do, from making a jump shot to sending astronauts into space.

When we use electricity in our home, the electrical power was probably generated by burning coal, by a nuclear reaction, or by a hydroelectric plant on a river, to name just a few sources. Therefore, coal, nuclear, and hydro are called energy sources. When we fill up a gas tank, the source might be petroleum refined from crude oil or ethanol made by growing and processing corn.

For example, the food you eat contains chemical energy, and your body stores this energy until you use it when you work or play.

Forms of Energy

We already learned that there are many forms of energy, but they can all be put into two categories:

- Potential energy
- Kinetic energy

Potential Energy

Potential energy is stored energy and the energy of position. There are several forms of potential energy.

- Chemical energy is energy stored in the bonds of atoms and molecules. Batteries, biomass, petroleum, natural gas, and coal are examples of stored chemical energy. Chemical energy is converted to thermal energy when we burn wood in a fireplace or burn gasoline in a car's engine.
- Mechanical energy is energy stored in objects by tension. Compressed springs and stretched rubber bands are examples of stored mechanical energy.

173

- Nuclear energy is energy stored in the nucleus of an atom—the energy that holds the nucleus together. Large amounts of energy can be released when the nuclei are combined or split apart. Nuclear power.
- Gravitational energy is energy stored in an object's height. The higher and heavier the object, the more gravitational energy is stored. When you ride a bicycle down a steep hill and pick up speed, the gravitational energy is being converted to motion energy. Hydropower is another example of gravitational energy, where the dam piles up water from a river into a reservoir.

Kinetic Energy

Kinetic energy is the motion of waves, electrons, atoms, molecules, substances, and objects.

- Radiant energy is electromagnetic energy that travels in transverse waves. Radiant energy includes visible light, x-rays, gamma rays and radio waves. Light is one type of radiant energy. Sunshine is radiant energy, which provides the fuel and warmth that make life on earth possible.
- Thermal energy, or heat, is the vibration and movement of the atoms and molecules within substances. As an object is heated up, its atoms and molecules move and collide faster. Geothermal energy is the thermal energy in the earth.
- Motion energy is energy stored in the movement of objects. The faster they move, the more energy is stored. It takes energy to get an object moving, and energy is released when an object slows down. Wind is an example of motion energy. A dramatic example of motion is a car crash, when the car comes to a total stop and releases all its motion energy at once in an uncontrolled instant.
- Sound is the movement of energy through substances in longitudinal (compression/rarefaction) waves. Sound is produced when a force causes an object or substance to vibrate. The energy is transferred through the substance in a wave. Typically, the energy in sound is far less than other forms of energy.
- Electrical energy is delivered by tiny charged particles called electrons, typically moving through a wire. Lightning is an example of electrical energy in nature.

Sources of Energy

We use many different energy sources to do work. Energy sources are classified as renewable or nonrenewable. Renewable and nonrenewable energy can be converted into secondary energy sources like electricity and hydrogen.

Most of Our Energy Is Nonrenewable

We get most of our energy from nonrenewable energy sources, which include fossil fuels (oil, natural gas, and coal). These energy sources are called fossil fuels because they were formed over millions of years by the action of heat from the earth's core and pressure from rock and soil on the remains (or fossils) of dead plants and creatures like microscopic diatoms. Another nonrenewable energy source is uranium, whose atoms can be split (through a process called nuclear fission) to create heat and eventually electricity.

We use renewable and nonrenewable energy sources to generate the electricity we need for our homes, businesses, schools, and factories. Electricity powers our computers, lights, refrigerators, washing machines, and heating and cooling systems.

Most of the gasoline used in cars and motorcycles, and the diesel fuel used in trucks, tractors, and buses are both made from crude oil and other hydrocarbon liquids that are nonrenewable resources. Natural gas, used to heat homes, dry clothes, and cook food, is also a nonrenewable resource.

Nonrenewable energy sources accounted for 90% of all energy used in the nation (Figures 11.1 and 11.2). Biomass, which includes wood, biofuels, and biomass waste, is the largest renewable source accounting for about half of all renewable energy and 5% of total energy consumption.

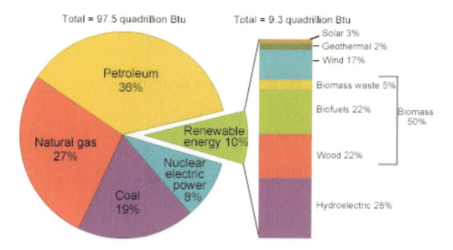

FIGURE 11.1   U.S energy consumption by energy source, 2013.

Biomass *renewable* Heating, electricity, transportation	4.6%	**Petroleum** *nonrenewable* Transportation, manufacturing	36.1%
Hydropower *renewable* Electricity	2.6%	**Natural gas** *nonrenewable* Heating, manufacturing, electricity	27.4%
Geothermal *renewable* Heating, electricity	0.2%	**Coal** *nonrenewable* Electricity, manufacturing	18.5%
Wind *renewable* Electricity	1.6%	**Nuclear/Uranium** *nonrenewable* Electricity	8.5%
Solar & other *renewable* Light, heating, electricity	0.3%		

FIGURE 11.2   U.S. energy consumption by source, 2013.

Renewable Energy

There are five main renewable energy sources:

- Solar energy from the sun, which can be turned into electricity and heat
- Wind energy
- Geothermal energy from heat inside the earth
- Biomass from plants, which includes firewood from trees, ethanol from corn, and biodiesel from vegetable oil
- Hydropower from hydroelectric turbines

Use of Renewable Energy Is Growing

Renewable energy sources include biomass, geothermal energy, hydropower, solar energy, and wind energy. They are called renewable energy sources because they are naturally replenished regularly. Day after day, the sun shines, the wind blows, and rivers flow. We use renewable energy sources to create electricity, for heat, and for transportation fuels.

How Are Secondary Sources of Energy Different from Primary Energy Sources?

Electricity and hydrogen are different than other energy sources because they are secondary sources of energy. Secondary sources of energy—energy carriers—are used to store, move, and deliver energy in an easily useable form. We have to use another energy source to make secondary sources of energy like electricity and hydrogen.

The Law of Conservation of Energy

To scientists, conservation of energy does not mean saving energy. Instead, *the law of conservation of energy* says that energy is neither created nor destroyed. When we use energy, it doesn't disappear. We change it from one form of energy into another form of energy (Figure 11.3).

A car engine burns gasoline, converting the chemical energy in gasoline into mechanical energy. Solar cells change radiant energy into electrical energy. Energy changes form, but the total amount of energy in the universe stays the same.

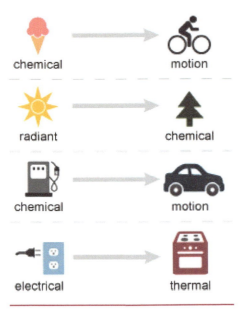

FIGURE 11.3 Energy transformations.

Converting One Form of Energy into Another

Energy efficiency is the amount of useful energy you get from any type of system. A perfectly energy-efficient machine would convert all the energy put into the machine to useful work. In reality, converting one form of energy into another form of energy always involves a loss of useable energy.

Most energy transformations are not very efficient. The human body is a good example. Your body is like a machine, and the fuel it requires is food. Food gives you energy to move, breathe, and think. But your body isn't very efficient at converting food into useful work. Your body is less than 5% efficient most of the time. The rest of the energy is lost as heat.

11.2 U.S Energy Facts

Americans Use Many Types of Energy

Petroleum (oil) is the largest share of U.S. primary energy consumption, followed by natural gas, coal, nuclear electric power, and renewable energy (including hydropower, wood, biofuels, biomass waste, wind, geothermal, and solar). Electricity is a secondary energy source that is generated from primary forms of energy.

Energy sources are commonly measured in different physical units to include barrels of oil, cubic feet of natural gas, tons of coal, and kilowatthours of electricity. In the United States, British thermal units (Btu), a measure of heat energy, is commonly used for comparing different types of energy. In 2013, total U.S. primary energy use was about 97.5 quadrillion (1015, or one thousand trillion) Btu.

The major energy users are residential and commercial buildings, industry, transportation, and electric power generators. The pattern of fuel use varies widely by sector. For example, petroleum oil provides 92% of the energy used for transportation, but only 1% of the energy used to generate electricity.

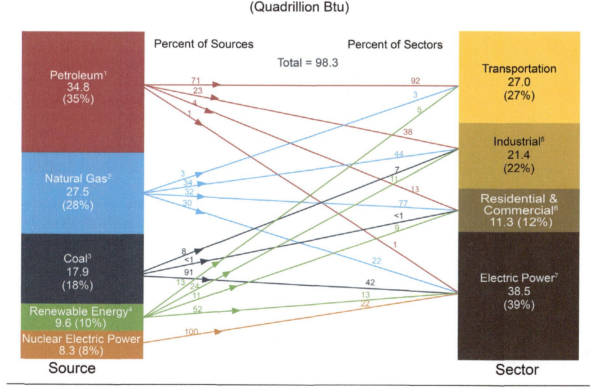

Primary Energy Consumption by Source and Sector, 2014
(Quadrillion Btu)

Courtesy of EIA

[1] Does not include biofuels that have been blended with petroleum—biofuels are included in "Renewable Energy."
[2] Excludes supplemental gaseous fuels.
[3] Includes less than -0.1 quadrillion Btu of coal coke net imports.
[4] Conventional hydroelectric power, geothermal, solar/photovoltaic, wind, and biomass.
[5] Includes industrial combined-heat-and-power (CHP) and industrial electricity-only plants.
[6] Includes commercial combined-heat-and-power (CHP) and commercial electricity-only plants.

[7] Electricity-only and combined-heat-and-power (CHP) plants whose primary business is to sell electricity, or electricity and heat, to the public. Includes 0.2 quadrillion Btu of electricity net imports not shown under "Source."
   Notes: Primary energy in the form that it is first accounted for in a statistical energy balance, before any transformation to secondary or tertiary forms of energy (for example, coal is used to generate electricity). • Sum of components may not equal total due to independent rounding.
   Sources: U.S. Energy Information Administration, Monthly Energy Review (March 2015), Tables 1.3, 2.1-2.6.

FIGURE 11.4   This graphic details where energy sources are consumed by different energy use sectors.

Domestic Energy Production Meets About 84% of U.S. Energy Demand

In 2013, energy produced in the United States provided about 84% of the nation's energy needs. The remaining energy was supplied mainly by imports of petroleum.

The three major fossil fuels—petroleum, natural gas, and coal—accounted for most of the nation's energy production in 2013:

- Natural gas—30%
- Coal—24%
- Petroleum (crude oil and natural gas plant liquids)—24%
- Renewable energy—11%
- Nuclear electric power—10%

The Mix of U.S. Energy Production Changes

The three major fossil fuels—petroleum, natural gas, and coal—have dominated the U.S. energy mix for more than 100 years. There are several recent changes in U.S. energy production:

- The share of coal produced from surface mines increased significantly from 25% in 1949 to 51% in 1971 to 66% in 2012. The remaining share was produced from underground mines.
- In 2013, natural gas production was higher than in any previous year. In recent years, more efficient and cost-effective drilling and production techniques have resulted in increased production of natural gas from shale formations.
- Total U.S. crude oil production generally decreased each year from a peak in 1970, but the trend reversed in 2010. In 2013, crude oil production was the highest since 1989. These increases were the result of increased use of horizontal drilling and hydraulic fracturing techniques, most notably in North Dakota and Texas.
- Natural gas plant liquids (NGPL) are hydrocarbons that are separated as liquids from natural gas at processing plants. They are important ingredients for manufacturing plastics and gasoline. Propane is the only NGPL that is widely used for heating and cooking. Production of NGPL fluctuates with natural gas production, but the NGPL share of total U.S. crude oil and petroleum field production increased from 8% in 1950 to 26% in 2013.

- In 2013, total renewable energy production and consumption reached record highs of about 9 quadrillion Btu each. Hydroelectric power production in 2013 was about 9% below the 50-year average, but increases in biofuels use and wind power generation increased the overall total contribution of renewable energy. Production of energy from wind and solar were at record highs in 2013.

11.3 Use of Energy

How We Use Energy

The United States is a highly developed and industrialized society. We use a lot of energy in our homes, in businesses, in industry, and for personal travel and transporting goods. There are four major sectors that consume energy at the point of end use (Figure 11.5).

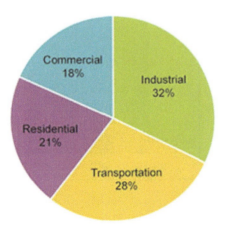

¹Includes electricity consumption.
Source: U.S. Energy Information Administration, *Monthly Energy Review*, Table 2.1 (April 2013) preliminary 2012 data.

FIGURE 11.5 Share of total energy consumed by major sectors of the economy, 2012.¹

The industrial sector includes facilities and equipment used for manufacturing, agriculture, mining, and construction.

The transportation sector comprises vehicles that transport people or goods, such as: cars, trucks, buses, motorcycles, trains, subways, aircraft, boats, barges, and even hot air balloons.

The residential sector consists of homes and apartments.

The commercial sector includes buildings such as offices, malls, stores, schools, hospitals, hotels, warehouses, restaurants, places of worship, and more.

Each end-use sector consumes electricity produced by the electric power sector.

Primary energy consumption in the United States was almost three times greater in 2012 as in 1949. In all but 18 of the years between 1949 and 2012, primary energy consumption increased over the previous year.

The year 2009 provided a sharp contrast to this historical trend, in part due to the economic recession. Real gross domestic product (GDP) fell 2% compared to 2008, and energy consumption declined by nearly 5%, the largest single-year decline since 1949. Decreases occurred in all four major end-use sectors: residential—3%, commercial—3%, industrial—9%, and transportation—3%. Consumption increased about 3% in 2010, but then declined slightly in 2011 and about 2% in 2012. Economic growth and other factors like weather and fuel prices impact consumption in each sector differently.

In Industry

The United States is a highly industrialized country. Industry accounts for about one-third of the energy used in the country.

There are many different uses and a variety of different energy sources used in the industrial sector. One main use is as boiler fuel, which means producing heat that is transferred to the boiler vessel to generate steam or hot water. Another use is as process heating, which is when energy is used directly to raise the temperature of products in the manufacturing process; examples are separating components of crude oil in petroleum refining, drying paint in automobile manufacturing, and cooking packaged foods. Energy sources are also used as feedstocks (see below) to make products.

"Other energy sources" account for 38% of the energy manufacturers' use of heat, power, and electricity generation. Among these sources are steam, pulping liquor from paper making, agricultural waste, tree wood, wood residues from mill processing, and wood-related and paper-related refuse.

In the manufacturing sector, the predominant energy sources are natural gas and electricity (a secondary source Figure 11.6).

For Transportation

America is a nation on the move. About 28% of all the energy we consume goes to transporting people and goods from one place to another.

Petroleum fuels are made from crude oil and liquids from natural gas processing. The petroleum fuels used for transportation include gasoline, diesel fuel, jet fuel, residual fuel oil, and liquid petroleum gases. In 2013, those fuels provided about 92% of the total energy used by the transportation sector

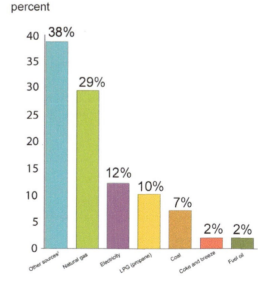

Source: U.S. Energy Information Administration, *Manufacturing Energy Consumption Survey 2010*, Table 1.2 (March 2013)

[1] Includes all use of energy and fuels; excludes shipments of energy sources produced onsite.

[2] Other sources include steam, pulping liquor from paper making, agricultural waste, tree wood, wood residues from mill processing, and wood-related and paper-related refuse.

FIGURE 11.6   Sources of energy used for industry and manufacturing, 2010.[1]

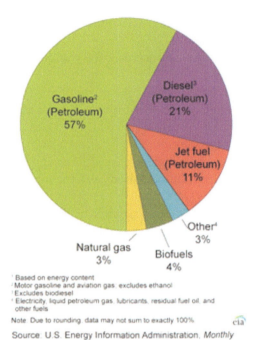

Based on energy content
Motor gasoline and aviation gas, excludes ethanol
Excludes biodiesel
Electricity, liquid petroleum gas, lubricants, residual fuel oil, and other fuels

Note: Due to rounding, data may not sum to exactly 100%.

Source: U.S. Energy Information Administration, *Monthly Energy Review June 2014*, tables 2.5 and 3.8c, preliminary data for 2013

FIGURE 11.7 Fuel used for U.S. transportation, 2013.[1]

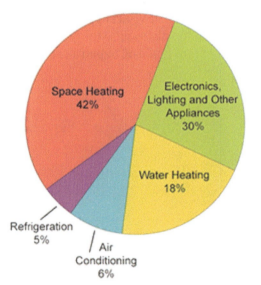

* 2009 is the most recent year for which data are available.

Source: U.S. Energy Information Administration, *Residential Energy Consumption Survey (RECS) 2009*

FIGURE 11.8 How energy is used in homes (2009).*

in the United States. Biofuels, such as ethanol and biodiesel, contributed about 5%, and natural gas contributed about 3%1. Electricity provided less than 1%.

Gasoline is the dominant transportation fuel used in the United States. Gasoline (excluding fuel ethanol) accounted for 57% of total U.S. transportation energy use in 2013 (Figure 11.7). When the ethanol that is blended into finished motor gasoline is included, gasoline's share was about 61%. Gasoline (including fuel ethanol) consumption for transportation averaged about 8.6 million barrels (363 million gallons) per day. About 6 million gallons per day of gasoline were consumed for uses other than for transportation. Gasoline is used mainly in cars, motorcycles, and light trucks. Diesel fuel (or distillate) is used mainly by heavy trucks, buses, and trains. Kerosene is used in jet airplanes, and residual fuel oil is used in ships. Natural gas and liquid petroleum gas are used in all types of vehicles, but they are used predominantly in heavy duty vehicles such as buses and other transportation fleets. Most of the electricity used for transportation is used by public mass transit systems.

Ethanol and biodiesel were actually some of the first fuels used in automobiles, but they were replaced by gasoline and diesel fuel. Today, most of the biofuels used in vehicles are added to gasoline and diesel fuel. Government incentives and mandates contributed to large increases in the use of biofuels in the United States over the past several decades. The amount of fuel ethanol added to motor gasoline consumed for transportation went from about 1.3 billion gallons in 1995 to about 13 billion gallons in 2013. Biodiesel consumption increased from 10 million gallons in 2001 to about 1.4 billion gallons in 2013.

In Homes

The amount of energy we use in our homes mainly depends on the climate where we live and the types and number of energy consuming devices we use. The pie chart on the right (Figure 11.8) shows the major energy uses in homes in 2009, when most energy use was for space heating (42%), followed by electronics, lighting and other appliances (30%), water heating (18%), air conditioning (6%), and refrigeration (5%).

The number and variety of ways we use energy in homes is changing rapidly. Energy use for air conditioning has doubled since 1980. U.S. households currently plug in more appliances and electronics at home than ever before. While refrigerators and cooking equipment have long been standard in homes, the ownership of appliances such as microwaves, dishwashers, and clothes washers and dryers has increased over the past 30 years.

It is increasingly common for homes to use multiple televisions and computers. Additionally, the home electronics market is constantly innovating, and new products such as DVRs, game systems, and rechargeable electronic devices are becoming ever more integral to our modern lifestyle. As a result of these changes, appliances and electronics (including refrigerators) now account for nearly one-third of all energy used in homes.

Types of Energy Used in Homes

Natural gas and electricity are the most-consumed energy sources in U.S. homes, followed by heating oil, and propane. Natural gas and heating oil (fuel oil) are used mainly for home/space heating. Space heating accounts for the largest share of the energy used in U.S. homes. Electricity, which is used for heating and cooling, also lights our homes and runs almost all of our appliances including refrigerators, toasters, and computers. Many homes in rural areas use propane for heating, while others use it to fuel their barbecue grills.

Regional Consumption Data Reflect Population Shifts and Climate

In the late 1990s, homes in the South Census Region surpassed the Midwest in consuming the most energy in the United States (Figure 11.9). This shift reflects the economic boom in the region, which stimulated U.S. migration to the South and the construction of more and larger homes. In 2009, homes in the South consumed 3.22 quadrillion Btu, about 3% of the country's total energy use and about 32% of energy used in homes.

Due to the longer heating seasons, the Northeast and Midwest regions still, consume the most energy per household, at 108 and 112 Million Btu per household in 2009, respectively.

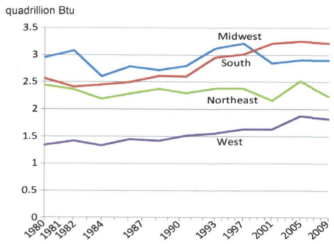

Source: U.S. Energy Information Administration, *Residential Energy Consumption Survey (RECS)* 1980-2009.

FIGURE 11.9   Total home energy use by region, 1980–2009.

In Commercial Buildings

Commercial buildings include a wide variety of building types—offices, hospitals, schools, police stations, places of worship, warehouses, hotels, and shopping malls. Different commercial activities all have unique energy needs but, as a whole, commercial buildings use more than half of their energy for heating and lighting (Figure 11.10).

Electricity and natural gas are the most common energy sources used in commercial buildings. Commercial buildings also use another source that you don't usually find in residential buildings—district energy. When there are many buildings close together, like on a college campus or in a big city, it is sometimes more efficient to have a central heating and cooling plant that distributes steam, hot water, or chilled water to all of the different buildings. This type of system (referred to as a district system) can reduce equipment and maintenance costs as well as save energy.

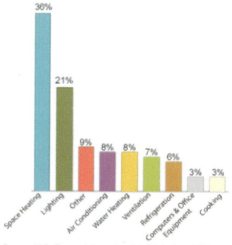

Source: U.S. Energy Information Administration, 2003 *Commercial Building Energy Consumption Survey*, Table E1A (September 2008).

FIGURE 11.10   Energy use in commercial buildings, 2003.

The types of buildings in the commercial sector are used for a mix of many different activities and uses. Retail and service buildings use the most total energy of all the commercial building types. This is not very surprising when you think about all the stores and service businesses there are all over the United States. Other commercial users of energy include offices, schools, health care and lodging facilities, food establishments, and many others.

Efficiency and Conservation

All of us use energy every day—for transportation, cooking, heating and cooling rooms, manufacturing, lighting, and entertainment. The choices we make about how we use energy—turning machines off when we're not using them or choosing to buy energy efficient appliances—impact our environment and our lives.

The terms energy conservation and energy efficiency have two distinct definitions. There are many things we can do to use less energy (conservation) and use it more wisely (efficiency).

Energy conservation is any behavior that results in the use of less energy. Turning the lights off when you leave the room and recycling aluminum cans are both ways of conserving energy.

Energy efficiency is the use of technology that requires less energy to perform the same function. A compact fluorescent light bulb that uses less energy than an incandescent bulb to produce the same amount of light is an example of energy efficiency. However, the decision to replace an incandescent light bulb with a compact fluorescent is an act of energy conservation.

11.4 Secondary Sources

What Are Secondary Energy Sources?

Secondary energy sources are also referred to as energy carriers, because they move energy in a useable form from one place to another. There are two well-known energy carriers:

- Electricity
- Hydrogen

We make electricity and hydrogen from the conversion of other sources of energy, such as coal, nuclear, or solar energy. These are called primary sources.

For many energy needs, it is much easier to use electricity or hydrogen than it is to use the primary energy sources themselves.

Electricity

Electricity is the flow of electrical power or charge. It is both a basic part of nature and one of our most widely used forms of energy.

Electricity is actually a secondary energy source, also referred to as an energy carrier. That means that we get electricity from the conversion of other sources of energy, such as coal, nuclear, or solar energy. These are called primary sources. The energy sources we use to make electricity can be renewable or non-renewable, but electricity itself is neither renewable or nonrenewable.

Before electricity became available over 100 years ago, houses were lit with kerosene lamps, food was cooled in iceboxes, and rooms were warmed by wood-burning or coal-burning stoves.

gillimar/Shutterstock

FIGURE 11.11 Energy efficient light bulb.

Many scientists and inventors have worked to decipher the principles of electricity since the 1600s. Some notable accomplishments were made by Benjamin Franklin, Thomas Edison, and Nikola Tesla. Benjamin Franklin demonstrated that lightning is electricity. Thomas Edison invented the first long-lasting incandescent light bulb.

Prior to 1879, direct current (DC) electricity had been used in arc lights for outdoor lighting. In the late 1800s, Nikola Tesla pioneered the generation, transmission, and use of alternating current (AC) electricity, which reduced the cost of transmitting electricity over long distances. Tesla's inventions used electricity to bring indoor lighting to our homes and to power industrial machines.

Despite its great importance in our daily lives, few of us probably stop to think what life would be like without electricity. Like air and water, we tend to take electricity for granted. But we use electricity to do many jobs for us every day—from lighting, heating, and cooling our homes to powering our televisions and computers.

Everything Is Made of Atoms

In order to understand electricity, we need to know something about atoms. Everything in the universe is made of atoms—every star, every tree, every animal. The human body is made of atoms. Air and water are, too. Atoms are the building blocks of the universe. Atoms are so small that millions of them would fit on the head of a pin.

Atoms Are Made of Even Smaller Particles

The center of an atom is called the nucleus. It is made of particles called protons and neutrons. The protons and neutrons are very small, but electrons are much, much smaller. Electrons spin around the nucleus in shells a great distance from the nucleus. If the nucleus were the size of a tennis ball, the atom would be the size of the Empire State Building. Atoms are mostly empty space.

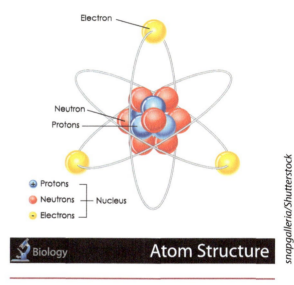

FIGURE 11.12

Graph of an Atom

If you could see an atom, it would look a little like a tiny center of balls surrounded by giant invisible bubbles (or shells). The electrons would be on the surface of the bubbles, constantly spinning and moving to stay as far away from each other as possible. Electrons are held in their shells by an electrical force.

The protons and electrons of an atom are attracted to each other. They both carry an electrical charge. Protons have a positive charge (+) and electrons have a negative charge (−). The positive charge of the protons is equal to the negative charge of the electrons. Opposite charges attract each other. An atom is in balance when it has an equal number of protons and electrons. The neutrons carry no charge and their number can vary.

The number of protons in an atom determines the kind of atom, or element, it is. An element is a substance consisting of one type of atom (the Periodic Table shows all the known elements), all with the same number of protons. Every atom of hydrogen, for example, has one proton, and every atom of carbon has six protons. The number of protons determines which element it is.

Electrons usually remain a constant distance from the nucleus in precise shells. The shell closest to the nucleus can hold two electrons. The next shell can hold up to eight. The outer shells can hold even more. Some atoms with many protons can have as many as seven shells with electrons in them. The electrons in the shells closest to the nucleus have a strong force of attraction to the protons. Sometimes, the electrons

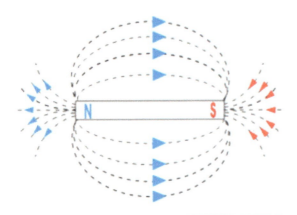

FIGURE 11.13   Magnets have magnetic fields.

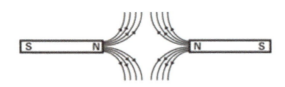

FIGURE 11.14   The same ends of magnets repel one another.

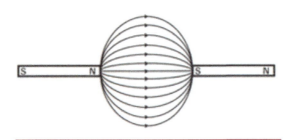

FIGURE 11.15   Opposite poles of magents (N-S) attract each other.

Source: National Energy Education Development Project (public domain)

in an atom's outermost shells do not. These electrons can be pushed out of their orbits. Applying a force can make them move from one atom to another. ***These moving electrons are electricity***.

Static Electricity

Lightning is a form of electricity. It is electrons moving from one cloud to another or jumping from a cloud to the ground. Have you ever felt a shock when you touched an object after walking across a carpet? A stream of electrons jumped to you from that object. This is called static electricity.

Have you ever made your hair stand straight up by rubbing a balloon on it? If so, you rubbed some electrons off the balloon. The electrons moved into your hair from the balloon. They tried to get far away from each other by moving to the ends of your hair. They pushed against each other and made your hair move—they repelled each other. Just as opposite charges attract each other, like charges repel each other.

Magnetic Fields & Electricity

The spinning of the electrons around the nucleus of an atom creates a tiny magnetic field. Most objects are not magnetic because their electrons spin in different, random directions, and cancel out each other. Magnets are different; the molecules in magnets are arranged so that their electrons spin in the same direction. This arrangement of atoms creates two poles in a magnet, a North-seeking pole and a South-seeking pole. The magnetic force in a magnet flows from the North pole to the South pole. This creates a magnetic field around a magnet (Figure 11.13).

Have you ever held two magnets close to each other? They don't act like most objects. If you try to push the South poles together, they repel each other. Two North poles also repel each other (Figure 11.14).

Turn one magnet around, and the North (N) and the South (S) poles are attracted to each other. Just like protons and electrons—opposites attract (Figure 11.15).

Electrochemical Energy

An electrochemical battery produces electricity using two different metals in a chemical solution. A chemical reaction between the metals and the chemicals frees more electrons in one metal than in the other. One end of the battery is attached to one of the metals, and the other end is attached to the other metal.

The end that frees more electrons develops a positive charge, and the other end develops a negative charge. If a wire is attached from one end of the battery to the other, electrons flow through the wire to balance the electrical charge.

A load is a device that does work or performs a job. If a load—such as an incandescent light bulb—is placed along the wire, the electricity can do work as it flows through the wire. Electrons flow from the negative end of the battery through the wire to the light bulb. The electricity flows through the wire in the light bulb and back to the positive end of the battery.

Electricity Travels in Circuits

Electricity must have a complete path before the electrons can move. If a circuit is open, the electrons cannot flow. When we flip on a light switch, we close a circuit (Figure 11.16). The electricity flows from an electric wire, through the light bulb, and back out another wire. The light bulb produces light as electricity flows through a tiny wire in the bulb, gets very hot, and glows.

When we flip the switch off, we open the circuit and no electricity flows to the light. The bulb burns out when the tiny wire breaks and the circuit is opened.

When we turn on a TV, electricity flows through wires inside the TV set, producing pictures and sound. Sometimes electricity runs motors like in washers or mixers. Electricity does a lot of work for us many times each day.

Transformers Help to Move Electricity Efficiently Over Long Distances

To solve the problem of sending electricity over long distances, William Stanley developed a device called a transformer. A transformer allows electricity to be efficiently transmitted over long distances. A transformer also makes it possible to supply electricity to homes and businesses located far from electric generating plants.

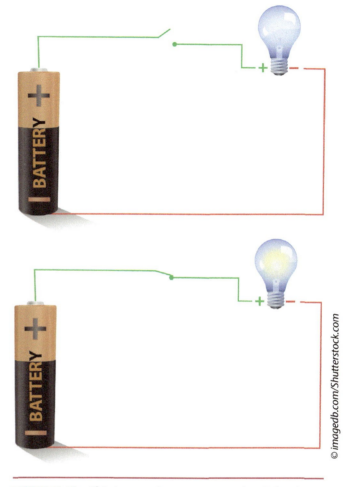

FIGURE 11.16   This image shows an open circuit (top) and a closed circuit (bottom) for a light bulb.

© imagedb.com/Shutterstock.com

A transformer changes the voltage of electricity in power lines. The electricity produced by a generator travels along cables to a transformer, which then changes electricity from low voltage to high voltage. Electricity can be moved long distances more efficiently using high voltage. Transmission lines are used to carry the electricity to a substation. Substations have transformers that change the high voltage electricity into lower voltage electricity. From the substation, distribution lines carry the electricity to homes, offices, and factories that all use low voltage electricity.

Electricity is measured in units of power called watts. It was named to honor James Watt, the inventor of the steam engine. One watt is a very small amount of power. It would require nearly 750 watts to equal one horsepower.

A kilowatt is the same as 1,000 watts. A kilowatthour (kWh) is equal to the energy of 1,000 watts working for one hour. The amount of electricity a power plant generates or a customer uses over a period of time is measured in kilowatthours (kWh). Kilowatthours are determined by multiplying the number of kilowatts required by the number of hours of use (Figure 11.17).

For example, if you use a 40-watt light bulb for 5 hours, you have used 200 watthours, or 0.2 kilowatthours, of electrical energy.

FIGURE 11.17   Picture of a residential electricity meter.

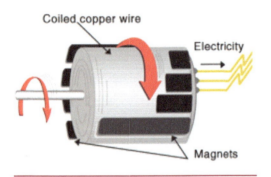

FIGURE 11.18   Diagram of a turbine generator—Spinning rotor turning coiled copper wir inside stationary magnets to generate electricity.

Source: Adapted from Energy For Keeps (Public Domain)

How Electricity Is Generated

A generator is a device that converts mechanical energy into electrical energy. The process is based on the relationship between magnetism and electricity. In 1831, scientist Michael Faraday discovered that when a magnet is moved inside a coil of wire, electrical current flows in the wire (Figure 11.18).

A typical generator at a power plant uses an electromagnet—a magnet produced by electricity—not a traditional magnet. The generator has a series of insulated coils of wire that form a stationary cylinder. This cylinder surrounds a rotary electromagnetic shaft. When the electromagnetic shaft rotates, it induces a small electric current in each section of the wire coil. Each section of the wire becomes a small, separate electric conductor. The small currents of individual sections are added together to form one large current. This current is the electric power that is transmitted from the power company to the consumer.

An electric utility power station uses either a turbine, engine, water wheel, or other similar machine to drive an electric generator—a device that converts mechanical or chemical energy to electricity. Steam turbines, internal-combustion engines, gas combustion turbines, water turbines, and wind turbines are the most common methods to generate electricity.

Steam turbine power plants powered by coal and nuclear energy produce about 70% of the electricity used in the United States. These plants are about 35% efficient. That means that for every 100 units of primary heat energy that go into a plant, only 35 units are converted to useable electrical energy.

Most of the electricity in the United States is produced using steam turbines.

A turbine converts the kinetic energy of a moving fluid (liquid or gas) to mechanical energy. In a steam turbine, steam is forced against a series of blades mounted on a shaft. This rotates the shaft connected to the generator. The generator, in turn, converts its mechanical energy to electrical energy based on the relationship between magnetism and electricity.

In steam turbines powered by fossil fuels, such as coal, petroleum (oil), and natural gas, the fuel is burned in a furnace to heat water in a boiler to produce steam.

Electricity Is Delivered to Consumers through a Complex Network

Electric power is generated at power plants and then moved to substations by transmission lines—large, high-voltage power lines. In the United States, the network of nearly 160,000 miles of high voltage transmission lines is known as the "grid" (Figure 11.19).

The utility, distribution company, or retail service provider selling you power may be a not-for-profit municipal entity, an electric co-operative owned by its members, a private, for-profit company owned by stockholders (often called an investor-owned utility), or a power marketer. Some Federally-owned authorities—including the Bonneville Power Administration and the Tennessee Valley Authority, among others—also buy, sell, and distribute power.

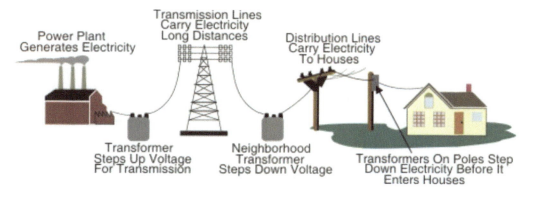

FIGURE 11.19   A flow diagram of power generation, transmission, and distribution from the power plant to residential houses.

Source: National Energy Education Development Project (Public Domain)

The origin of the electricity you consume may vary. Utilities may generate all the electricity they sell using just the power plants they own. Utilities may also purchase some of their supply on the wholesale market from other utilities, power marketers, independent power producers, or from a market based on membership in a regional transmission reliability organization.

Most of the existing grid was built during a highly structured, highly regulated era designed to ensure that everyone in the United States had reasonable access to electricity service. Utility customers, through fees authorized and regulated by State regulatory commissions, generally paid for developing and maintaining the grid. Many local grids are interconnected for reliability and commercial purposes, forming larger, more dependable networks to maximize coordination and planning. These networks extend throughout many States.

The North American Electric Reliability Corporation (NERC) was established to ensure that the grid in the United States was reliable, adequate, and secure (Figure 11.20). Some NERC members have formed regional organizations with similar missions. These organizations are referred to as Independent System Operators (ISOs) and Regional Transmission Organizations (RTOs). They are part of a national standard design advocated by the Federal Energy

FIGURE 11.20   NERC map.

Source: Energy Information Adminsitration. Map shows NERC (North American Electric Reliability Corporation) regions.

Regulatory Commission (FERC). Some have members who connect to lines in Canada or Mexico. Most, depending on the location and the utility, are indirectly connected to dozens and often hundreds of power plants. Some power consumed in the United States is imported from Canada and Mexico.

Electricity is an essential part of modern life. In our homes we use it for lighting, running appliances and electronics, and often for heating and cooling. Most consumers don't think much about their electricity until a power outage, or when they get a high utility bill.

Consumers Depend on Reliable Electricity

Fortunately in the United States, power outages are relatively infrequent and short in duration mainly because we have one of the most advanced, reliable, and well-maintained electricity generation, transmission, and distribution systems in the world. We plug in an appliance or turn on a switch without giving it a second thought. To ensure that continuous flow of electricity is there to meet our needs, electric power producers operate several types of large generators.

Did You Know

Electricity was first sold in the United States in 1879 by the California Electric Light Company in San Francisco, which produced and sold only enough electricity to power 21 electric lights (Brush arc light lamps).

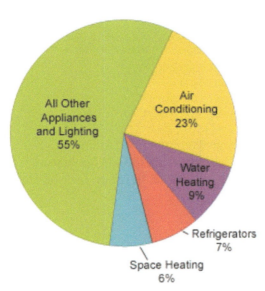

Source: U.S. Energy Information Administration, *Annual Energy Outlook 2012, Early Release*, Table 4.

FIGURE 11.21   How electricity is used in homes, 2011.

Source: Energy Information Adminsitration. Map shows NERC (North American Electric Reliability Corporation) regions.

U.S. electricity use in 2011 was more than 13 times greater than electricity use in 1950. Share of electricity use by major consuming sectors:

- Residential—37%
- Commercial—34%
- Industrial—26% (includes "direct use")
- Transportation—Only a small percentage of electricity is used in the transportation sector, mostly for trains and plug-in electric cars

Direct use is electricity generated mainly by manufacturing that is used by the facilities. Most of the electricity used in the residential sector is for air conditioning, refrigerators, space and water heating, lighting, and powering appliances and equipment (Figure 11.21).

Electricity and the Environment

Although electricity is a clean and relatively safe form of energy to use, there are environmental impacts associated with the production and transmission of electricity. **Nearly all types of electric power plants have some impacts or effects on the environment, some more than others**.

The United States has laws to reduce these impacts. Perhaps the most important such law is the Clean Air Act, which established regulations for the control of air emissions from most power plants. The Environmental Protection Agency (EPA) administers the Act and sets emissions standards for power plants through various programs, such as the Acid Rain Program. The Act has resulted in a substantial reduction of emissions of some of the major types of air pollutants in the United States (see Figure 11.22 that shows air emissions from a coal-powered electricity plant).

Gary Whitton/Shutterstock

FIGURE 11.22   Hunter Power Plant, a Coal-Fired Power Plant South of Castle Dale, Utah.

The Impact of Power Plants on the Landscape

All power generators or plants have a physical footprint (the area where they are placed or located). Some can be located inside, on, or next to an existing building, so the impact of their footprint is very small. Most large power plants require clearing land to locate the power plant and any necessary fuel storage areas (in the case of hydro-power dams, a reservoir forms behind the dam). Some plants may also require the construction of access roads, rail, pipelines, transmission lines, and access to cooling water supplies or reservoirs.

Besides the physical footprint, many power plants are large physical structures that have impacts on the visual landscape. Some people may not like this especially where the landscape is relatively natural or pristine.

In general, the larger the area disturbed, the greater the real and potential impacts on the landscape.

Electric Power Lines Also Have a Footprint

Power transmission and distribution lines carry electricity from power plants to customers. Most transmission lines are strung above ground on large towers. The towers and lines impact the visual landscape, especially when they pass through pristine natural areas. Trees near the wires may be disturbed and have to be managed to keep them from touching the wires and these activities can affect native plant populations and wildlife. Power lines can be placed under the ground, but this is more expensive and may result in a greater disturbance of the landscape than overhead lines.

Hydrogen

What Is Hydrogen?

> **Did You Know**
>
> Hydrogen is the lightest element and is a gas at normal temperature and pressure. Hydrogen condenses to a liquid at temperatures of −423° Fahrenheit (−253° Celsius).

Hydrogen is the simplest element. Each atom of hydrogen has only one proton. It is also the most plentiful gas in the universe. Stars like the sun are made primarily of hydrogen. The sun is basically a giant ball

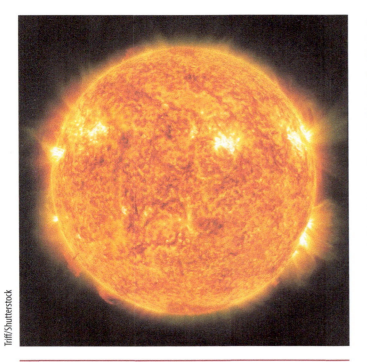

FIGURE 11.23 The sun.

of hydrogen and helium gases (Figure 11.23). In the sun's core, hydrogen atoms combine to form helium atoms. This process— alled fusion—gives off radiant energy. This radiant energy sustains life on Earth. It gives us light and makes plants grow. It makes the wind blow and rain fall. It is stored as chemical energy in fossil fuels. Most of the energy we use today originally came from the sun's radiant energy.

Hydrogen gas is so much lighter than air that it rises fast and is quickly ejected from the atmosphere. This is why hydrogen as a gas (H_2) is not found by itself on Earth. It is found only in compound form with other elements. Hydrogen combined with oxygen, is water (H_2O). Hydrogen combined with carbon forms different compounds, including methane (CH_4), coal, and petroleum. Hydrogen is also found in all growing things—for example, biomass. It is also an abundant element in the Earth's crust.

Hydrogen has the highest energy content of any common fuel by weight (about three times more than gasoline), but the lowest energy content by volume (about four times less than gasoline).

Hydrogen Is an Energy Carrier

Energy carriers move energy in a useable form from one place to another. Electricity is the most well-known energy carrier. We use electricity to move the energy in coal, uranium, and other energy sources from power plants to homes and businesses. We also use electricity to move the energy in flowing water from hydropower dams to consumers. For many energy needs, it is much easier to use electricity than the energy sources themselves.

Like electricity, hydrogen is an energy carrier and must be produced from another substance. Hydrogen is not currently widely used, but it has potential as an energy carrier in the future. Hydrogen can be produced from a variety of resources (water, fossil fuels, or biomass) and is a byproduct of other chemical processes.

How Is Hydrogen Made?

Because hydrogen doesn't exist on Earth as a gas, it must be separated from other elements. Hydrogen atoms can be separated from water, biomass, or natural gas molecules. The two most common methods for producing hydrogen are steam reforming and electrolysis (water splitting). Scientists have discovered that even some algae and bacteria give off hydrogen.

Steam reforming is currently the least expensive method of producing hydrogen and accounts for about 95% of the hydrogen produced in the United States. This method is used in industries to separate hydrogen atoms from carbon atoms in methane (CH_4). But the steam reforming process results in greenhouse gas emissions that are linked with global warming.

1. Electrolysis Creates No Emissions but Is Costly
2. Electrolysis is a process that splits hydrogen from water. It results in no emissions, but it is currently an expensive process. New technologies are currently being developed.
3. Hydrogen can be produced at large central facilities or at small plants for local use.

How Much Hydrogen Is Produced in the United States?

About 9 million metric tons of hydrogen are produced in the United States annually, enough to power 20-30 million cars or 5-8 million homes. Most of this hydrogen is produced in three states: California, Louisiana, and Texas.

Nearly all of hydrogen consumed in the United States is used by industry for refining, treating metals, and processing foods.

The National Aeronautics and Space Administration (NASA) is the primary user of hydrogen as an energy fuel; it has used hydrogen for years in the space program. Liquid hydrogen fuel lifts NASA's space shuttles into orbit (Figure 11.24). Hydrogen batteries, called fuel cells, power the shuttle's electrical systems. The only by-product is pure water, which the crew uses as drinking water.

FIGURE 11.24   NASA space shuttle enterprise.

Hydrogen Fuel Cells Produce Electricity

Hydrogen fuel cells make electricity (Figure 11.25). They are very efficient, but expensive to build. Small fuel cells can power electric cars. Large fuel cells can provide electricity in remote places with no power lines. Because of the high cost to build fuel cells, large hydrogen power plants won't be built for a while. However, fuel cells are being used in some places as a source of emergency power, from hospitals to wilderness locations. Portable fuel cells are being sold to provide longer power for laptop computers, cell phones, and military applications.

Today, there are more than 300 hydrogen-fueled vehicles in the United States. Most of these vehicles are buses and automobiles powered by electric motors. They store hydrogen gas or liquid on board and convert the hydrogen into electricity for the motor using a fuel cell. Only a few of these vehicles burn the hydrogen directly (producing almost no pollution).

The present cost of fuel cell vehicles greatly exceeds that of conventional vehicles in large part due to the expense of producing fuel cells.

Hydrogen vehicles are starting to move from the laboratory to the road. Hydrogen vehicles are in use by a few state agencies and a few private entities. Currently, there are 68 hydrogen refueling stations in the United States, nearly a third of which are located in California. There are so-called "chicken and egg" questions that hydrogen developers are working hard to solve, including: who will buy hydrogen cars if there are no refueling stations? And who will pay to build a refueling station if there are no cars and customers?

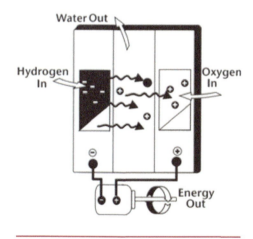

FIGURE 11.25   Hydrogen fuel cell.

Source: The National Energy Education Project (Public Domain)

CHAPTER 12

Fossil Fuels

It's deceptive to lump all forms of fossil fuels into 1 chapter, because these sources of energy are such a big part of the U.S. and global energy equation. They are discussed together because they share many key characteristics. In terms of environmental science, a primary characteristic is that they all give off greenhouse gases when combusted (burned), thereby contributing to global climate change.

12.1 Oil and Petroleum Products

Nonrenewable Sources

Recall that energy sources are considered nonrenewable if they cannot be replenished (made again) in a short period of time. On the other hand, renewable energy sources such as solar and wind can be replenished naturally in a short period of time.

The four nonrenewable energy sources used most often are:

Oil and petroleum products—including gasoline, diesel fuel, heating oil, and propane
Natural gas
Coal
Uranium (nuclear energy) – NOT a fossil fuel

Nonrenewable energy sources come out of the ground as liquids, gases, and solids. Crude oil (petroleum) is the only commercial nonrenewable fuel that is naturally in liquid form. Natural gas and propane are normally gases, and coal is a solid.

Coal, petroleum, natural gas, and propane are all considered fossil fuels because they were formed from the buried remains of plants and animals that lived millions of years ago. Conversely, uranium ore, a solid, is mined and converted to a fuel used at nuclear power plants. Uranium is not a fossil fuel, but is a nonrenewable fuel.

Oil and Petroleum Products

Oil was formed from the remains of animals and plants (diatoms) that lived millions of years ago in a marine (water) environment before the dinosaurs. Over millions of years, the remains of these animals and plants were covered by layers of sand and silt. Heat and pressure from these layers helped the remains turn into what we today call crude oil (Figure 12.1). The word petroleum means rock oil or oil from the earth.

Did you know crude oil can be sweet or sour? Crude oil is called sweet when it contains only a small amount of sulfur and sour if it contains a lot of sulfur. Crude oil is also classified by the weight of its molecules. Light crude oil flows freely like water, while heavy crude oil is thick like tar.

After crude oil is removed from the ground, it is sent to a refinery by pipeline, ship, barge, or rail. At a refinery, different parts of the crude oil are separated into useable petroleum products. Crude oil is measured in barrels.

All materials courtesy of the National Oceanic and Atmospheric Administration

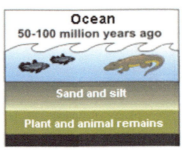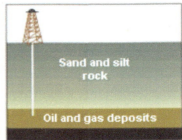

FIGURE 12.1 Petroleum and natural gass formation.

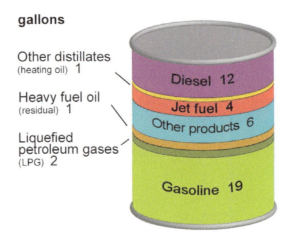

FIGURE 12.2 Products made from barrel of crude oil, 2013.

Refining Crude Oil

A refinery is a factory. Just as a paper mill turns lumber into paper, a refinery takes crude oil and other feedstocks and turns them into gasoline and many other petroleum products. A refinery runs 24 hours a day, 365 days a year and requires a large number of employees to run it. A refinery can occupy as much land as several hundred football fields. Workers often ride bicycles to move from place to place inside the complex. One barrel of crude oil, when refined, produces about 19 gallons of finished motor gasoline, and 10 gallons of diesel, as well as other petroleum products. Most petroleum products are used to produce energy. For instance, many people across the United States use propane to heat their homes. A 42 U.S. gallon barrel of crude oil yields about 45 gallons of petroleum products. This gain from processing the crude oil is similar to what happens to popcorn, which gets bigger after it is popped.

There are many other products made from petroleum, some of those products include:

- Ink
- Crayons
- Dishwashing liquids
- Deodorant
- Eyeglasses

- CDs and DVDs
- Tires
- Ammonia
- Heart valves and other medical equipment
- Computers
- Tires

Where Our Oil Comes From

Crude oil is produced in 31 states and U.S. coastal waters. In 2012, 61% of U.S. crude oil production came from five states (Figure 12.3):

- Texas (31%)
- North Dakota (10%)
- California (8%)
- Alaska (8%)
- Oklahoma (4%)

As of 2012, about 20% of U.S. crude oil was produced from wells located offshore in federally administered waters of the Gulf of Mexico.

Although total U.S. crude oil production generally declined between 1985 and 2008, it has been increasing since 2008. More cost effective drilling technology has helped boost production, especially in North Dakota, Texas, and the offshore Gulf of Mexico.

Natural gas plant liquids (NGPL) are liquids that are separated from natural gas at processing plants and are important ingredients for manufacturing plastics and gasoline. Propane is the only NGPL that is widely used for heating and cooking. Production of NGPL fluctuates with natural gas production, but their share of total U.S. petroleum production has increased from 8% in 1950 to 27% in 2012.

In 2012, the United States relied on net imports (imports minus exports) for about 40% of the petroleum that we used.

About 100 countries produce crude oil and NGPL. The top five producing countries in 2012 and their share of total world production were (Figure 12.4):

- Saudi Arabia (13%)
- Russia (12%)
- United States (12%)
- China (5%)
- Canada (4%)

After the fall of the Union of Soviet Socialist Republics (USSR) in 1991, Saudi Arabia became the world's top petroleum producer.

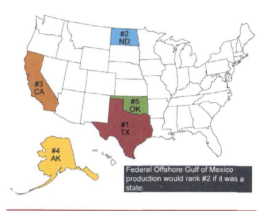

FIGURE 12.3 Top 5 crude producing states in 2012 from most to least include; Texas (#1), North Dakota, California, Alaska, and Oklahoma (#5).

Source: U.S. Energy Information Administration, *Petroleum Supply Monthly.* Table 26 (June 2013).

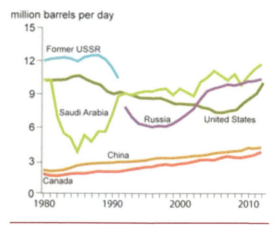

FIGURE 12.4 Top oil* producing countries, 1980–2012.

* Crude oil and natural gas plant liquids (NGPL).

Source: U.S. Energy Information Administration, International Energy Statistics (as of 9/30/13), China does not include NGPL.

Offshore Oil and Gas

When you are at your favorite beach in Florida or California, you are not at the very edge of the country. Although it might seem like the ocean is the border of the United States, the border is actually 200 miles out from the land. This 200-mile-wide band around the country is called the Exclusive Economic Zone (EEZ). In 1983, President Reagan claimed the area of the EEZ in the name of the United States (Figure 12.6). In 1994, all countries were granted an EEZ of 200 miles from their coastline according to the International Law of the Sea.

There is a lot of activity just beyond the beach. The beach extends from the shore into the ocean on a continental shelf that gradually descends to a sharp drop, called the continental slope. This continental shelf can be as narrow as 20 kilometers or as wide as 400 kilometers. The water on the continental shelf is shallow, rarely more than 150 to 200 meters deep. The EEZ is part of the United States. The federal government manages the land under the sea on behalf of the American people.

The U.S. Bureau of Ocean Energy Management, Regulation, and Enforcement (BOEMRE) leases the land under the ocean to producers. These companies pay BOEMRE rental fees and royalties on all the minerals they extract from the ocean floor. Individual states control the waters off their coasts out to three miles for most states and between nine and 12 for Florida, Texas, and some other states.

The continental shelf drops off at the continental slope, ending in abyssal plains that are three to five kilometers below sea level. Many of the plains are flat, while others have jagged mountain ridges, deep canyons, and valleys. The tops of some of these mountain ridges form islands where they extend above the water.

Most of the energy we get from the ocean is extracted from the ground. Oil, natural gas, and minerals come from the ocean floor.

People are working on other new ways to use the ocean. Solar and wind energy have been used on land, and now they are also being used at sea. Other energy sources that are being explored in the ocean are wave energy, tidal energy, methane hydrates, and ocean thermal energy conversion.

Ksenia Ragozina/Shutterstock

FIGURE 12.5 Oil and natural gas are frequently pumped from reservoirs under the ocean floor.

Use of Oil

Crude oil and other liquids produced from fossil fuels are refined into petroleum products that we use for many different purposes. Biofuels, such as ethanol and biodiesel, are also used as petroleum products, mainly in mixtures with gasoline and diesel fuel.

Did You Know

In 2013, nearly three-fourths of total U.S. petroleum consumption was in the transportation sector.

The United States consumes more energy from petroleum than from any other energy source. In 2013, total U.S. petroleum consumption was 18.9 million barrels per day (bbl/d), or 36% of all the energy we consumed.

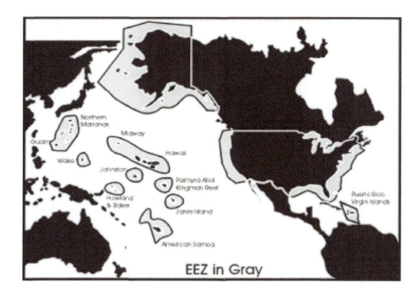

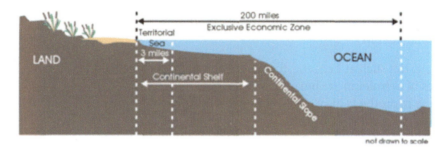

FIGURE 12.6   The Exclusive Economic Zone (EEZ) extends 200 miles from the shoreline.

When petroleum products are burned to produce energy, they may be used to propel a vehicle, to heat a building, or to produce electric power in a generator. In addition, petroleum may be used as a raw material (a feedstock) to create products such as plastics, polyurethane, solvents, asphalt, and hundreds of other intermediate and end-user goods.

Gasoline is the main petroleum product consumed in the United States. In 2007, gasoline consumption reached a record high of 9.3 million bbl/d (or 391 million gallons per day).

Distillate fuel oil includes diesel fuel and heating oil. Diesel fuel is used in the diesel engines of heavy construction equipment, trucks, buses, tractors, boats, trains, some automobiles, and electricity generators. Heating or fuel oil is used in boilers and furnaces to heat homes and buildings, for industrial heating, and for producing electricity. Total distillate fuel oil consumption in 2013 was more than 3.8 million bbl/d, or 20% of total petroleum consumption.

Liquefied petroleum gases (LPGs) include propane, ethane, butane, and other gases that are produced at natural gas processing plants and oil refineries. LPG consumption in 2013 was nearly 2.4 million bbl/d.

Propane, a heavily consumed LPG, is used in homes for space and water heating, clothes drying, cooking, as a transportation fuel, for heating greenhouses and livestock housing, and for drying crops. In addition, the chemical industry uses about half of all the propane consumed in the United States as a raw material for making plastics, nylon, and other materials.

Jet fuel consumption in 2013 was 1.4 million bbl/d.

Worldwide consumption of petroleum was 89.4 million bbl/d in 2012. The three largest petroleum consuming countries in 2012 included the United States, China, and Japan:

- United States (18.5 million barrels per day)
- China (10.3 million barrels per day)
- Japan (4.7 million barrels per day)

The U.S. Energy Information Administration projects that most petroleum-based and nonpetroleum-based liquid fuels—including those derived from fuels such as coal, biomass, and natural gas—will continue to be used for transportation over the next two decades.

Projections to 2040 show consumption of petroleum-based motor gasoline declining while biofuel consumption grows as a result of government mandates. Diesel fuel consumption expands, primarily for use in heavy-duty vehicles. Consumption of petroleum and other liquids increases in the industrial sector but decreases in all the other end-use sectors (residential, commercial, and transportation).

Oil and the Environment

Products from oil (petroleum products) help us do many things. We use them to fuel our airplanes, cars, and trucks, to heat our homes, and to make products like medicines and plastics. Even though petroleum products make life easier—finding, producing, moving, and using them can harm the environment through air and water pollution.

Petroleum products give off the following emissions when they are burned as fuel:

- Carbon dioxide (CO_2)
- Carbon monoxide (CO)
- Sulfur dioxide (SO_2)
- Nitrogen oxides (NO_X) and Volatile Organic Compounds (VOC)
- Particulate matter (PM)

Lead and various air toxics such as benzene, formaldehyde, acetaldehyde, and 1,3-butadiene may also be emitted when some types of petroleum are burned.

Nearly all of these byproducts have negative impacts on the environment and human health:

- Carbon dioxide is a greenhouse gas and a source of global warming.[1]
- SO_2 causes acid rain, which is harmful to plants and to animals that live in water, and it worsens or causes respiratory illnesses and heart diseases, particularly in children and the elderly.

- NO_X and VOCs contribute to ground-level ozone, which irritates and damages the lungs.
- PM results in hazy conditions in cites and scenic areas, and, along with ozone, contributes to asthma and chronic bronchitis, especially in children and the elderly. Very small, or "fine PM" is also thought to cause emphysema and lung cancer.
- Lead can have severe health impacts, especially for children, and air toxics are known or probable carcinogens.

Over the years, new technologies and laws have helped to reduce problems related to petroleum products. As with any industry, the Government monitors how oil is produced, refined, stored, and sent to market to reduce the impact on the environment. Since 1990, fuels like gasoline and diesel fuel have also been improved so that they produce less pollution when we use them. Because a lot of air pollution comes from cars and trucks, many environmental laws have been

Virunja/Shutterstock

FIGURE 12.7 Signs like these help to discourage people from dumping used motor oil (for example) into storm drains, which eventually drain to natural waterways.

aimed at changing the make-up of gasoline and diesel fuel so that they produce fewer emissions. These "reformulated fuels" are much cleaner-burning than gasoline and diesel fuel were in 1990.

Exploring and drilling for oil may disturb land and ocean habitats. New technologies have greatly reduced the number and size of areas disturbed by drilling, sometimes called "footprints."[2] Satellites, global positioning systems, remote sensing devices, and 3-D and 4-D seismic technologies make it possible to discover oil reserves while drilling fewer wells.

The use of horizontal and directional drilling makes it possible for a single well to produce oil from a much bigger area. Today's production footprints are also smaller those 30 years ago because of the development of movable drilling rigs and smaller "slimhole" drilling rigs.

When the oil in a well becomes uneconomic to produce, the well must be plugged below ground, making it hard to tell that it was ever there. As part of the "rigs-to-reefs" program, some old offshore rigs are tipped over and left on the sea floor to become artificial reefs that attract fish and other marine life. Within six months to a year after a rig is toppled, it becomes covered with barnacles, coral, sponges, clams, and other sea creatures.

Oil Spills

While oil spills from ships and offshore platforms are the most well-known source of oil in ocean water, a lot of oil actually gets into water from natural oil seeps coming from the ocean floor. The natural seeps may be a "major" source of oil that enters the environment globally, but they are slow, small, and spread out over large areas, and the ecosystem has adapted to them, versus the catastrophic impact that a tanker or well spill has on the areas affected.

Leaks also happen when we use petroleum products on land. For example, gasoline sometimes drips onto the ground when people are filling their gas tanks, when motor oil gets thrown away after an oil change, or when fuel escapes from a leaky storage tank. When it rains, the spilled products get washed into the gutter and eventually flow to rivers and into the ocean. Another way that oil sometimes gets into water is when fuel is leaked from motorboats and jet skis.

When a leak in a storage tank or pipeline occurs, petroleum products can also get into the ground, and the ground must be cleaned up. To prevent leaks from underground storage tanks, all buried tanks are supposed to be replaced by tanks with a double lining.

If oil is spilled into rivers or oceans, it can harm wildlife. When we talk about "oil spills," people usually think about oil that leaks from a ship that is involved in an accident. The amount of oil spilled from ships dropped significantly during the 1990s partly because new ships were required to have a "double-hull" lining to protect against spills.

Case Study: Deepwater Horizon Oil Spill

On April 20, 2010, an explosion on the Deepwater Horizon MC252 drilling platform in the Gulf of Mexico caused the rig to sink and oil began leaking into the Gulf. Before it was finally capped in mid-July, almost 5 million barrels of oil were released into the Gulf. The magnitude of this spill is something our nation has not seen before, causing significant impacts to wildlife and the fishing community along the large coastal areas of Louisiana, Mississippi, Texas, Alabama, and Florida. Although it will be months before the full extent of the damage will be known, NOAA (National Oceanic and Atmospheric Administration) acted quickly to begin preliminary assessments and plan for restoration along the coast.

With 20 years of experience restoring habitat impacted by oil, including restoration activities conducted in the Gulf prior to the spill, NOAA is prepared to lead the effort to develop a range of restoration strategies. We will be involved in both short-term and long-term restoration efforts to return the Gulf to pre-spill conditions.

Key Facts

- NOAA responds to as many as 150 oil spills every year.
- In response to oil spills, NOAA has restored thousands of acres of coastal habitat in the past 20 years.
- NOAA and the other trustees involved hold the responsible party accountable for assessment and restoration costs.

Potential Pathways for Oil to Reach Bottom Sediments

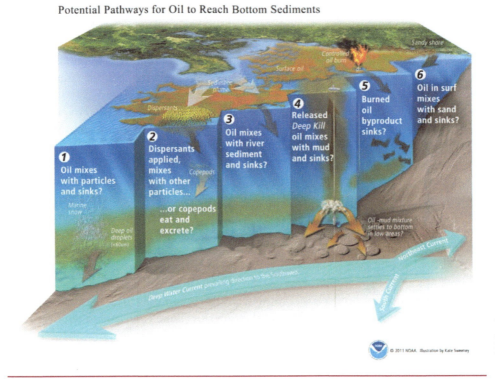

NOAA/Kate Sweeney.

FIGURE 12.8 Illustration showing the potential pathways of spilled oil following the 2010 Deepwater Horizon oil spill in the Gulf of Mexico.

Visualizing the Deepwater Horizon Oil Spill

After the Deepwater Horizon spill, oil moved through the water column in a variety of ways. We knew there were several possible scenarios for how it might move into the sediments at the bottom of the ocean.

Using mapping data and discussing the concepts with NOAA scientists, medical and scientific illustrator Kate Sweeney developed a single, striking graphic illustration that clearly encompassed all the most likely possibilities (Figure 12.8):

Another of Kate's images was used as part of the **Programmatic Environmental Impact Statement** public scoping process to illustrate the Gulf ecosystem and potential oil impacts (Figure 12.9):

Gasoline

Gasoline is a nonrenewable fuel made from petroleum. Refineries in the United States produce about 19 gallons of gasoline from every 42-gallon barrel of crude oil that is refined. The rest of the barrel gets turned into other petroleum products like diesel fuel, heating oil, jet fuel, and propane.

Did You Know
Gasoline changes with the seasons.

Gasoline changes with the seasons.

The main difference between winter- and summer-grade gasoline is their vapor pressure. Gasoline vapor pressure is important for an automobile engine to work properly. During cold winter months, vapor pressure must be high enough for the engine to start easily.

Gasoline evaporates more easily in warm weather, releasing more volatile organic compounds that contribute to health problems and the formation of ground-level ozone and smog. In order to cut down on pollution, the U.S. Environmental Protection Agency requires petroleum refiners to reduce the vapor pressure of gasoline used during the summer months.

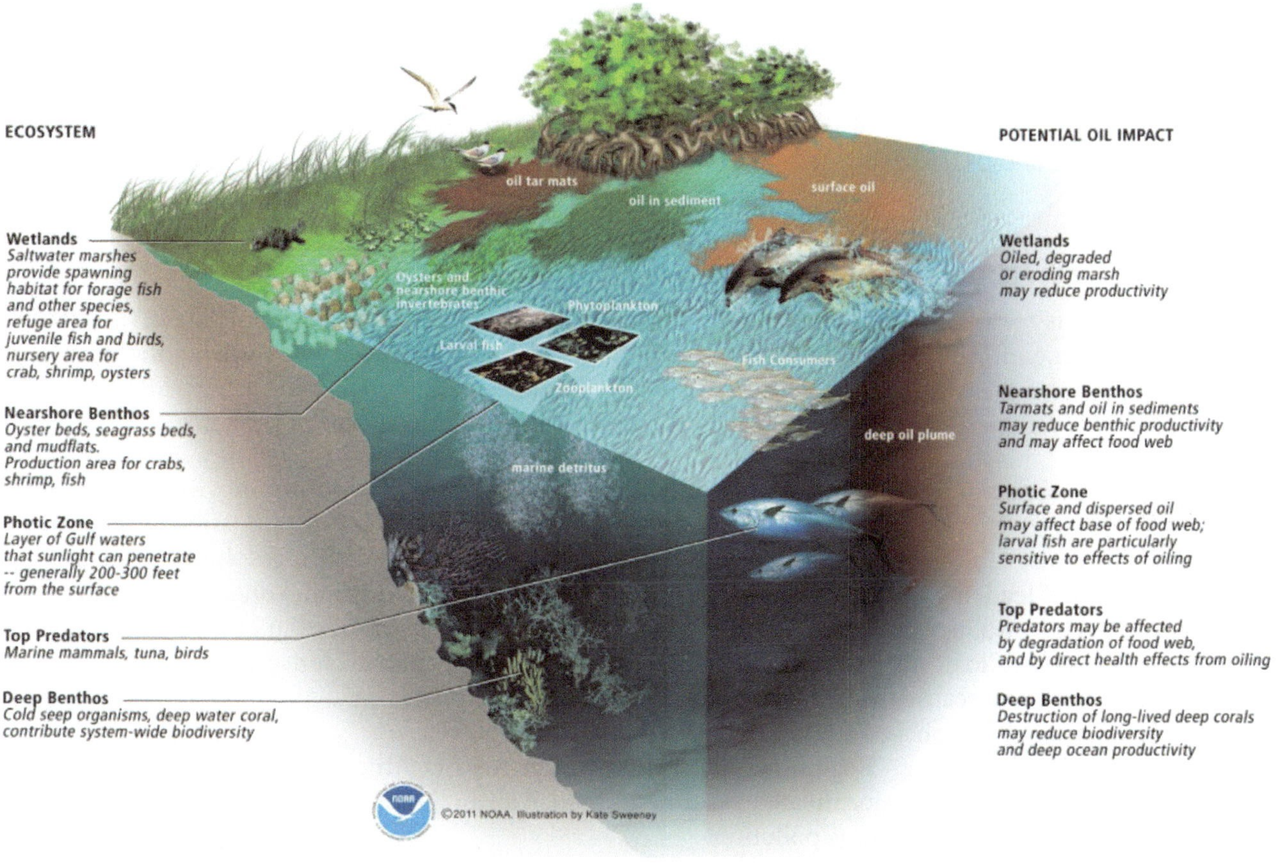

FIGURE 12.9 Illustration showing the potential impact of oil on the Gulf ecosystem following the 2010 Deepwater Horizon oil spill in the Gulf of Mexico.

Americans used about 366 million gallons per day of gasoline in 2012. With about 305 million people in the United States, that equals more than a gallon of gasoline every day for each man, woman, and child. The United States does not produce enough crude oil to make all of the gasoline used by U.S. motorists. Only about 40% of the crude oil used by U.S. refineries is produced in the United States. The rest is imported from other countries. Gasoline accounts for slightly more than 66% of all the energy used for transportation, 47% of all petroleum consumption, and 18% of total U.S. energy consumption.

While gasoline is produced year-round, extra volumes are made and imported to meet higher demand in the summer. Gasoline is delivered from oil refineries mainly through pipelines to an extensive distribution chain serving about 162,000 retail gasoline stations in the United States.

Today, gasoline is the fuel used by most passenger vehicles in the United States. There are about 254 million vehicles that use gasoline, and they each travel more than 11,618 miles per year. There are about 162,000 fueling stations that provide convenient refueling for consumers.

Most gasoline is used in cars and light trucks. It also fuels boats, recreational vehicles, and farm, construction, and landscaping equipment.

Gasoline & the Environment

Is a toxic and highly flammable liquid. The vapors given off when it evaporates and the substances produced when it is burned (carbon monoxide, nitrogen oxides, particulate matter, and unburned hydrocarbons) contribute to air pollution. Also, burning gasoline produces carbon dioxide, a greenhouse gas linked to global climate change.

Did You Know

Burning a gallon of gasoline (that does not contain ethanol) produces about 19.6 pounds of carbon dioxide. In 2013, total U.S. carbon dioxide emissions from gasoline combustion were about 1,110 million metric tons, approximately 21% of total U.S. energy-related carbon dioxide emissions.

Laws Like the Clean Air Act Reduce Environmental Impacts

The Clean Air Act is a law that seeks to reduce air pollution in the United States. The Clean Air Act (first passed in 1970) and its amendments have aimed to reduce pollution from gasoline use by requiring the use of less polluting engines and fuels [2], among other items. The U.S. Environmental Protection Agency (EPA) put these goals into action by implementing several new requirements following the passage of the act:

- **Emissions control devices and cleaner burning engines**—Emissions control devices on passenger vehicles were required beginning in 1976. In the 1990s, the EPA established emissions standards for other types of vehicles and for engines used in offroad equipment [3] that burn gasoline.
- **Removal of leaded gasoline**—Lead in gasoline proved to be a public health concern. The move away from leaded gasoline began in 1976 when catalytic converters were installed in new vehicles to reduce the emission of toxic air pollutants. Vehicles equipped with a catalytic converter cannot operate on leaded gasoline because the presence of lead in the fuel damages the catalytic converter. Leaded gasoline was completely phased out of the U.S. fuel system by 1986.
- **Reformulated gasoline**—Beginning in 1995, the Clean Air Act Amendments of 1990 required the use of cleaner burning reformulated gasoline to reduce air pollution in metropolitan areas that had significant ground-level ozone pollution.
- **Low sulfur gasoline**—Since 2006, refiners have been required to supply gasoline with 90% less sulfur content than they made in 2004. More reductions in gasoline sulfur content are planned to begin in 2017. Gasoline with lower sulfur content reduces emissions from old and new vehicles, and it is necessary for advanced vehicle emission control devices to work properly.
- **Reduced risk of gasoline leaks**—Gasoline leaks happen at gas stations every day. As we fill up our gas tanks, gasoline drips from the nozzle onto the ground and vapors leak from the open gas tank into the air. Gasoline leaks can also happen in pipelines or underground storage tanks [4] where they can't be seen. Beginning in 1990, all underground storage tanks had to be replaced by tanks with a double lining that provide an additional safeguard for preventing leaks.
- **Methyl tertiary butyl ether (MTBE)**—one of the chemicals added to gasoline to help it burn cleaner, is toxic, and a number of states started banning the use of MTBE in gasoline in the late 1990s. By 2007, the U.S. refining industry had voluntarily stopped using MTBE when making reformulated gasoline for sale in the United States. MTBE was replaced with ethanol, which is not toxic.

Diesel Fuel

Diesel fuel is used in the diesel engines found in most freight trucks, trains, buses, boats, and farm and construction vehicles (Figure 12.10). Diesel fuel powers the vehicles that we use to produce and transport nearly all of our food and all of the other products we make and buy. Some cars and small trucks and boats also have diesel engines.

Diesel fuel is also used in diesel engine-generators to generate electricity. Many industrial facilities, large buildings, institutional facilities, hospitals, and electric utilities have diesel generators for backup and emergency power supply. Most remote villages in Alaska use diesel generators for their electricity.

Heating oil and diesel fuel are closely related products called distillates. The main difference between the two fuels is that diesel fuel contains less sulfur than heating oil. Approximately 12 gallons of distillate are produced from each 42-gallon barrel of crude oil. Of these 12 gallons of distillate, less than 2 gallons have a high-sulfur content, which can only be sold as heating oil.

In the past, diesel fuel contained high quantities of sulfur, which is considered harmful to the environment when burned through combustion. Because diesel fuel requires additional processing to remove sulfur, it is more costly to produce than heating oil.

Diesel & the Environment

Diesel Fuel and Engines Are Getting Cleaner

Diesel fuel produces many harmful emissions when it is burned, and diesel fueled vehicles are major sources of harmful pollutants such as ground level ozone and particulate matter. The U.S. Environmental Protection Agency (EPA) is working to address this problem with standards for the sulfur content of diesel fuel and emission control technologies for new diesel engines.

FIGURE 12.10 A diesel powered tanker truck.

EPA standards require a major reduction in the sulfur content of diesel fuels. To meet the EPA standards, the petroleum industry is producing Ultra Low Sulfur Diesel (ULSD) fuel, a cleaner-burning diesel fuel containing a maximum 15 parts-per-million (ppm) sulfur. By December 1, 2010, all highway diesel fuel had to be ULSD fuel, and in 2014 all diesel fuel sold for all uses must be ULSD.

The EPA also established emissions control standards for diesel-powered highway vehicles for model year 2007 and later. These engines are designed to operate only with ULSD fuel. Using ULSD sulfur diesel fuel and advanced exhaust control systems can reduce vehicle particulate emissions by up to 90% and nitrogen compounds (NO_x) by 25–50%. ULSD helps to reduce emissions in older engines too.

Even with these advances, diesel fuel will still contribute significantly to air pollution in the U.S. because it will take a long time for new cleaner-burning diesel vehicles to replace older ones.

Carbon Dioxide Emissions

About 22.4 pounds of carbon dioxide (CO_2) are produced when a gallon of diesel fuel is burned. Carbon dioxide is a greenhouse gas that is linked to global climate change. Diesel engines get better fuel economy than gasoline powered engines, so the amount of CO_2 produced for each mile traveled may be lower in a vehicle with a diesel engine.

Heating Oil

Heating oil is a petroleum product used by many Americans, especially in the Northeast, to heat their homes. At refineries, crude oil is separated into different fuels including gasoline, jet fuel/kerosene, lubricating oil, heating oil, and diesel.

Heating oil and diesel fuel are closely related products called distillates. The main difference between the two fuels is that heating oil is allowed to contain more sulfur than diesel fuel. However, a number of states in the Northeast have begun to reduce the level of sulfur allowed in heating oil.

Historically, heating oil prices have fluctuated from year to year and month to month. They are generally higher during the winter months when demand for heating oil is higher.

Propane

Propane is an energy-rich gas, C_3H_8 (Figure 12.11). It is one of the liquefied petroleum gases (LP-gases or LPGs) that are found mixed with natural gas and oil. Propane and other liquefied gases, including ethane and butane, are separated from natural gas at natural gas processing plants, or from crude oil at refineries. The amount of propane produced from natural gas and from oil is roughly equal.

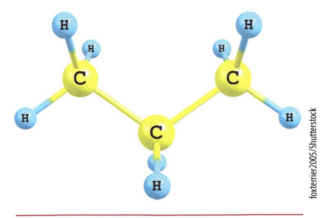

foxterrier2005/Shutterstock

FIGURE 12.11   A ball and stick model of a propane molecule.

Propane naturally occurs as a gas. However, at higher pressure or lower temperatures, it becomes a liquid. Because propane is 270 times more compact as a liquid than as a gas, it is transported and stored in its liquid state. Propane becomes a gas again when a valve is opened to release it from its pressurized container. When returned to normal pressure, propane becomes a gas so that we can use it.

Liquefied petroleum gases are mixtures of propane, ethane, butane, and other gases that are produced at natural gas processing plants and refineries. Fractionation plants then separate the liquids from each other. LP-gases were discovered in 1912 when a U.S. scientist, Dr. Walter Snelling, discovered that these gases could be changed into liquids and stored under moderate pressure. The LP-gas industry got its start shortly before World War I when a problem in the natural gas distribution process occurred. A section of the pipeline in one natural gas field ran under a cold stream, and the coldness led to a lot of liquids building up in the pipeline, sometimes to the point of blocking the entire pipeline. Soon, engineers figured out a solution: facilities were built to cool and compress natural gas, and to separate the gases that could be turned into liquids (including propane and butane).

Propane & the Environment

Propane is a nonrenewable fossil fuel, like the natural gas and oil it is produced from. Like natural gas (methane), propane is colorless and odorless. Although propane is nontoxic and odorless, foul-smelling mercaptan is added to it to make gas leaks easy to detect.

Propane is a clean burning fossil fuel, which is why it is often chosen to fuel indoor equipment such as fork lifts. Its clean burning properties and its portability also make it popular as an alternative transportation fuel.

Propane-fueled engines produce much fewer emissions of carbon monoxide and hydrocarbons compared to gasoline engines. Like all fossil fuels, propane emits water vapor and carbon dioxide, a greenhouse gas.

12.2 Natural Gas

How was natural gas formed?

The main ingredient in natural gas is methane, a gas (or compound) composed of one carbon atom and four hydrogen atoms. As we have already learned, millions of years ago, the remains of plants and animals (diatoms) decayed and built up in thick layers. This decayed matter from plants and animals is called organic material—it was once alive. Over time, the sand and silt changed to rock, covered the organic material, and trapped it beneath the rock. Pressure and heat changed some of this organic material into coal, some into oil (petroleum), and some into natural gas—tiny bubbles of odorless gas.

The search for natural gas begins with geologists, who study the structure and processes of the Earth. They locate the types of rock that are likely to contain gas and oil deposits. Today, geologists' tools include seismic surveys that are used to find the right places to drill wells. Seismic surveys use echoes from a vibration source at the Earth's surface (usually a vibrating pad under a truck built for this purpose) to collect information about the rocks beneath. Sometimes it is necessary to use small amounts of dynamite to provide the vibration that is needed. Scientists and engineers explore a chosen area by studying rock samples from the earth and taking measurements. If the site seems promising, drilling begins. Some of these areas are on land but many are offshore, deep in the ocean. Once the gas is found, it flows up through the well to the surface of the ground and into large pipelines.

Some of the gases that are produced along with methane, such as butane and propane (also known as "by-products"), are separated and cleaned at a gas processing plant. The by-products, once removed, are used in a number of ways. For example, propane can be used for cooking on gas grills. Natural gas withdrawn from a well may contain liquid hydrocarbons and nonhydrocarbon gases. This is called "wet" natural gas. The natural gas is separated from these components near the site of the well or at a natural gas processing plant. The gas is then considered "dry" and is sent through pipelines to a local distribution company, and, ultimately, to the consumer.

Dry natural gas is also known as consumer-grade natural gas. In addition to natural gas production, the U.S. gas supply is augmented by imports, withdrawals from storage, and by supplemental gaseous fuels.

Most of the natural gas consumed in the United States is produced in the United States. Some is imported from Canada and shipped to the United States in pipelines. A small amount of natural gas is shipped to the United States as liquefied natural gas (LNG).

We can also use machines called "digesters" that turn today's organic material (plants, animal wastes, etc.) into natural gas. This process replaces waiting for millions of years for the gas to form naturally.

Storing Natural Gas for Times of Peak Demand

Underground natural gas storage provides pipelines, local distribution companies, producers, and pipeline shippers with an inventory management tool, seasonal supply backup, and access to natural gas needed to avoid imbalances between receipts and deliveries on a pipeline network. Underground storage facilities are usually hollowed-out salt domes, geological reservoirs (depleted oil or gas fields) or water-bearing sands (called aquifers) topped by an impermeable cap rock.

There are three main types of natural gas underground storage facilities used in the United States today:

Depleted natural gas or oil fields
Most existing gas storage in the United States is in depleted natural gas or oil fields that are close to consumption centers.

Salt caverns
Salt caverns provide high withdrawal and injection rates relative to their working gas capacity. Base gas requirements are relatively low. The large majority of salt cavern storage facilities have been developed in salt dome formations located in the Gulf Coast states. Salt caverns have also been leached from bedded salt formations in states in the Midwest, Northeast, and Southwest.

Aquifers
In some areas, most notably in the Midwest, natural aquifers have been converted to gas storage reservoirs. An aquifer is suitable for gas storage if the water bearing sedimentary rock formation is overlaid with an impermeable cap rock.

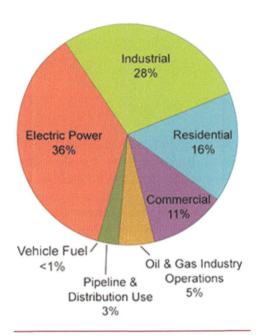

FIGURE 12.12 Natural Gas Use, 2012.

Source: U.S. Energy Information Administration, Natural Gas Monthly (March 29, 2013).

How Natural Gas Is Used

Natural Gas is a major energy source for the United States.

About 25% of energy used in the United States came from natural gas in 2012. The United States used 25.46 trillion cubic feet (Tcf) of natural gas in 2012. Natural gas is used to produce steel, glass, paper, clothing, brick, electricity, and as an essential raw material for many common products. Some products that use natural gas as a raw material are: paints, fertilizer, plastics, antifreeze, dyes, photographic film, medicines, and explosives.

Slightly more than half of the homes in the United States use natural gas as their main heating fuel. Natural gas is also used in homes to fuel stoves, water heaters, clothes dryers, and other household appliances.

The major consumers of natural gas in the United States in 2012 (Figure 12.12) included:

- Electric power sector—9.1 trillion cubic feet (Tcf)
- Industrial sector—7.1 Tcf
- Residential sector—4.2 Tcf
- Commercial sector—2.9 Tcf

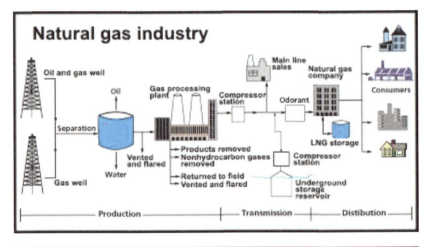

FIGURE 12.13 Transporting natural gas from the producing well.

Delivery & Storage

Transporting natural gas from the wellhead to the final customer requires many infrastructure assets and involves several physical transfers of custody and multiple processing steps. Natural gas delivery infrastructure can be grouped into three categories; production, transmission and distribution (Figure 12.13).

The natural gas transmission line is a wide-diameter, often-times long-distance, portion of a natural gas pipeline system, located between the gathering system (production area), natural gas processing plant, and other receipt points and the principal customer service area(s). There are three types of transmission pipelines:

1. Interstate natural gas pipelines operate and transport natural gas across state borders.
2. Intrastate natural gas pipelines operate and transport natural gas within a state border.
3. Hinshaw natural gas pipelines receive natural gas from interstate pipelines and deliver it to consumers for consumption within a state border.

When the natural gas gets to the communities where it will be used (usually through large pipelines), it flows into smaller pipelines called mains. Very small lines, called services, connect to the mains and go directly to homes or buildings where the natural gas will be used.

Natural Gas Pipelines

The U.S. natural gas pipeline network is a highly integrated grid that can move natural gas to nearly any location in the continental United States. It is an intricate transportation system made up of about 1.5 million miles of mainline and other pipelines links production areas and natural gas markets. The U.S. natural gas transportation network delivered more than 24 trillion cubic feet (Tcf) of natural gas during 2012 to about 72 million customers.

Did You Know
Customers in more than half of the Lower 48 states are totally dependent on the interstate natural gas pipeline system to supply their natural gas needs.

How did this transmission and distribution network become so large? About 142,000 miles (Figure 12.14) of the current 306,000 miles of the mainline natural gas transmission network were installed in the 1950s and 1960s as consumer demand for low-priced natural gas more than doubled following World War II. In fact, about half of the natural gas pipeline mileage currently installed in Texas and Louisiana, two of the largest natural gas production areas in the United States, was constructed between 1950 and 1969. By the close of 1969, marketed natural gas production exceeded 20 Tcf for the first time.

A large portion of the 1.2 million miles of local distribution pipelines that receive natural gas from the mainline transmission grid and deliver it to consumers was also installed between 1950 and 1969. However, the period of greatest local distribution pipeline growth happened more recently. In the 1990s, more than 225,000 miles of new local distribution lines were installed to provide service to the many new commercial facilities and housing developments that wanted access to natural gas supplies.

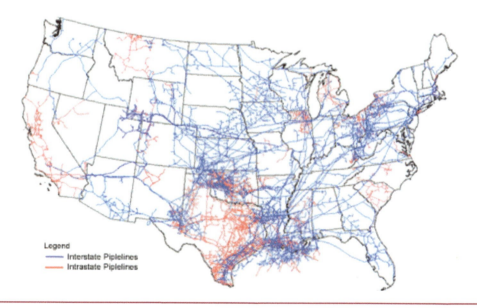

Legend
— Interstate Pipelines
— Intrastate Pipelines

FIGURE 12.14   The national natural mainline transmission grid is made up of approximately 217,000 miles of interstate pipelines and 89,000 miles of intrastate pipeline.

Source: Energy Information Administration, Natural Gas Transportation information System, Natural Gas Pipeline Maps Database (December 2008).

INSAGO/Shutterstock

FIGURE 12.15   Natural gas travels through many pipelines on its journey to the consumer.

Natural gas prices, along with oil prices, increased substantially between 2003 and 2008. Higher prices provided natural gas producers an incentive to expand development of existing fields and begin exploration of previously undeveloped natural gas fields. Consequently, new pipelines have been constructed and others are being built to link these expanded and new production sources to the existing mainline transmission network.

How Much Gas Is Left and Where Is It?

A "reservoir" is a place where large volumes of methane, the major component of natural gas, can be trapped in the subsurface of the Earth at places where the right geological conditions occurred at the right times. Reservoirs are made up of porous and permeable rocks that can hold significant amounts of oil and gas within their pore spaces.

What are proved reserves?

Did You Know
In 1821, William Hart dug the first well specifically to produce natural gas in the United States in the Village of Fredonia on the banks of Canadaway Creek in Chautauqua County, New York. It was 27 feet deep, excavated with shovels by hand, and its gas pipeline was hollowed-out logs sealed with tar and rags.

Proved reserves of natural gas are estimated quantities that analyses of geological and engineering data have demonstrated to be economically recoverable in future years from known reservoirs.

Proved reserves are added each year with successful exploratory wells and as more is learned about fields where current wells are producing. For this reason those reserves constantly change and should not be considered a finite amount of resources available. Application of new technologies can convert categories of previously uneconomic natural gas resources into proved reserves. U.S. proved reserves of natural gas have increased in every year since 1999, a trend accelerated by shale gas drilling (Figure 12.16).

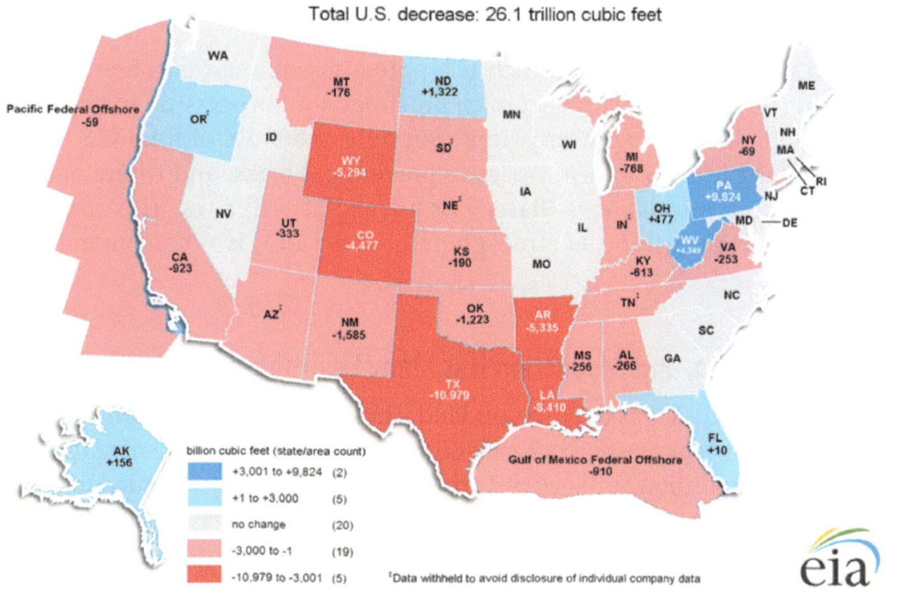

FIGURE 12.16   Change in wet natural gas proved reserves by state/area, 2012.

Source: U.S. Energy Information Administration, Form EIA-23, "Annual Survey of Domestic Oil and Gas Reserves."

How much natural gas reserves are in the United States?

In 2011, U.S. natural gas proved reserves, estimated as "wet" gas which includes natural gas plant liquids, increased by 10% to 349 trillion cubic feet (Tcf). Major improvements in shale gas exploration and production technologies drove the increase in U.S. natural gas proved reserves.

What are undiscovered technically recoverable resources?

In addition to proved natural gas reserves, there are large volumes of natural gas classified as undiscovered technically recoverable resources. Undiscovered technically recoverable resources are expected to exist because the geologic settings are favorable despite the relative uncertainty of their specific location. Undiscovered technically recoverable resources are also assumed to be producible over some time period using existing recovery technology.

Natural Gas & the Environment

Natural gas has many qualities that make it an efficient, relatively clean, and economical energy source. There are, however, environmental and safety issues with its production and use. Many of the areas that are now being explored and developed for natural gas production are relatively pristine and or are wilderness areas, and development of these areas have large impacts on the area's environment, wildlife, and human populations.

Natural Gas Is a Relatively Clean Burning Fossil Fuel

Burning natural gas for energy results in much fewer emissions of nearly all types of air pollutants and carbon dioxide (CO_2) per unit of heat produced than coal or refined petroleum products. About 117 pounds of carbon dioxide are produced per million Btu equivalent of natural gas compared to over 200 pounds of CO_2 per million Btu of coal and over 160 pounds per million Btu of distillate fuel oil. These clean burning properties have contributed to an increase in natural gas use for electricity generation and as a transportation fuel for fleet vehicles in the United States.

Natural Gas Is Mainly Methane—A Strong Greenhouse Gas

Natural gas is made up mostly of methane, which is a potent greenhouse gas. Some natural gas leaks into the atmosphere from oil and gas wells, storage tanks, pipelines, and processing plants. These leaks were the source of about 25% of total U.S. methane emissions, but only about 3% of total U.S. greenhouse gas emissions in 2011. The oil and natural gas industry tries to prevent gas leaks, and where natural gas is produced but can't be transported economically, it is "flared" or burned at well sites. This is considered to be safer and better than releasing methane into the atmosphere because CO_2 is not as potent a greenhouse gas as methane.

Natural Gas Exploration, Drilling, and Production Has Many Environmental Impacts

When geologists explore for natural gas deposits on land, they may have to disturb vegetation and soils with their vehicles. A gas well on land may require a road and clearing and leveling an area to make a drill pad. Well drilling activities produce air pollution and may disturb wildlife. Pipelines are needed to transport the gas from the wells, and this usually requires clearing land to bury the pipe. Natural gas production can also result in the production of large volumes of contaminated water. This water has to be properly handled, stored, and treated so that it does not pollute land and water.

While the natural gas that we use as a fuel is processed so that it is mainly methane, unprocessed gas from a well may contain many other compounds, including hydrogen sulfide, a very toxic gas. Natural gas with high concentrations of hydrogen sulfide is usually flared. Natural gas flaring produces CO_2, carbon monoxide, sulfur dioxide, nitrogen oxides, and many other compounds depending on the chemical composition of the natural gas and how well the gas burns in the flare. Natural gas wells and pipelines often have engines to run equipment and compressors, which produce additional air pollutants and noise.

Advances in Drilling and Production Technologies Have Positive and Negative Impacts

New drilling and gas recovery technologies have greatly reduced the amount of area that has to be disturbed to produce each cubic foot of natural gas. Horizontal and directional drilling techniques make it possible to produce more gas from a single well than in the past, so fewer wells are needed to develop a gas field. Hydraulic fracturing (commonly called "hydrofracking," or "fracking," or "fracing") of shale rock formations is opening up large reserves of gas that were previously too expensive to develop. Hydrofracking involves pumping liquids under high pressure into a well to fracture the rock and allow gas to escape from tiny pockets in the rock. However, there are some potential environmental concerns that are also associated with the production of shale gas.

The fracturing of wells requires large amounts of water. In some areas of the country, significant use of water for shale gas production may affect the availability of water for other uses, and can affect aquatic habitats. If mismanaged, hydraulic fracturing fluid—which may contain potentially hazardous chemicals—can be released by spills, leaks, faulty well construction, or other exposure pathways. Any such releases can contaminate surrounding areas.

Hydrofracturing also produces large amounts of wastewater, which may contain dissolved chemicals and other contaminants that require treatment before disposal or reuse. Because of the quantities of water used and the complexities inherent in treating some of the wastewater components, treatment and disposal is an important and challenging issue.

According to the United States Geological Survey, hydraulic fracturing "causes small earthquakes, but they are almost always too small to be a safety concern. In addition to natural gas, fracking fluids and formation waters are returned to the surface. These wastewaters are frequently disposed of by injection into deep wells. The injection of wastewater into the subsurface can cause earthquakes that are large enough to be felt and may cause damage."

Natural gas may be released to the atmosphere during and after well drilling, the amounts of which are being investigated.

Strict Safety Regulations and Standards Are Required for Natural Gas

Because a natural gas leak can cause an explosion, there are very strict government regulations and industry standards in place to ensure the safe transportation, storing, distribution, and use of natural gas. Because natural gas has no odor, natural gas companies add a strong smelling substance called mercaptan to it so that people will know if there is a leak. If you have a natural gas stove, you may have smelled this "rotten egg" smell of natural gas when the pilot light has gone out.

12.3 Coal

Coal takes millions of years to form. Coal is a combustible black or brownish-black sedimentary rock composed mostly of carbon and hydrocarbons. It is the most abundant fossil fuel produced in the United States. Coal is a nonrenewable energy source because it takes millions of years to develop. The energy in coal comes from the energy stored by plants that lived hundreds of millions of years ago, when the earth was partly covered with swampy forests.

For millions of years, a layer of dead plants at the bottom of the swamps was covered by layers of water and dirt, trapping the energy of the dead plants. The heat and pressure from the top layers helped the plant remains develop into what we now call coal (Figure 12.17).

Types of Coal

Coal is classified into four main types, or ranks (anthracite, bituminous, subbituminous, and lignite). Classification depends on the amounts and types of carbon the coal contains and on the amount of heat energy the coal can produce. The rank of a deposit of coal depends on the pressure and heat that acted on the plant debris as it sank deeper and deeper over millions of years. For the most part, the higher ranks of coal contain more heat-producing energy.

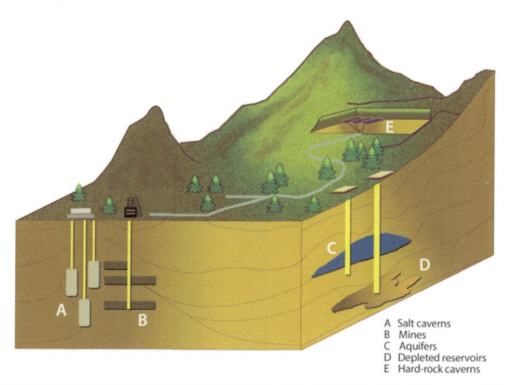

A Salt caverns
B Mines
C Aquifers
D Depleted reservoirs
E Hard-rock caverns

FIGURE 12.17 Underground storage locations for natural gas.
Source: PB-KBB, inc., enhanced by EIA.

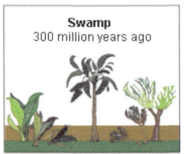

Swamp
300 million years ago

Before the dinosaurs, many
giant plants died in swamps.

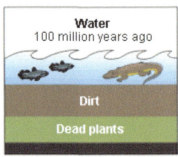

Water
100 million years ago

Dirt

Dead plants

Over millions of years, the plants
were buried under water and dirt.

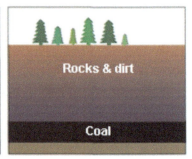

Rocks & dirt

Coal

Heat and pressure turned
the dead plants into coal.

FIGURE 12.18   How coal was formed.

Anthracite contains 86%–97% carbon, and generally has a heating value slightly higher than bituminous coal. It accounts for 0.2% of the coal mined in the United States. All of the anthracite mines in the United States are located in northeastern Pennsylvania.

Bituminous coal contains 45%–86% carbon. Bituminous coal was formed under high heat and pressure. Bituminous coal in the United States is between 100 and 300 million years old. It is the most abundant rank of coal found in the United States, accounting for nearly half of U.S. coal production. Bituminous coal is used to generate electricity and is an important fuel and raw material for the steel and iron industries. In 2012, 48% of the U.S. coal produced was bituminous. West Virginia, Kentucky, and Pennsylvania are the largest producers of bituminous coal.

Subbituminous coal has a lower heating value than bituminous coal. Subbituminous coal typically contains 35%–45% carbon. Most subbituminous coal in the United States is at least 100 million years old. About 44% of the coal produced in the United States is subbituminous. Wyoming is the leading source of subbituminous coal.

Lignite is the lowest rank of coal with the lowest energy content. Lignite coal deposits tend to be relatively young coal deposits that were not subjected to extreme heat or pressure, containing 25%–35% carbon. Lignite is crumbly and has high moisture content. There are 20 lignite mines in the United States, producing about 7% of U.S. coal. Most lignite is mined in Texas and North Dakota. Lignite is mainly burned at power plants to generate electricity.

Mining Coal and Getting It to Market

Coal miners use large machines to remove coal from the ground. Many U.S. coal beds are near the ground's surface, and about two-thirds of U.S. coal production comes from surface mines. Modern mining methods allow coal miners to easily reach most of the nation's coal reserves. As a result of growth in surface mining and improved mining technology, the amount of coal produced by one miner in one hour has more than tripled since 1978.

Coal miners use two primary methods to remove coal:

■ Surface mining (including mountain top removal) is used to produce most of the coal in the United States because it is less expensive than underground mining. Surface mining can be used when the coal is buried fewer than 200 feet underground. In surface mining, large machines remove the top soil and layers of rock known as overburden to expose the coal seam (Figure 12.19). Once the mining is finished, the dirt and rock are returned to the pit, the topsoil is replaced, and the area is replanted.

Top soil
Overburden

Shallow coal seam

FIGURE 12.19   Surface mining.

- Underground mining, sometimes called deep mining, is used when the coal is buried several hundred feet below the surface (Figure 12.20). Some underground mines are 1,000 feet deep. To remove coal in these underground mines, miners ride elevators down deep mine shafts where they run machines that dig out the coal.

Processing Coal

After coal is removed from the ground, it typically goes on a conveyor belt to a preparation plant located at the mining site. The plant cleans and processes coal to remove other rocks and dirt, ash, sulfur, and unwanted materials. This increases the heating value of the coal.

Transporting Coal

After coal is mined and processed, it is ready to be shipped to market. The cost of shipping coal can be more expensive than the cost of mining it. Nearly 70% of coal delivered in the United States is transported, for at least part of its trip to market, by train. Coal can also be transported by barge, ship, truck, and even pipeline. It is often cheaper to transport coal on river barges, but barges are unable to take coal everywhere it's needed. If the coal is used near the coal mine, it can be moved by trucks and conveyors. Coal can also be crushed, mixed with water, and sent through a slurry pipeline. Sometimes, coal-fired electric power plants are built near coal mines to lower transportation costs.

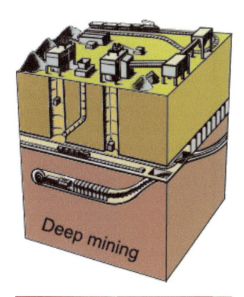

FIGURE 12.20   Underground mining requires a lot of equipment and tunneling, and is only economically feasible if the coal is far below the surface.

Did You Know

The Black Thunder Mine and North Antelope Rochelle Mine in Wyoming produce almost as much coal as West Virginia, the nation's second-largest coal-producing state.

In 2012, 1,016.4 million short tons of coal were mined in 25 states. Wyoming mined the most coal, followed by West Virginia, Kentucky, Pennsylvania, and Illinois (Figure 12.21).

Coal is mainly found in three large regions, the Appalachian coal region, the Interior coal region, and the Western coal region (includes the Powder River Basin).

Almost 29% of the coal produced in the United States comes from the Appalachian coal region.

West Virginia is the largest coal-producing state in the region, and the second-largest coal-producing state in the United States.

Coal mined in the Appalachian coal region is primarily used for steam generation for electricity, coke production, and for export.

As of January 1, 2013, the recoverable reserves at producing mines were 18.7 billion short tons. However, the amount of coal reserves at producing mines is a small portion of the total amount of coal that exists in the United States. So how much coal is there? It is impossible to know exactly how much coal exists in the United States because

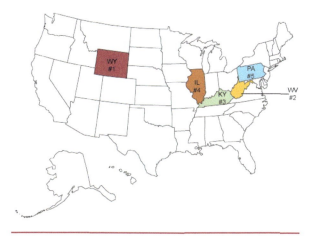

FIGURE 12.21   Top coal producinig states, 2012.

Source: U.S. Energy Information Administration, Quarterly Coal Report (March 2013), preliminary 2012 data.

it is buried underground, but it is possible to make estimates. Total resources is our best estimate of the total amount of coal (including undiscovered coal) in the United States. Currently, total resources are estimated to be about 4 trillion short tons.[1] Total resources includes several categories of coal with various degrees of geologic assurance and data reliability.

The Demonstrated Reserve Base (DRB2) is the sum of coal in both measured and indicated resource categories of reliability. This represents 100% of the in-place coal that could be mined commercially at a given time. EIA estimates that the DRB in 2012 was 481.4 billion short tons.

Estimated recoverable reserves include only the coal that can be mined with today's mining technology, after accessibility constraints and recovery factors are taken into consideration. EIA estimates there are 257.6 billion short tons of U.S. recoverable coal reserves, about 54% of the Demonstrated Reserve Base.

Based on U.S. coal production for 2012, the U.S. estimated recoverable coal reserves represent enough coal to last 253 years. However, EIA projects in the most recent Annual Energy Outlook (April 2014) that U.S. coal production will increase by approximately 0.3% per year from 2012–2040. If that growth rate continues into the future, U.S. estimated recoverable coal reserves would be exhausted in about 180 years if no new reserves are added.

The Distribution of World Coal Reserves Varies from Oil and Natural Gas

Coal is widely distributed across the continents in comparison to all other fossil fuels. As of December 31, 2011, total recoverable reserves of coal **around the world** were estimated at 979.8 billion short tons. There are significant coal reserves in the United States and Russia, but not in the Middle East. In fact, the United States and Russia account for nearly half of global coal reserves. Oil reserves are predominantly found in Venezuela, the Middle East, and Canada while more than half of the world's natural gas reserves are in Russia, Iran, and Qatar (Figure 12.22).

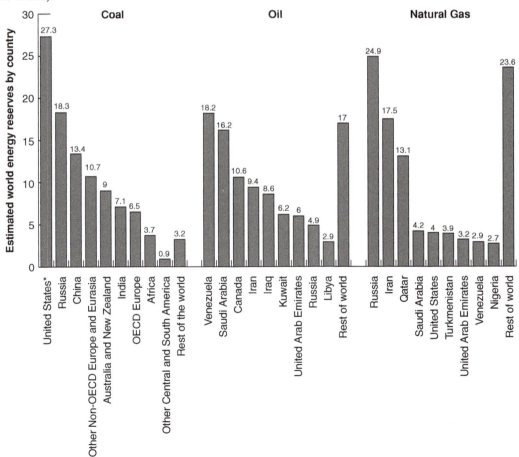

FIGURE 12.22 Estimated world energy reserves by country.

1. The most comprehensive national assessment of U.S. coal resources was published by the U.S. Geological Survey (USGS) in 1975, which indicated that as of January 1, 1974 coal resources in the United States totaled 4 trillion short tons. While more recent regional assessments of U.S. coal resources have been conducted by the U.S. Geological Survey, a new national level assessment of U.S. coal resources has not been conducted.

Primary Uses for Coal: Electricity Generation & Industry

Coal is used to create about 39% of all the electricity generated in the United States. Power plants can make steam by burning coal. The steam then turns turbines (machines for generating rotary mechanical power) to generate electricity. In addition to companies in the electric power sector, industries and businesses with their own power plants use coal to generate electricity.

Did You Know
A pound of coal supplies enough electricity to power ten 100-watt light bulbs for about an hour.

Many industries use coal and coal byproducts. Methanol and ethylene can be produced from coal and can be used to make plastics, tar, synthetic fibers, fertilizers, and medicines. Coal is used indirectly to make steel. First, coal is baked in hot furnaces to make coke, and then coke is used to smelt iron ore into iron to make steel. The high temperatures created by burning coke give steel the strength and flexibility needed for bridges, buildings, and automobiles. The concrete and paper industries also use large amounts of coal for heat.

Coal is an abundant fuel that is relatively inexpensive to produce and convert to useful energy. However, producing and using coal has many impacts on the environment.

FIGURE 12.23   Blast furnace in a modern steel works.

chinahbzyd/Shutterstock

Coal and the Environment

Impacts of Coal Mining

Surface, or strip mines, are the source of about 70% of the coal that is mined in the U.S. These mining operations remove the soil and rock above coal deposits, or "seams," disturbing land at its surface. The amount of coal produced at a surface mine is not only determined by the area of land being mined at the surface but the thickness of the coal deposit. For example, in Wyoming's Powder River Basin, where coal deposits may run 70 feet deep, one acre of land may produce over 100,000 tons of coal.

One surface mining technique that has affected large areas of the Appalachian Mountains in West Virginia and Kentucky is mountain top removal and valley fill mining, where the tops of mountains have been removed using a combination of explosives and mining equipment and deposited into nearby valleys. As a result, the landscape is changed, and streams may be covered with a mixture of rock and dirt. The water draining from these filled valleys may contain pollutants that can harm aquatic wildlife downstream. While mountain-top mining has been around since the 1970s, its use became more widespread and controversial since the 1990s.

Did You Know

Some electric power plants use "scrubbers," also known as flue gas desulfurization equipment, to reduce the amount of sulfur going out their smokestacks.

U.S. laws require that dust and water runoff from the affected area has to be controlled, and that the area has to be "reclaimed" close to its original condition. Many surface mines have been reclaimed so well that it can be hard to tell that there was a surface mine in the area. However, there are areas that have not been reclaimed as successfully.

Underground mines have less overall impact on the environment than surface mines. The most serious impact of underground mining may be the methane gas that has to be vented out of mines to make the mines safe to work in. Methane is a strong greenhouse gas, meaning that on an equal-weight basis its global warming potential is much higher than for that of other greenhouse gases. In 2011, the most recent year for which estimates are available, methane emissions from underground mines accounted for about 7% of total U.S. methane emissions and 1% of total U.S. greenhouse gas emissions (based on global warming potential). Some mines capture and use or sell the methane extracted from mines. Surface mines contributed about 2% of total U.S. methane emissions. Learn more about greenhouse gas emissions.

The ground above mine tunnels can collapse, and acidic water can drain from abandoned underground mines. Underground coal mining is a dangerous profession, and coal miners can be injured or killed in mining accidents, especially in countries without strict enforcement of safety regulations and procedures. Miners can also get black lung disease from the coal dust in the mines.

Emissions from Burning Coal

In the United States, most coal is used as a fuel to generate electricity. Burning coal produces numerous emissions that adversely affect the environment and human health.

The principal emissions resulting from coal combustion are:

- Sulfur dioxide (SO_2), which contributes to acid rain and respiratory illnesses
- Nitrogen oxides (NO_x), which contributes to smog and respiratory illnesses
- Particulates, which contribute to smog, haze, and respiratory illnesses and lung disease
- Carbon dioxide (CO_2), which is the primary greenhouse gas emission from the burning of fossil fuels (coal, oil, and natural gas (Figure 12.24))

Mercury and other heavy metals, which has been linked with both neurological and developmental damage in humans and other animals. Mercury concentrations in the air usually are low and of little direct concern. However, when mercury enters water—either directly or through deposition from the air—biological processes transform it into methylmercury, a highly toxic chemical that accumulates in fish and the animals (including humans) that eat fish.

Fly ash and bottom ash are residues created when coal is burned at power plants. In the past, fly ash was released into the air through the smokestack, but by law much of it now must be captured by pollution control devices, like scrubbers. In the United States, fly ash is generally stored at coal power plants or placed in landfills. Pollution leaching from ash storage and landfills into groundwater and the rupture of several large impoundments of ash have emerged as new environmental concerns.

Reducing the Impacts of Coal Use

The Clean Air Act and the Clean Water Act require industries to reduce pollutants released into the air and the water. Industry has found several ways to reduce sulfur, nitrogen oxides (NO_x), and other impurities from coal. They have found more effective ways of cleaning coal after it is mined, and coal consumers have shifted towards greater use of low sulfur coal.

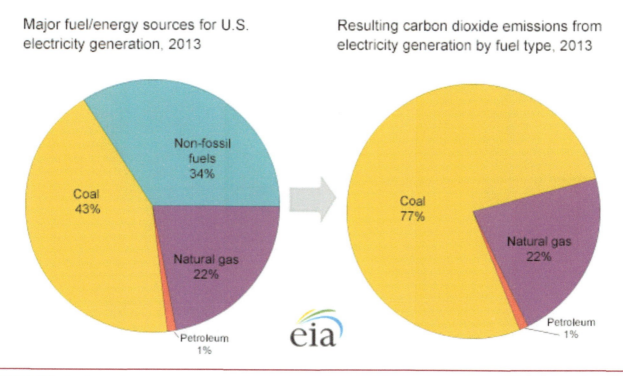

Major fuel/energy sources for U.S. electricity generation, 2013

Resulting carbon dioxide emissions from electricity generation by fuel type, 2013

FIGURE 12.24   Fuels used to generate electricity and their resulting carbon dioxide emissions.

Source: U.S. Energy Information Administration, Monthly Energy Review. Table 2.6 (May 2014). preliminary.
Source: U.S. Energy Information Administration, Monthly Energy Review. Table 12.6 (May 2014). preliminary.

Power plants use flue gas desulfurization equipment, also known as "scrubbers," to clean sulfur from the smoke before it leaves their smokestacks. In addition, industry and government have cooperated to develop technologies that can remove impurities from coal or that make coal more energy-efficient so less needs to be burned.

Equipment intended mainly to reduce SO_2 (such as scrubbers), NO_x (such as catalytic converters), and particulate matter (such as electrostatic precipitators and baghouses) is also able to reduce mercury emissions from some types of coal. Scientists are also working on new ways to reduce mercury emissions from coal-burning power plants.

Research is underway to address emissions of carbon dioxide from coal combustion. Carbon capture separates CO_2 from emissions sources and recovers it in a concentrated stream. The CO_2 can then be sequestered, which puts CO_2 into storage, possibly underground, in such a way that it will remain there permanently.

Reuse and recycling can also reduce coal's environmental impact. Land that was previously used for coal mining can be reclaimed for uses like airports, landfills, and golf courses. Waste products captured by scrubbers can be used to produce products like cement and synthetic gypsum for wallboard.

CHAPTER 13

Global Climate Change

13.1 A Brief History of the Climate Change Debate

Most people would assume that concern over global warming or global climate change was a fairly recent phenomenon. However, as this brief article by Stephan Harding of Schumacher College in England points out, scientists as far back as the 1820s took note of possible links between human activities and the climate system.

These scientists developed a basic understanding of what is known as the **greenhouse effect***, which will be described in more detail in the following section. At a basic level, the greenhouse effect involves certain gases that are naturally present in the atmosphere that trap infrared or heat energy as it escapes from the Earth's surface. In essence, these gases act like the glass panes on a greenhouse. They allow sunlight in, but block heat from escaping, which results in warmer temperatures. The greenhouse effect is a natural phenomenon. In fact, without it, the Earth's average surface temperature would be a frigid 0 degrees Fahrenheit (°F) rather than its actual average temperature of around 59°F.*

> **greenhouse effect**: the global warming of our atmosphere caused by the presence of carbon dioxide and other greenhouse gases, which trap the sun's radiation

What got the attention of the scientists discussed in this article was the increased concentrations of greenhouse gases in the atmosphere due to human activities. In particular, the burning of fossil fuels such as coal and oil, which release carbon dioxide (CO_2), the most common of the greenhouse gases after water vapor. The scientists hypothesized that adding more CO_2 to the atmosphere could increase global temperatures and change climate. It was not until decades later, however, that scientific instrumentation advanced enough to determine that, in fact, atmospheric CO_2 concentrations were increasing and that this was due almost entirely to human activities.

Today it is widely acknowledged by scientists across the world that global average temperatures have increased since the Industrial Revolution and the widespread use of fossil fuels. However, scientists do not accept the simple presence of a correlation (higher CO_2 concentrations and higher temperatures) as proof of a cause-effect relationship. Instead, other possible causes of increased temperatures (such as variations in solar energy output from the sun) also must be considered. After examining the many different factors that influence climate, the Intergovernmental Panel on Climate Change (IPCC), a working group that includes thousands of scientists from around the world, has concluded that increased CO_2 concentrations, due mainly to fossil fuel combustion, are primarily responsible for the observed increase in global temperatures in recent decades.

> **attribution**: assigning some quality or character to a person or thing

By *Stephan Harding*

Our understanding of climate change began with intense debates amongst 19th century scientists about whether northern Europe had been covered by ice thousands of years ago. In the 1820s Jean Baptiste Joseph Fourier [a French mathematician and physicist] discovered that "greenhouse gasses" trap heat radiated from the Earth's surface after it has absorbed energy from the sun. In 1859 John Tyndall [a British physicist] suggested that ice ages were caused by a decrease in the amount of atmospheric carbon dioxide. In 1896 Svente Arrhenius [a Swedish physicist and chemist] showed that doubling the carbon dioxide content of the air would gradually raise global temperatures by 5–6°C—a remarkably prescient result that was virtually ignored by scientists obsessed with explaining the ice ages.

The idea of global warming languished until 1938, when Guy S Callender [a British engineer and inventor] suggested that the warming trend revealed in the 19th century had been caused by a 10% increase in atmospheric carbon dioxide from the burning of fossil fuels. At this point scientists were not alarmed, as they were confident that most of the carbon dioxide emitted by humans had dissolved safely in the oceans. However, this notion was dispelled in 1957 by Hans Suess [an Austrian physical chemist and nuclear physicist] and Roger Revelle [an American oceanographer], who discovered a complex chemical buffering system which prevents sea water from holding on to much atmospheric carbon dioxide.

The possibility that humans could contribute to global warming was now being taken seriously by scientists, and by the early 1960s some had begun to raise the spectre of severe climate change within a century. They had started to collect evidence to test the idea that global temperatures were increasing alongside greenhouse gas emissions, and to construct mathematical models to predict future climates.

In 1958 Charles Keeling [an American geochemist and oceanographer] began long-term measurements of atmospheric carbon dioxide at the Mauna Loa observatory in Hawaii. Looked at now, the figures show an indisputable annual increase, with roughly 30% more of the gas relative to pre-industrial levels in today's atmosphere—higher than at any time in the last 700,000 years. Temperature readings reveal an average warming of 0.5–0.6°C over the last 150 years.

Climate change skeptics have pointed out that these records could have been due to creeping urbanisation around weather stations, but it is now widely accepted that this 'urban heat island effect' is relatively unimportant and that it doesn't explain why most of the warming has been detected far away from cities, over the oceans and the poles.

The Case for Global Warming

Since the 1960s, evidence of global warming has continued to accumulate. In 1998 Michael Mann [professor and director of the Earth System Science Center at Penn State University] and colleagues published a detailed analysis of global average temperature over the last millennium known as the "hockey stick graph," revealing a rapid temperature increase since the industrial revolution. Despite concerted efforts to find fault with Mann's methodology, his basic result is now accepted as sound. Then, in 2005, just as the **Kyoto Protocol** for limiting greenhouse gas emissions was ratified, James Hansen [head of the NASA Goddard Institute for Space Studies] and his team detected a dramatic warming of the world's oceans—just as expected in a warming world.

> **Kyoto Protocol**: an international treaty that stipulates highly developed countries must cut their emissions of carbon dioxide and other gases that cause climate warming by an average of 5.2% by 2012

There is now little doubt that the temperature increase over the last 150 years is real, but debate still surrounds the causes. We know that the warming during the first half of the last century was almost certainly due to a more vigorous output of solar energy, and some scientists have suggested that increased solar activity and greater volcanic emissions of carbon dioxide are responsible for all of the increase. But others point out that during the last 50 years the sun and volcanoes have been less active and could not have caused the warming over that period.

By 2005 a widespread scientific consensus had emerged that serious, large-scale disruption could occur around 2050, once average global temperature increase exceeds about 2°C, leading to abrupt and irreversible changes. These include the melting of a large proportion of the Greenland ice cap (now already under way), the reconfiguration of the global oceanic circulation, the disappearance of the Amazon forest, the emission of methane from permafrost and undersea methane hydrates, and the release of carbon dioxide from soils.

This new theory of "abrupt climate change" has overturned earlier predictions of gradual change and has prompted some scientists to warn that unmitigated climate change could lead to the complete collapse of civilisation. Fears have been fuelled by the possibility that smoke, hazes and particles from burning vegetation and fossil fuels could be masking global warming by bouncing solar energy back to space. This "global dimming" effect is diminishing as we clean up air pollution. As a result global average temperature could rise by as much as 10 degrees Celsius [approximately 18°F] by the close of the century—a catastrophic increase.

A more conservative assessment by the Intergovernmental Panel on Climate Change (IPCC) in 2001 indicated that with unabated carbon emissions, global temperature could rise gradually to around 5.8°C [roughly 10°F] by 2100. An increase of this nature would still threaten the lives of millions of people, particularly in the global south, due to sea level rise and extreme weather events.

Although some people still deny that climate change is a problem we can do something about, last year the UK government indicated that it was on board. The Stern Review showed that without immediate and relatively inexpensive action, climate change would lead to severe and permanent global economic depression by 2050. There is now a strong scientific and economic consensus about the severity of the climate crisis.

Adapted from: Harding, S. 2007. The Long Road to Enlightenment. The Guardian, January 8. Available online at: http://www.guardian.co.uk/environment/2007/jan/08/climatechange.climatechangeenvironment/print.

13.2   The Evidence & Observations Behind Our Current Understanding of Global Climate Change

Introduction

"The perched bowlders which are found in the Alpine valleys . . . occupy at times positions so extraordinary that they excite in a high degree the curiosity of those who see them. For instance, when one sees an angular stone perched upon the top of an isolated pyramid, or resting in some way in a very steep locality, the first inquiry of the mind is, When and how have these stones been placed in such positions, where the least shock would seem to turn them over?"

Louis Agassiz, Etudes sur les Glaciers, 1840.

Jean-Pierre Perraudin wasn't a scientist, but he knew that glaciers had carved out the alpine landscape around his home. He had seen the strange, giant boulders perched high in the Val de Bagnes as he hiked and hunted through the Swiss Alps near his home. Those granite rocks were different from their surroundings—they did not belong where he had seen them scattered across the valley. Perraudin had also noticed that long scratches marked the exposed rocks that lined the narrow valley. Only one thing in his experience could explain the rocks and the marks: glaciers. High in the southern portion of the valley, he had seen the large sheets of ice and the stripes they left on the land. He could picture how the ice might carry large boulders to the valley below. Perraudin concluded that glaciers must have once extended far down into the valley.

Most scientists of those times considered the scratches and misplaced boulders, called erratics, firm evidence of the great Biblical flood. In Perraudin's experience, giant rocks did not float. But when he presented his idea to naturalist Jean de Charpentier in 1815, the reception was cool. Charpentier wrote, "I found his hypothesis so extraordinary and even so extravagant that I considered it as not worth examining or even considering" (quoted in Imbrie, p. 22).

Despite this reception, Perraudin stood by his theory that glaciers had once extended far down into the Val de Bagnes. "I am ready to demonstrate this fact to incredulous people by the obvious proof of comparing these marks with those uncovered by glaciers at present," he wrote defiantly (Imbrie, p. 22). He soon got his chance to prove his idea to another naturalist, Ignace Venetz, who came to the area for work. Venetz eventually convinced Charpentier that the glacier theory had merit, and he in turn converted the influential scientist Louis Agassiz.

Balanced precariously above the road, these boulders were left behind by retreating glaciers. The boulders protected the soil underneath from erosion while the surrounding material washed away. Rocks transported by glaciers and deposited on different types of rock are called glacial erratics. (Photograph from the National Snow and Ice Data Center Glacier Photo Collection Figure 13.1).

Image courtesy of NASA

FIGURE 13.1 **Photograph of glacial erratics perched on top of a morraine in the Alps.**

Long parallel grooves on exposed rock faces, called striations, are gouged out by rocks and debris carried on the bottom surface of a flowing glacier. Jean-Pierre Perraudin inferred that striations he saw in valleys far from existing glaciers were caused by glaciers that had since disappeared.

Agassiz quickly developed a broad theory of what he termed the "Ice Age," a period in which giant glaciers extended down from the North Pole to cover Europe and much of North America. When he presented the theory on July 24, 1837, in a meeting of the Swiss Society of Natural Sciences in Neuchatel, Switzerland, he was met with sheer rage from a scientific community well-established in its views. Agassiz was undeterred. He enthusiastically showed leading scientists evidence of glaciation, and captured the public imagination with his descriptions of an icy Earth. While most scientists disagreed at first, his fervor caught the attention of the community, and by the 1870s, the Ice Age theory was widely accepted.

Louis Agassiz (1807–1873) found evidence for a past ice age in the mountains of Switzerland. Although controversial at first, his ideas were eventually accepted by the scientific community. Modern scientists have extended his techniques, and developed many new ones, to explore past climates (both warm and cold) in detail.

As scientists came to accept that the erratics pointed to an Ice Age, the realization triggered more questions. What had caused the Ice Age, and why had it ended? As fossils of a temperate forest were discovered sandwiched between Ice Age soil layers, scientists' questions became increasingly more complicated. Could the Ice Ages return? Might our climate change again?

Ironically, one of the theories proposed to explain what had caused the Ice Ages to come and go would become the key argument supporting an opposite, and equally controversial, kind of climate change: global warming. In 1895 Svante Arrhenius proposed that the drop in temperatures that occurred during the Ice Age could have been produced by a drop in the concentration of atmospheric carbon dioxide compared to the then-current levels. He even proposed that industrial emissions could raise Earth's temperature in coming centuries.

Despite his theory, most scientists didn't believe that humans could affect climate until the 1960s, when evidence of human impacts on the natural world were becoming increasingly obvious. Once again, climate change became controversial, as the theory of global warming became one of the biggest scientific debates of the late twentieth century. The questions of the past generation took on a new urgency. Could climate change be a threat to humankind? And could our own actions increase the threat?

These questions have compelled scientists to scour the Earth for signs of past climate change. Their search has developed into an entire scientific field: **paleoclimatology**, the study of past climates. In determining what has triggered climate change in the past, scientists hope to learn how natural and human triggers might change our climate in the future.

Most scientists who study the past have fossils and artifacts that help them reconstruct history. But without thermometer readings, how can they know how cold it was when the ice from the last Ice Age began to retreat? Scientists who want to reconstruct past climates gather clues buried in the Earth in much the same way that an archeologist reveals past culture by looking at artifacts or a detective reconstructs a crime scene using multiple bits of evidence.

Agassiz used detailed drawings of mountainous terrain, glacial erratics, and other glacial features to bolster his theories. The boulder in the left foreground protects the column of ice underneath it from melting, while being carried by the glacier (Figure 13.1). Eventually, the boulder will be deposited in a moraine, or left behind as an isolated erratic.

Climate leaves an imprint on the planet, in the chemical and physical structure of its oceans, life, and land. Some of these artifacts, known as climate proxies, reveal general climate patterns over the entire Earth, while other proxies reveal seasonal change in specific regions. By reading the signs of past climate, scientists reconstructed the history of Earth's climate over hundreds of thousands—in some cases millions—of years. When combined with observations of Earth's modern climate into computer models, paleoclimate data help scientists to predict future climate change.

Written in the Earth

By *Holli Riebeek · design by Robert Simmon · June 28, 2005*

FIGURE 13.2   Loess stratigraphy in a road cut.

The first pieces of evidence for climate change came from the land itself, from the misplaced boulders scattered across much of the Northern Hemisphere, though there were other signs as well. A homogeneous, fine yellow soil covered more than one million square miles of Europe, Asia, and North America. The soil was as thick as 3 meters (10 feet) in some places, and nearly nonexistent in others. Scientists eventually recognized that the soil, called loess (rhymes with "bus"), was made of rock that had been ground into powder under the weight and movement of the glaciers. As the ice melted, water swept the dust out from under the glaciers into streams along the edge of the ice. When water levels dropped, the dust blew across the land, leaving an uneven layer of fine, homogeneous soil.

By mapping the loess and the trail of rock debris left by the glaciers, scientists determined that the ice sheets had once stretched down over the familiar bowls of the Great Lakes in North America and across the British Isles and Scandinavia. Thick glaciers, far larger than those that currently cap the mountain peaks, covered the Alps. In time, geologists discovered other layers of similar soil from earlier times—a sign that the climate had not changed just once, but at least three or four times.

Loess deposits, composed of fine wind-blown dust produced by the grinding action of glaciers, indicate the former presence of ice sheets in locations around the Northern Hemisphere (Figure 13.2).

The story of glaciation and climate change may have been etched into rocks and brushed from the soil like a giant sand painting, but then Nature left the book out in the open. Exposed to erosion and weathering for thousands, even millions of years, some parts of the original story have been revised or become difficult to read. Buried in the Earth, however, is another version of the Earth's climate story, and scientists have begun to leaf through its pages.

Speleothems: Cave Rocks

A deep cavern dips into the New Mexico desert, shielding spiky icicle-like rocks that hang from the ceiling and the rounded columns that grow from the floor in a myriad of shapes. One of the world's largest underground chambers, the Carlsbad Cavern is resplendent in the intricate finery of the rock formations that form there (Figure 13.3). Beyond their breath-taking beauty, the formations in Carlsbad and the more than 100 other caves in the area provide a record of rainfall in the southwestern United States.

Spectacular rock formations in Carlsbad Cavern hold information on past rainfall and temperatures.

Tucked away inside the Earth, the rocks are protected from the weathering and large-scale erosion that taints other land-based climate records. As water runs through the ground, it picks up minerals, the most common of which is calcium carbonate. When the mineral-rich water drips into caves, it leaves behind solid mineral deposits—the same solid material that forms white spots on water faucets or glass dishes. The mineral deposits accumulate in the well-known icicle-shaped rock on the ceiling, a stalactite, and in a mound on the floor where the drip lands, a stalagmite. Less well known, water deposits can also dry in a flat slab called a flowstone.

Geologists refer to the mineral formations in caves as "speleothems." While the water flows, the speleothems grow in thin, shiny layers. The amount of growth is an indicator of how much ground water dripped into the cave. Little growth might indicate a drought, just as rapid growth could point to heavy precipitation. When the speleothems stop growing, the outside becomes dirty and eroded in places, giving it a dull appearance. A growing speleothem looks smooth and wet.

Minerals dissolved in groundwater can crystallize on exposed surfaces of underground passages. The deposited minerals form stalactites, stalagmites, and other formations, collectively named speleothems. Soda straw stalactites are delicate tubes of calcium carbonate (limestone).

John Blanton/Shutterstock

FIGURE 13.3 **Photograph of chandelier formation, Carlsbad Caverns.**

Scientists can date the layers in the speleothem by measuring how much uranium, a radioactive element, has decayed. Uranium from the surrounding bedrock seeps into the water and forms a carbonate that becomes part of each layer of the speleothem as it forms. Uranium decays into thorium, which sticks to the clay in the bedrock instead of seeping into ground water and from there into the speleothem. As a result, the newest layers of a growing speleothem typically contain no thorium.

Over time uranium predictably turns into thorium, so scientists can tell how old a layer is by measuring the ratio of uranium to thorium. Once the layers have been dated, scientists can create a rough record of how ground water levels changed over the lifetime of the formation. Because speleothem growth is influenced by geography, ground water chemistry, and other factors, the record from one cave cannot serve as a record of climate change. Scientists must look for similar patterns of growth in speleothems in caves over a broad area to infer that the climate changed.

The cross section of a stalagmite reveals a sequence of layers, laid down over time (Figure 13.4). Researchers determine the age of the rings using Uranium-Thorium radioisotopic dating, and examine ring thickness and oxygen isotopes to determine past climate.

Scientists are trying to glean more climate information from speleothems. The rocks could provide a climate record through the oxygen isotope ratios. Oxygen in water comes in two important varieties for paleoclimate research: heavy and light. The ratio of these different types of oxygen in water vary based on air temperature, the total amount of ice in the world, and the amount of local precipitation—all important pieces of the climate puzzle. To an extent, the ratio is preserved in the cave rocks, and scientists can use this clue to learn about the climate at the time the rock formed.

Scientists can use the oxygen isotope ratio to track changes in the amount of rainfall (heavy rain results in more light oxygen) or changes in where the rain came from—the ocean or inland sources. "In southeast Brazil, for example, winter rain comes from the nearby Atlantic Ocean, but summer rain comes from the Amazon

Sliim Sepp/Shutterstock

FIGURE 13.4   Photograph of a cross section of a stalagmite.

Basin," says Stephen J. Burns, a geochemist at the University of Massachusetts. "The two have quite different oxygen isotope ratios." By examining the ratios in the rocks, Burns and others can track seasonal changes and rainfall patterns. Since caves exist all over the Earth, speleothems have the potential to become a pivotal land-based climate record.

Other land features help fill in the picture of climate in the past. Water marks trace the shorelines of ancient lakes where none exist now. Beaches record changes in sea level as ice sheets formed, and then melted. Dust deposits reveal where winds blew across ancient deserts. Fossils of animals and plants—even microscopic fossilized pollen grains—give clues about the climate where and when those organisms lived.

Paleoclimatologists analyze the growth rate of stalactites and stalagmites to reveal patterns of past rainfall. This graph shows the thickness of near-annual growth rings for the past 450 years from a stalagmite in Carlsbad Cavern. Thick rings indicate a relatively wet climate, while thin rings indicate a dry climate.

While these artifacts in the Earth are valuable to scientists reconstructing large-scale climate patterns on land, most have been disturbed by subsequent weathering. For a global picture of past climate, scientists also need a consistent record that covers a broad section of the Earth. For that, they turn to the oceans.

A Record from the Deep

By Holli Riebeek · design by Robert Simmon · September 27, 2005

Clad in a hard hat and steel-toe boots, paleoceanographer Jerry McManus strides onto the deck of the JOIDES Resolution (Figure 13.5), staring through the steel rigging that supports the ship's drilling equipment at the brilliant star-studded sky. Here, in the middle of the ocean, city lights do not dim the night sky, and the clear view is spectacular. McManus, an associate scientist at the Woods Hole Oceanographic Institution, has just completed another 12-hour shift in one of the ship's six science labs, where he has been analyzing samples of the sea floor to glean bits of evidence about past climates.

The ship cruises the globe sampling sediments from the bottom of the world's oceans. The ship is capable of drilling holes over 2,100 meters (6,890 feet) below the sea floor in water up to 8,000 meters (26,000 feet) deep.

Even now, in the dead of the night, it is not quiet. The ship's twelve powerful thrusters whine constantly as their 750-horsepower engines struggle against the ocean currents to keep the ship in one place while the

FIGURE 13.5   Photograph of the ocean drilling ship, JOIDES Resolution.

drilling crew pulls long sections of mud from the sea floor. The science labs continue to bustle as another crew replaces those who are leaving for the night, and caterers and housekeepers move through the ship to support the science and drilling teams. McManus pauses to scan the surface of the ocean for signs of whales or other sea life. He saw a manta ray jumping once, but not tonight. He returns to his room in the forecastle deck. Two sets of bunk beds accommodate the four people who share the room, but he rarely sees his roommates. They are on other shifts in this all-too-brief voyage to coax climate secrets out of the ocean depths.

Year after year, a steady rain of dust, plants, and animal skeletons settles on the ocean floor. As new materials pile on top of old materials, layers of sediment form a vertical timeline extending millions of years into the past. McManus and his colleagues on the Resolution are drilling long cores of the ocean floor to read the timeline. The 470-foot-long research vessel is specially equipped to pull cores of mud from the sea floor. Much of the equipment, and the ship itself, is adapted from tools the oil industry uses to drill at sea, and, as a result, the Resolution resembles an oil rig with its steel drill tower and deck-top cranes.

In the center of the ship, long sections of pipe snake down to the sea floor where a drill is fitted on the outside of the pipe. A solid piston inside the pipe moves up as the pipe plunges into the mud so that the pipe fills with mud as it sinks. The goal is to pull up a column of sediment without disturbing it. Stirring the sediment would destroy the timeline preserved in the layers. The pipe draws up 10-meter segments of earth at a time. A cone with a homing device rests over the drill hole so the pipe can be lowered into the same location to retrieve the next 10 meters until the drill hits the solid rock of the sea floor.

The Resolution is perhaps the most advanced scientific ocean drilling ship, and an international consortium of ocean researchers called the International Ocean Drilling Program is responsible for it. Though the technology is vastly different, the idea of a science-dedicated ocean exploring vessel isn't too far off from the first explorations in the 1870s. On December 21, 1872, a three-masted, square-rigged wooden ship set sail from Portsmouth, England, to start a three-and-a-half-year voyage that would take the HMS Challenger from the North Atlantic to Antarctica and around the world. The ship's crew and teams of physicists, biologists, and chemists from around the world sounded out the depths of the ocean, collected samples of plants and animals and ocean water, and recorded sea temperature at various depths. They published their results in a 50-volume report, each volume containing 29,500 pages. The voyage of the Challenger became the basis of modern oceanography.

Scientists on the Challenger dredged the ocean floor with large bags to collect plant and animal samples. They found that the ocean was covered in fine sediment that contained the fossils of sea animals. What was more, the fossils were different in cold areas verses warm areas. The finding thrilled paleoclimatologists, who wanted to use the fossils to determine how cold the oceans had been in the past. Scientists almost immediately began to devise systems of hollow pipes that could be used to bring a column of the sea floor to the surface.

These records from the deep yielded many important insights to the Earth's past climates. Each layer within the core holds fossils of the tiny plants and animals that dominate the ocean, as well as grains of dust and minerals that can tell about wind and current patterns. Like land fossils, marine fossils offer clues about conditions in the ocean when the plant or animal lived. The cores are carefully labeled ("this way up" is a crucial designation for the vertical time lines) and divided into smaller sections for analysis. The 19th-century voyage of the HMS Challenger set the standard for subsequent ocean research vessels. Among other discoveries, scientists on the Challenger realized that fossils retrieved from samples dredged from the sea floor revealed past climates.

The Ice Core Record

By *Holli Riebeek · design by Robert Simmon · December 19, 2005*

Richard Alley might have envied paleoceanographer Jerry McManus' warm, ship-board lab. (See previous installment: "A Record from the Deep.") One of the researchers in the Greenland Ice Sheet Project 2 (GISP2), Alley huddled in a narrow lab cut into the Greenland Ice Sheet, where "the temperature stayed at a 'comfortable' twenty below [Fahrenheit]," he wrote in his book about his research, The Two-Mile Time Machine. An assembly line of science equipment lined the twenty-foot-deep trench that served

James Jones Jr/Shutterstock

FIGURE 13.6   Drilling on board the Resolution continues day and night.
Image: drilling derrick, drill bit, Engraving of the HMS Challenger.

as a makeshift lab. For six weeks every summer between 1989 and 1993, Alley and other scientists pushed columns of ice along the science assembly line, labeling and analyzing the snow for information about past climate, then packaging it to be sent for further analysis and cold storage at the National Ice Core Laboratory in Denver, Colorado. Nearby, a specially built drill bored into the thick ice sheet twenty-four hours a day under the perpetual Arctic sun. Essentially a sharpened pipe rotating on a long, loose cable, the drill pulled up cores of ice from which Alley and others would glean climate information.

Throughout each year, layers of snow fall over the ice sheets in Greenland and Antarctica. Each layer of snow is different in chemistry and texture, summer snow differing from winter snow. Summer brings 24 hours of sunlight to the polar regions, and the top layer of the snow changes in texture—not melting exactly, but changing enough to be different from the snow it covers. The season turns cold and dark again, and more snow falls, forming the next layers of snow. Each layer gives scientists a treasure trove of information about the climate each year. Like marine sediment cores, an ice core provides a vertical timeline of past climates stored in ice sheets and mountain glaciers.

Ice sheets contain a record of hundreds of thousands of years of past climate, trapped in the seasonal snow layers are easiest to see in snow pits, writes Alley, the Evan Pugh Professor in the Environment Institute and Department of Geosciences at Pennsylvania State University. To see the layers, scientists dig two pits separated by a thin wall of snow. One pit is covered, and the other is left open to sunlight. By standing in the covered pit, scientists can study the annual snow layers in the snow wall as the sunlight filters through the other side. "I have stood in snow pits with dozens of people—drillers, journalists, and others—and so far, every visitor has been impressed. The snow is blue, something like the blue seen by deep sea divers, an indescribable, almost achingly beautiful blue," writes Alley. "The next thing most people notice is the layering."

To pry climate clues out of the ice, scientists began to drill long cores out of the ice sheets in Greenland and Antarctica in the late 1960s. By the time Alley and the GISP2 project finished in the early 1990s, they had pulled a nearly 2-mile-long core (3,053.44 meters) from the Greenland ice sheet, providing a record of at least the past 110,000 years. Even older records going back about 750,000 years have come out of Antarctica. Scientists have also taken cores from thick mountain glaciers in places such as the Andes Mountains in Peru and Bolivia, Mount Kilimanjaro in Tanzania, and the Himalayas in Asia.

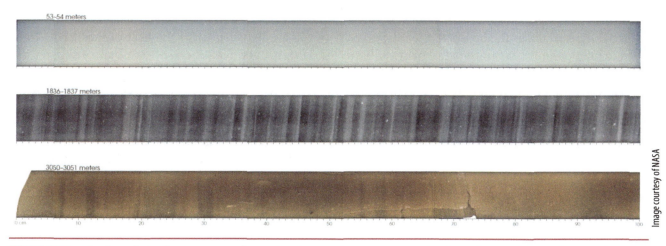

Image courtesy of NASA

FIGURE 13.7   **The bands in the ice cores show the change in seasons, and colors change depending on the depth of the core.**

The gradually increasing weight of overlying layers compresses deeply buried snow into ice, but annual bands remain. Relatively young and shallow snow becomes packed into coarse and granular crystals called firn (top: 53 meters deep). Older and deeper snow is compacted further (middle: 1,836 meters). At the bottom of a core (lower: 3,050 meters), rocks, sand, and silt discolor the ice. (Photographs courtesy U.S. National Ice Core Laboratory Figure 13.7).

The ice cores can provide an annual record of temperature, precipitation, atmospheric composition, volcanic activity, and wind patterns. In a general sense, the thickness of each annual layer tells how much snow accumulated at that location during the year. Differences in cores taken from the same area can reveal local wind patterns by showing where the snow drifted. More importantly, the make-up of the snow itself can tell scientists about past temperatures. As with marine fossils, the ratio of oxygen isotopes in the snow reveals temperature, though in this case, the ratio tells how cold the air was at the time the snow fell. In snow, colder temperatures result in higher concentrations of light oxygen.

Scientists can confirm these chemistry-based temperature measurements by observing the temperature of the ice sheet directly. The ice sheet's thickness makes its temperature much more resistant to change than the six inches of snow that might fall on your driveway during a winter snowstorm. As Alley explained to the Earth Observatory, the ice sheet can be compared to a frozen roast that is put directly into the oven. The outside heats up quickly, but the center remains cold, close to the temperature of the freezer, for a long time. Similarly, the ice sheet has warmed somewhat since the Ice Age, but not completely. The top has warmed as global temperatures have warmed, while the bottom has been warmed by heat flow from deep inside the Earth. But in the middle of an ice sheet, the ice remains close to the Ice Age temperatures at which it formed. "Because we understand how heat moves in ice, [and] we know how cold the ice is today, we can calculate how cold the ice was during the Ice Age," says Alley.

When scientists lower an ultra-precise thermometer into a hole in the ice, they can detect the temperature variations that have occurred since the Ice Age. The near-surface ice temperature, like the atmosphere today, is warm, and then the temperature drops in the layers formed roughly between AD 1450 and 1850, a period known as the Little Ice Age, one of several cold snaps that briefly interrupted the overall warming trend ongoing since the end of the Ice Age. As the thermometer goes deeper into the ice sheet, the temperature warms again, and then plummets to the temperatures indicative of the Ice Age. Finally, the bottom layers of the ice sheet are warmed by heat coming from the Earth. These directly measured temperatures represent a rough average—a record of trends, not variable, daily temperatures—but climatologists can compare the thermometer temperatures with the oxygen isotope record as a way to calibrate those results.

Scientists measure the temperature of an ice sheet directly by lowering a thermometer into the borehole that was drilled to retrieve the ice core. Like an insulated thermos, snow and ice preserve the temperature of each successive layer of snow, which reflects general atmospheric temperatures when the layer accumulated. Close to the surface of the bedrock, the lowest layers of the ice are warmed by the heat of the Earth. These physical temperature measurements help calibrate the temperature record scientists obtain from oxygen isotopes.

As valuable as the temperature record may be, the real treasure buried in the ice is a record of the atmosphere's characteristics. When snow forms, it crystallizes around tiny particles in the atmosphere, which fall to the ground with the snow. The type and amount of trapped particles, such as dust, volcanic ash, smoke, or pollen, tell scientists about the climate and environmental conditions when the snow formed. As the snow settles on the ice, air fills the space between the ice crystals. When the snow gets packed down by subsequent layers, the space between the crystals is eventually sealed off, trapping a small sample of the atmosphere in newly formed ice. These bubbles tell scientists what gases were in the atmosphere, and based on the bubble's location in the ice core, what the climate was at the time it was sealed. Records of methane levels, for example, indicate how much of the Earth wetlands covered because the abundance of life in wetlands gives rise to anaerobic bacteria that release methane as they decompose organic material. Scientists can also use the ice cores to correlate the concentration of carbon dioxide in the atmosphere with climate change—a measurement that has emphasized the role of carbon dioxide in global warming.

Finally, anything that settles on the ice tends to remain fixed in the layer it landed on. Of particular interest are wind-blown dust and volcanic ash. As with dust found in sea sediments, dust in ice can be analyzed chemically to find out where it came from. The amount and location of dust tells scientists about wind patterns and strength at the time the particles were deposited. Volcanic ash can also indicate wind patterns. Additionally, volcanoes pump sulfates into the atmosphere, and these tiny particles also end up in the ice cores. This evidence is important because volcanic activity can contribute to climate change, and the ash layers can often be dated to help calibrate the timeline in the layers of ice.

Air bubbles trapped in the ice cores provide a record of past atmospheric composition. Ice core records prove that current levels of carbon dioxide and methane, both important greenhouse gases, are higher than any previous level in the past 400,000 years.

Though ice cores have proven to be one of the most valuable climate records to date, they only provide direct evidence about temperature and rainfall where ice still exists, though they hint at global conditions. Marine sediment cores cover a broader area—nearly 70 percent of the Earth is covered in oceans—but they only give tiny hints about the climate over the land. Soil and rocks on the Earth's surface reveal the advance and retreat of glaciers over the land surface, and fossilized pollen traces out rough boundaries of where the climate conditions were right for different species of plants and trees to live. Unique water and rock formations in caves harbor a climate record of their own. To understand the Earth's climate history, scientists must bring together all of these scattered threads into a single, seamless story.

Living Things with a Climate "Memory"

By *Holli Riebeek · design by Robert Simmon · December 22, 2005*

"This history in trees tells us the climatic story of the Southwest with amazing accuracy. When a real theory of climate has been developed and we can predict drought and flood over a period of years, this Arizona story in tree rings will have played a creditable part in developing that climatic foresight which is perhaps the most valuable economic advantage yet lying beyond our reach."

—Andrew Ellicott Douglass, 1929

While cave rocks and ice cores provide a long-term, annual record of past climate (see "Written in the Earth" and "The Ice Core Record" in this series), some other climate proxies can offer a detailed record of seasonal temperature or rainfall changes. As they grow from season to season, coral reefs in the oceans and trees on the land both record small variations in the climate. These records can tell scientists about growing conditions in the oceans or on the land, but the record only stretches across the collective lifetimes of the organisms that have been preserved through the centuries. Thus, even though their records are more detailed, reefs and trees cannot provide records that are as long or continuous as ice or sediment core records.

Tree Rings

Squat and gnarled, Methuselah clings to the rocky slopes of the White Mountains in Southern California as it has for the past 4,770 years. When the ancient bristlecone pine took root, the earliest Greek civilization was being established and the Egyptians were just beginning construction of the Pyramids of Giza. Thousands of years later, both those civilizations are long gone, but Methuselah lives on. It is one of the Earth's oldest known living organisms.

The barren limestone soil around the tree is bare of grass or other plants, supporting only a widespread grove of scraggly Bristlecone Pine trees, at least one of which is even older than Methuselah (Figure 13.8). At about 11,000 feet above the arid Great Basin Desert, the trees receive precious little water—hardly a location hospitable to any life, let alone the oldest of living organisms. Ironically, it is the barrenness of the location that has allowed the trees to live so long. With no surrounding fuel, lightning-ignited forest fires can't engulf the grove.

The Methuselah Walk, high in the White Mountains of California, winds among the oldest known trees in the world. The ancient and twisted bristlecone pines grow extremely slowly, preserving a history of climate in their annual growth rings. The bristlecone climate record goes back 9,000 years, contained in living and dead wood as old as the last ice age.

The inhospitable environment has also made the trees excellent recorders of rainfall. Each year, the trees grow wider, adding another ring to their girth. Most people have counted the number of rings in a tree stump to find out how old the tree was, but the rings also tell about growing conditions the year it formed. High in the Great Basin Desert, where water is scarce, the growing conditions are most directly influenced by rainfall.

Variation in the closely spaced rings of a bristlecone pine correspond to annual changes in rainfall and temperature.

In the 1890s, a young astronomer at the Lowell Observatory in Flagstaff, Arizona, was trying to understand how sun spot cycles might affect plant growth. In his research, Andrew Ellicott Douglass noticed that the thickness of each ring in the pines and Douglas firs in the region depended on how much rain fell during the year (Figure 13.9). He wrote, "Through long-past ages and with unbroken regularity, trees have

FIGURE 13.8   **Photograph of bristlecone pines along the Methuselah Trail, Inyo National Forest.**

Dani Vincek/Shutterstock

FIGURE 13.9 **Close-up photograph of bristlecone pin tree rings.**

jotted down a record at the close of each fading year—a memorandum as to how they passed the time; whether enriched by added rainfall or injured by lightning and fire. . . . So, in the rings of the talkative pines we find lean years and fat years recorded. The same succession of drought and plenty appears throughout the forest." Because all of the trees in the area exhibited the same pattern of thick and thin rings, Douglass was able to construct a tree calendar going back to AD 700 by piecing together the tree ring patterns of living trees and patterns found in wood preserved in Native American Pueblo villages.

Andrew Ellicott Douglass and Edmund Schulman pioneered dendrochronology—the science of dating past events using tree rings.

In the 1950s one of Douglass' former students and a respected tree researcher in his own right, Edmund Schulman, headed into the White Mountains to look at the trees rumored to be very old. He discovered Methuselah and the old bristlecone pines surrounding it. Around the trees, even older dead trees remained on the ground. Together, they gave a climate record of the Southwest United States that extends back 9,000 years, the longest record for a single tree species. In Europe scientists have combined the ring-records from various trees to piece together the past 11,000 years of Europe's climate history.

Douglas' rings tell about rainfall in the southwestern United States, but trees also respond to changes in sunlight, temperature, and wind, as well as non-climate factors like the amount of nutrients in the soil and disease. By observing how these factors combine to affect tree rings in a region today, scientists can guess how they worked in the past. For example, rainfall in the southwestern United States is the factor that affects tree growth most, but in places where water is plentiful, like the Pacific Northwest, the key factor affecting tree ring growth may be temperature. Once scientists know how these factors affect tree ring formation, scientists can drill a small core from several trees in an area (a process that does not harm the tree) and determine what the climate was in previous years. The trees may also record things like forest fires by bearing a scar in a ring.

Short- and long-term variability of rainfall along the eastern margin of the Sierra Nevada is recorded in bristlecone tree rings (Figure 13.10). Several long and intense droughts that appear in the tree-rings are also found in sediments in nearby Mono Lake. (Graph derived from Hughes 1996)

Individual events such as forest fires are recorded in tree-rings. The dark arcs that interrupt the sequence of rings in this sample were caused by fires in the 19th century.

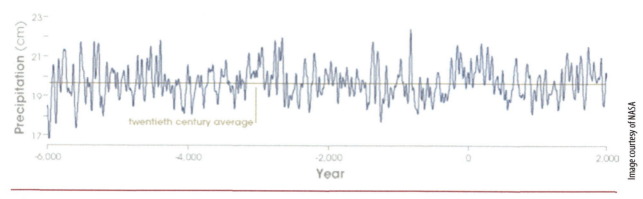

FIGURE 13.10 **Rainfall pattern recorded in the bristlecone pine trees.**

Coral Reefs

The warm, shallow ocean waters of the tropics have talkative "forests" of their own. Brightly colored mounds of coral grow in the warm ocean waters, quickly when nutrients are plentiful and more slowly when they are not. Like their land-based counterparts, corals add seasonal layers, which appear as bands in their hard calcium-carbonate shells. Corals respond to small changes in temperature, rainfall, and water clarity in a matter of months, making them a uniquely sensitive climate record. From a small core from the coral, scientists can put together a very detailed picture of climate in the Tropics—significant because much of Earth's weather is controlled by conditions in the Tropics.

The bands in the coral's shell can change in thickness with changes in temperature, water clarity, or nutrient availability, so while each band can record the season's climate, the interpretation of the record depends on how the three factors are related. Cool water rising from the ocean floor brings extra nutrients in many areas, so the shells are often thicker when the water is cool. In other areas, the cold may slow growth. Scientists have to couple their observations of patterns in the seasonal bands to other measurements, including modern observations of coral growth, to determine what the bands say about climate change.

Vibrant coral reefs harbor diverse communities of life in the tropical oceans. Like trees, corals produce annual rings that store a record of past conditions. Chemical analyses reveal details about past temperature, nutrient availability, salinity, and other information.

One of the most significant clues to climate in coral comes from the chemistry of the bands. The chemicals in each layer reflect conditions in the ocean when the layer formed. Like the scaly coverings of foraminifera and other marine organisms, the ratio of heavy and light oxygen in coral growth bands provide a record of temperature and rainfall during the growing season. Both more rain and higher temperatures result in a higher concentration of light oxygen in the ocean. The concentrations of other chemicals can help scientists separate the temperature and rainfall records implied by the oxygen ratio. In coral, the balance between strontium and calcium is largely determined by temperature. By comparing this ratio to the heavy-to-light-oxygen ratio, scientists can more accurately determine whether changes in coral skeletons are because of climate change involving temperature, or ocean salinity, which changes with rainfall, or a combination of both.

Coral can also tell scientists when heavy rains or floods carried extra sediment into the ocean. Sediment in the water can change the color of the coral as it absorbs elements from the land. Further, reef coral has a symbiotic relationship with algae that use photosynthesis to produce energy. When the water is clouded with sediment, the algae, and therefore the coral, cannot grow as quickly because it doesn't receive as much sunlight. This slow-down in growth appears in the growth layers pulled from core samples just as drought shows up in the growth rings of trees.

The climate record left in coral reefs is detailed, but limited. First, coral reefs don't exist everywhere in the world. They can only tell scientists about climate in warm, tropical waters. Scientists have discovered some deep water coral that may yield a detailed climate record of other regions, but the work is still in its early stages. Second, coral are living things that die. The record they preserve only covers the lifetime of the

Vlad61/Shutterstock

FIGURE 13.11 **Coral reefs are beautiful to look at, and they are also wonderful tools for studying climate change due to their sensitivity to changes in temperature, rainfall and the clarity of the surrounding water.**

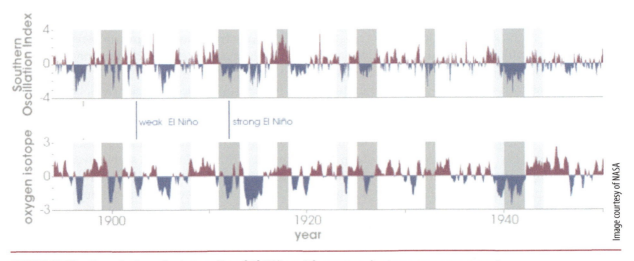

Image courtesy of NASA

FIGURE 13.12 **Correlating the intensity of El Niño with oxygen isotope measurements.**

individual—a few hundred years, then an older coral from the reef has to be found to stretch the record further back. Piecing together a continuous record can be very difficult and requires numerous samples from both living and fossil corals.

Scientists use coral cores to study cyclical events like El Niño. The upper graph shows the Southern Oscillation Index—a meteorological measurement of the intensity of El Niño. Low values correspond to El Niño events, high values to La Niña events. The lower graph shows change in oxygen-18 isotopes measured in coral cores on Tarawa Island. The Southern Oscillation Index and the coral oxygen isotope measurements rise and fall together, and they generally match historical records of weak (light gray bars) and strong (dark gray bars) El Niños. (Graphs in Figure 13.12 adapted from Cole, 1993).

Scientists can place the coral reef record in the timeframe recorded by other climate proxies once they know when the reef lived. They can date coral by measuring how much thorium and uranium it contains. Like speleothems, coral contains a large amount of uranium when it forms. Over time, the uranium decays into thorium until there are roughly equal amounts of uranium and thorium—a process that takes about three to four hundred thousand years. From the time the coral forms until the uranium decay evens out a few hundred thousand years later, scientists can tell exactly how old the coral is by measuring how much thorium it contains. This gives them a timeframe to relate to other climate records.

Explaining the Evidence

By *Holli Riebeek · design by Robert Simmon · May 9, 2006*

From the oceans' depths to the polar ice caps, clues to the Earth's past climates are engraved on our planet. Sea sediments reveal how much ice existed in the world and hint at past temperatures and weather patterns. Ice cores also provide a glimpse of past temperatures and preserve tiny bubbles of ancient atmosphere. Coral, tree rings, and cave rocks record cycles of drought and rainfall. Each piece of this complex puzzle must be put together to give us a picture of Earth's climate history. Scientists' efforts to explain the paleoclimate evidence—not just the when and where of climate change, but the how and why—have produced some of the most significant theories of how the Earth's climate system works.

The Earth's Shifting Orbit

From the scratched rocks strewn haphazardly across the landscape and the thin layer of soil left behind by retreating glaciers, scientists learned that **the Earth had gone through at least three or four ice ages. Noticing that the ice came and went cyclically, they began to suspect that the ice ages were connected to variations in the Earth's orbit.**

The Earth circles the Sun in a flat plane. It is as if the spinning Earth is also rolling around the edge of a giant, flat plate, with the Sun in the center. The shape of the Earth's orbit—the plate—changes from a nearly perfect circle to an oval shape on a 100,000-year cycle (eccentricity). Also, if you drew a line from the plate up through the Earth's North and South Poles—Earth's axis—the line would not rise straight up from the plate. Instead the axis is tilted, and the angle of the tilt varies between 22 and 24 degrees every 41,000 years (obliquity). Finally, the Earth wobbles on its axis as it spins. Like the handle of a toy top that wobbles toward you and away from you as the toy winds down, the "handle" of the Earth, the axis, wobbles toward and away from the Sun over the span of 19,000 to 23,000 years (precession). **These small variations in Earth-Sun geometry change how much sunlight each hemisphere receives during the Earth's year-long trek around the Sun, where in the orbit (the time of year) the seasons occur, and how extreme the seasonal changes are.**

In the early 1900s, a Serbian mathematician named Milutin Milankovitch meticulously calculated the amount of sunlight each latitude received in every phase of Earth's orbital variations. His work culminated in the 1930 publication of Mathematical Climatology and the Astronomical Theory of Climate Change. He theorized that the ice ages occurred when orbital variations caused the Northern Hemisphere around the latitude of the Hudson Bay and northern Europe to receive less sunshine in the summer. Short, cool summers failed to melt all of the winter's snow. The snow would slowly accumulate from year to year, and its shiny, white surface would reflect more radiation back into space. Temperatures would drop even further, and eventually, an ice age would be in full swing. Based on the orbital variations, Milankovitch predicted that the ice ages would peak every 100,000 and 41,000 years, with additional "blips" every 19,000 to 23,000 years.

Three variables of the Earth's orbit—eccentricity, obliquity, and precession—affect global climate. Changes in eccentricity (the amount the orbit diverges from a perfect circle) vary the distance of Earth from the Sun. Changes in obliquity (tilt of Earth's axis) vary the strength of the seasons. Precession (wobble in Earth's axis) varies the timing of the seasons.

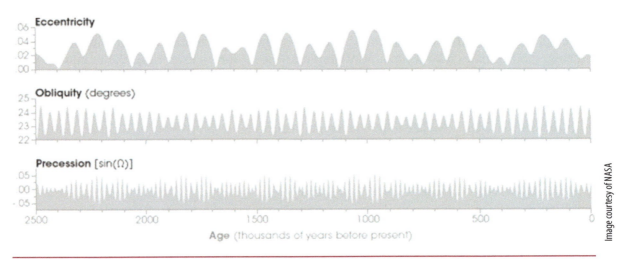

Image courtesy of NASA

FIGURE 13.13   Changes in these three variables above impact the amount of light energy that comes to Earth from the Sun, and where this energy hits Earth's surface most directly.

The Earth's orbit varies over tens and hundreds of thousands of years. **Combined changes in eccentricity, obliquity, and precession alter the strength and location of sunlight falling on the Earth's surface.** (Graphs by Robert Simmon based on data from Berger 1992, Figure 13.13).

The paleoclimate record shows peaks at exactly those intervals. Ocean cores showed that the Earth passed through regular ice ages—not just the 3 or 4 recorded on land by misplaced boulders and glacial loess deposits—but 10 in the last million years, and around 100 in the last 2.5 million years.

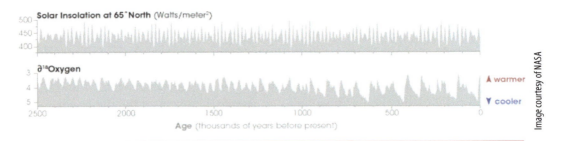

Image courtesy of NASA

FIGURE 13.14   (Upper graph) The rise and fall of the intensity of sunlight (insolation) in the far North during the summer—determined by the Earth's orbit—drives ice ages. Weak summer sunlight year after year allows snow to accumulate and glaciers to advance. The reflective ice sheets further cool the Earth's surface, resulting in global ice ages. When the Northern Hemisphere receives more sunlight, the snow melts, ice sheets retreat, and Earth warms. (Lower graph) Oxygen isotopes trapped in ocean sediments record cycles of ice ages millions of years into Earth's past. This climate record matches the frequency of orbital changes, although tangled feedbacks make the relationship complex. Dips represent ice ages, and spikes represent interglacials. (Graph by Robert Simmon, based on data from Berger 1992 and Lisiecki 2005.)

Evidence supporting Milankovitch's theory of the precise timing of the ice ages first came from a series of fossil coral reefs that formed on a shallow ocean bench in the South Pacific during warm interglacial periods. As the ice ages came, more and more water froze into polar ice caps and the ocean levels dropped, leaving the reef exposed. When the ice melted, the ocean rose and warmed, and another reef formed. At the same time, the peninsula on which the reefs formed was steadily being pushed up by the motion of the Earth's shifting tectonic plates. Today, the reefs form a visible series of steps along the shore of Papua New Guinea. The reefs, the age of which was well-defined because of the decaying uranium in the coral, measured out the millennia between ice ages. They also defined the maximum length of each ice age. **The intervals fell exactly where Milankovitch said they would.**

Explaining Rapid Climate Change: Tales from the Ice

When scientists started to analyze the paleoclimate evidence in the Greenland and Antarctic ice cores, they found that the record also supported Milankovitch's theory of when ice ages should occur. But they also found something that required additional explanation: some climate change appeared to have occurred very rapidly. Because Milankovitch's theory tied climate change to the slow and regular variations in Earth's orbit, the scientific community expected that climate change would also be slow and gradual. But the ice cores showed that while it took nearly 10,000 years for the Earth to totally emerge from the last ice age and warm to today's balmy climate, one-third to one-half of the warming—about 15 degrees Fahrenheit—occurred in about 10 years, at least in Greenland. A closer look at marine sediments confirmed this finding. Although the overall timing of the ice ages was clearly tied to variations in the Earth's orbit, other factors must have contributed to climate change as well. Something else made temperatures change very quickly, but what?

Rapid changes between ice ages and warm periods (called interglacials) are recorded in the Greenland ice sheet. Occurring over one or two decades, the warming of the Earth at the end of the last ice age happened much faster than the rate of change of the Earth's orbit. The last cool period (stadial), immediately before the current interglacial, began and ended suddenly, and was likely caused by changes in the deep ocean circulation. (Figure 13.15 by Robert Simmon, based on data provided by Alley 2004.)

FIGURE 13.15   Graph of Temperatures from 20,000 Years Ago to Present, Based on Data from the GISP2 Ice Core.

Greenhouse Gases

Scientists are now exploring a few possibilities. First, greenhouse gases probably influenced past climates. Ice cores record past greenhouse gas levels. In the past, when the climate warmed, the change was accompanied by an increase in greenhouse gases, particularly carbon dioxide. When scientists tried to build climate models, they could not get the models to simulate past climate change unless they also added changes in carbon dioxide levels. Though scientists aren't sure why carbon dioxide levels changed, almost all believe that the shift contributed to altering the climate. Ice cores also revealed that carbon dioxide levels are much higher today than at any time recorded in the past 750,000 years.

Global Conveyor Belt

Another possible trigger for rapid climate change is ocean circulation. Today, warm water from the equator is carried towards the poles on ocean surface currents. Because of the arrangement of the continents, warm water is carried far into the North Atlantic, moderating the climate in Northern Europe. As the warm surface water reaches the cold air in the north, it cools. The salty Atlantic water becomes very dense as it gets cold. The cold, salty water sinks to the bottom of the ocean before it can freeze, where it is pulled southward toward the equator. More warm water from the equator flows north to replace the sinking water, setting up a global oceanic "conveyor belt."

This pattern helps keep Northern Europe far warmer than other locations at the same latitude. The key to keeping the belt moving is the saltiness of the water, which increases the water's density and causes it to sink. Many scientists believe that if too much fresh water enters the ocean, for example, from melting Arctic glaciers and sea ice, the water will be diluted. Fresh water freezes at a higher temperature than salty water, so the cooling surface water would freeze before it could become dense enough to sink toward the bottom. If the water in the north does not sink, the water at the equator will not move north to replace it. The currents would eventually stop moving warm water northward, leaving Northern Europe cold and dry within a single decade.

This theory of rapid climate change is called the "conveyor belt theory." Recent paleoclimate studies have shown that when heat circulation in the North Atlantic Ocean slowed in the past, the climate changed in Northern Europe. Although the last ice age peaked about 20,000 years ago, the warming trend was interrupted at various points by cold spells. In a paper published in Nature on April 22, 2004, McManus and colleagues Roger Francois, Jeanne Gherardi, Lloyd Keigwin and Susan Brown-Leger at Woods Hole Oceanographic Institute and the Laboratoire des Sciences du Climat et de l'environnement in France showed that cold periods in Europe 17,500 and 12,700 years ago happened just after melting ice diluted the salty North Atlantic water, and the ocean "conveyor belt" slowed. The evidence, which they took from radioactive elements in ocean cores, is beginning to support the theory, but McManus cautions that there are still pieces to fill in before we fully understand what role the conveyor belt played in past climate change and what role it might play in the future.

The large-scale movement of water through the oceans, called the thermohaline circulation, plays a large role in the duration of ice ages. Dense, very salty (saline) water sinks in the North Atlantic, pulling the "conveyor belt" of currents behind it. The conveyor belt carries heat from the equator towards the poles, and raises Arctic temperatures, discouraging the growth of ice sheets. Influxes of fresh water from the lands that surround the North Atlantic can slow or shut down the circulation, cooling the Northern Hemisphere.

This map shows the general location and direction of the warm surface (red) and cold deep water (blue) currents of the thermohaline circulation. Salinity is represented by color in units of the Practical Salinity Scale. Low values (blue) are less saline, while high values (orange) are more saline. (Map by Robert Simmon, adapted from the IPCC 2001 and Rahmstorf 2002.)

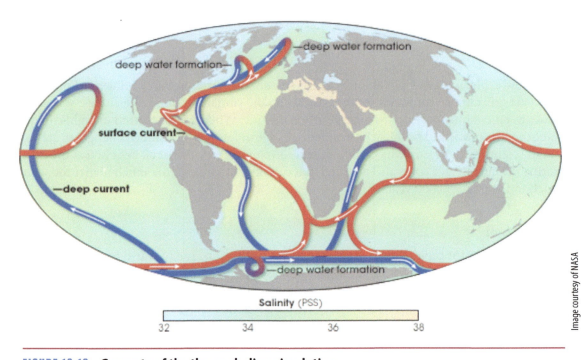

FIGURE 13.16   Currents of the thermohaline circulation.

Understanding the Past to Predict the Future

By *Holli Riebeek* · *design by Robert Simmon* · *November 14, 2006*

Scientists are using the theories they have constructed to explain the paleoclimate data record (see Part 5 in this series: Explaining the Evidence) to understand the modern climate and to predict how we can expect it to change in coming years or decades. To test climate theories, scientists at NASA's Goddard Institute for Space Studies (GISS) and elsewhere are building powerful mathematical models that can show how changing conditions on the Earth and in the atmosphere affect climate. The most sophisticated models might represent the Earth as a three-dimensional grid with the atmosphere split into ten different grid layers each containing 65,000 reference points. Scientists might have the computer model calculate what the effect of increasing carbon dioxide and air pollution would be at each of those 65,000 reference points. These highly detailed models of the Earth's climate system must be run on supercomputers, some of which perform more than 80 million calculations every hour.

13.3   Update: Recent Conclusions from the IPCC

The Intergovernmental Panel on Climate Change is an international organization established by the United Nations Environment Program (UNEP) and the World Meteorological Organization (WMO) in 1988. The group is made up of thousands of scientists from across the world who are trained in many different science disciplines (such as geology, ecology, physics, chemistry, atmospheric sciences and more). These scientists are divided into 3 Working Groups and a Task Force. The working groups compile scientific findings related to climate research, summarize these findings, and write reports about the status of the global climate and predictions for the future. It is important to note that the IPCC not only summarizes findings from many different scientists and studies. This group examines the studies to look for a consensus. For example, do the findings of one study support the findings of other studies, and if not, how does the new information fit into the climate puzzle? The IPCC won the Nobel Prize in 2007 for its work.

In 2013, the Working Group I published a report entitled "Climate Change 2013: The Physical Science Basis" in which they updated the evidence supporting global climate change. You may access their full report at http:// www.ipcc.ch/report/ar5/wg1/ or the Summary for Policymakers (which has a summary of their findings) at http:// www.ipcc.ch/pdf/assessment-report/ar5/wg1/WG1AR5_SPM_FINAL.pdf

Interpreting the Report

Please note the following if you read the report at the given links (which is highly recommended).

Many folks are unused to reading scientific reports, and as such, there is confusion when the reports cite the "degree of certainty" and the "level of confidence" of a conclusion. Scientists may seem to be very most wishy-washy people, because they will never report a conclusion with 100% confidence. This goes back to math, statistics, and the nature of science. Unless you can sample 100% of all possible cases and/or outcomes, you cannot make a conclusion about a hypothesis with 100% certainty. There is always a possibility that an observed trend is simply due to chance, even if it is a very small chance. Because science cannot conclude anything with 100% certainty, people assume that scientific conclusions are not really final conclusions. This is true.

What is NOT true is that a vast majority of scientific conclusions about global climate change are under dispute in the scientific community. The report from the IPCC looks at studies from all different fields of science, analyzes them to see where they overlap, and then calculates a "degree of certainty" that the agreement between these findings is not by chance. If the degree of certainty is 80%, it does not mean that 20% of studies are in opposition to the other studies. It just means that due to "noise" in the numbers (and climate data tends to be very "noisy"), there is 20% possibility that an observed trend is by chance, and is not really an observed trend.

The Summary for Policymakers from the "Climate Change 2013: The Physical Science Basis" reported the following basic conclusions:

- The global warming of air and water systems on Earth is "unequivocal".
- The last 30 years have been the warmest 30-year period in 1400 years. Furthermore, each decade within the most recent 30 years has been warmer than the decade before it (in other words, 2003–2012 was warmer than 1993–2002, which was warmer than 1983–1992).
- Most (90%) of the increase in heat stored on Earth is in the oceans, and the oceans have warmed measurably in the past 40 years on a global scale.
- There are **less** extreme (i.e., record-breaking) **cold** days and nights being reported, and **more extreme warm** days and nights. Additionally, there are more events of heavy precipitation (i.e., flood-causing rains or snowfalls) in North America and Europe.
- Glaciated areas have seen a reduction in the amount of ice and snow worldwide. The ice sheets in Greenland, Antarctica and the Arctic have lost, and continue to lose, mass. The Arctic summer sea ice has lost between 9.4% and 13.6% of it's surface extent **per decade**, and the Antarctic has lost 1.2% to 1.8% **per decade**.
- The sea level has risen by 19 cm (7.5 inches) since 1901, and the current rate of rise is higher than it has been in the past 2000 years. The rise in sea level is due to expansion of the warming ocean waters and due to the melting of the ice sheets in Antarctica, Greenland and the Arctic.
- Ice core analysis shows that the concentration of the greenhouse gasses carbon dioxide, methane and nitrous oxide are at the highest level in 800,000 years. Carbon dioxide increases are due to combustion of fossil fuels and deforestation (i.e., there are less trees taking up carbon dioxide, and they are combusted instead which releases carbon dioxide) and other land use changes.
- Mathematical models demonstrate that the largest contributing factor to global increases in temperature is the emission of greenhouse gasses. Changes due to volcanic aerosols and in the amount of incoming solar radiation made only a small contribution to global temperature changes.
- The average global temperature increase is expected to be 2 degrees C (as compared to the 1850–1900 average temperatures) by the year 2100, and it will continue to increase. Changes to precipitation, as a result, will not be consistent. There will be more extreme wet and dry seasons.

13.4 Looking Ahead and Responding to Global Climate Change

Computer climate models are constantly being updated and improved. These models incorporate the predominant factors that influence the climate system, and are calibrated using past climate data. These models are critical for our society as planning tools for reacting to future climate conditions.

The more GHGs humans release into the atmosphere, the stronger the enhanced greenhouse effect will become. Scenarios in which GHGs continue to be added to the atmosphere by human activities could cause additional warming of 2 to 11.5°F over the next century, depending on how much more GHGs are emitted and how strongly the climate system responds to them. Although the range of uncertainty for future temperatures is large, even the lower end of the range is likely to have many undesirable effects on natural and human systems.

Land areas warm more rapidly than oceans, and higher latitudes warm more quickly than lower latitudes. Therefore, regional temperature increases may be greater or less than global averages, depending on location. For example, the United States is projected to experience more warming than average, and the Arctic is expected to experience the most warming.

The future climate depends largely on the actions taken in the next few decades to reduce and eventually eliminate human-induced CO_2 emissions. In 2005, the U.S. National Academy of Sciences joined with 10 other science academies from around the world in a statement calling on world leaders to take "prompt action" on climate change. The statement was explicit about our ability to limit climate change: "Action taken now to reduce significantly the build-up of greenhouse gases in the atmosphere will lessen the magnitude and rate of climate change."

Changing Climate: Theory to Reality

Although "climate change" and "global warming" are often used interchangeably, rising temperatures are just one aspect of climate change. To understand why, it is important to distinguish between "weather" and "climate." The climate is the average weather over a long period of time. A simple way to think of this is: *weather* is what determines if you will use an umbrella today; *climate* determines whether you own an umbrella. Thus, when looking at climate change and its impacts, it is important to consider more than just global temperature trends. Changes in the climate other than average temperatures have more direct impacts on nature and society.

The USGCRP report says, "Climate changes are underway in the United States and are projected to grow," and "Widespread climate-related impacts are occurring now and are expected to increase." Sea level rise, the loss of sea ice, changes in weather patterns, more drought and heavy rainfall, and changes in river flows are among the documented changes in the United States. Climate change also threatens ecosystems and public health.

Dr. Jane Lubchencko, the Administrator of the National Atmospheric and Oceanographic Administration, has said, "Climate change is happening now and it's happening in our own backyards and it affects the kinds of things people care about."

More Extreme Weather

Extreme weather events have become more common in recent years, and this trend will continue in the future. Climate change has a significant effect on local weather patterns and, in turn, these changes can have serious impacts on human societies and the natural world.

Stronger Hurricanes

Scientists have confirmed that hurricanes are becoming more intense. Since hurricanes draw their strength from the heat in ocean surface waters, hurricanes have the potential to become more powerful as the water warms. A recent peer-reviewed assessment of the link between hurricanes and climate change concluded that "higher resolution modeling studies typically project substantial increases in the frequency of the most intense cyclones, and increases of the order of 20% in the precipitation rate within 100 km of the storm centre [*sic*]."

This trend toward stronger hurricanes is noteworthy because of the vulnerability of coastal communities to these extreme events. The USGCRP report says, "Sea-level rise and storm surge place many U.S. coastal areas at increasing risk of erosion and flooding Energy and transportation infrastructure and other property in coastal areas are very likely to be adversely affected." In recent years the massive destruction caused by Hurricane Katrina in the United States and by Cyclone Nargis, which devastated Burma in 2008, provide painful reminders of this vulnerability.

Hotter, Wetter Extremes

Average temperatures are rising, but extreme temperatures are rising even more: in recent decades, hot days and nights have grown more frequent and cold days and nights less frequent. There have been more frequent heat waves and hotter high temperature extremes.

In the United States, the USGCRP report says, "Many types of extreme weather events, such as heat waves and regional droughts, have become more frequent and intense during the past 40 to 50 years." More rain is falling in extreme events now compared to 50 years ago, resulting in more frequent flash flooding. In 1994 and 2008, the U.S. Midwest experienced flooding so severe that each event was considered a 500-year flood—a level of flooding so rare that it would not be expected to occur more than once in five centuries! In May 2010, the city of Nashville, Tennessee, experienced the worst flooding in its history, enduring what the U.S. Army Corps of Engineers declared a 1,000-year flood. Nearly the entire central city was underwater for the first time. *The Tennessean*—Nashville's principal daily newspaper—reported that the flood cost the city a year's worth of economic productivity. Individually, these events might be random occurrences, but they are part of a clear, long-term trend of increasing very heavy rainfall in the United States over the past 50 years.

In 2003, Europe experienced a heatwave so hot and so long that scientists estimated that such an extreme event had not occurred there in at least 500 years. That heat wave caused more than 30,000 excess deaths throughout southern and central Europe. A similarly historic heat wave struck Russia and other parts of Eastern Europe in the summer of 2010, killing thousands of people and destroying a large fraction of Russia's wheat crop. Since Russia is a large grain exporter, its crop losses drove up food prices globally.

Although there is no way to determine whether an individual weather event was caused by human-induced climate change, the types of events discussed here are the types of events that scientists have predicted will become more common in a warmer climate. Therefore, the events that actually occur are useful indicators of our vulnerabilities to project impacts and can teach us about the likely effects of climate change on our lives.

Too Much or Too Little: Effects on Water

Climate change will alter the quantity and quality of available fresh water and increase the frequency and duration of floods, droughts, and heavy precipitation events. Although climate change will affect different regions in different ways, it is generally expected that dry regions of the world will get drier and wet regions will get wetter.

More Floods and Droughts

A number of factors are expected to contribute to more frequent floods. More frequent heavy rain events will result in more flooding. Coastal regions will also be at risk from sea level rise and increased storm intensity. While some regions will suffer from having too much water, others will suffer from having too little. Diminished water resources are expected in semi-arid regions, like the western United States, where water shortages often already pose challenges. Areas affected by drought are also expected to increase. As the atmosphere becomes warmer, it can hold more water, increasing the length of time between rain events and the amount of rainfall in an individual event. As a result, areas where the average annual rainfall increases may also experience more frequent and longer droughts.

Altered Availability and Quality

Warmer temperatures threaten the water supplies of hundreds of millions of people who depend on water from the seasonal melting of mountain ice and snow in several ways: by increasing the amount of seasonal melt from glaciers and snowpack, by increasing the amount of precipitation that falls as rain instead of snow, and by altering the timing of snowmelt. In the near term, the melting of mountain ice and snow may cause flooding; in the long term, the loss of these frozen water reserves will significantly reduce the water available for humans, agriculture, and energy production. Earlier snowmelt brings other impacts. Western states have experienced a six-fold increase in the amount of land burned by wildfires over the past three decades because snowmelt has occurred earlier and summers are longer and drier.

Climate change will affect the quality of drinking water and impact public health. As sea level rises, saltwater will infiltrate coastal freshwater resources. Flooding and heavy rainfall may overwhelm local water infrastructure and increase the level of sediment and contaminants in the water supply. Increased rainfall could also wash more agricultural fertilizer and municipal sewage into coastal waters, creating more low-oxygen "dead zones" in the Chesapeake Bay and the Gulf of Mexico.

Effects on Human Health

Climate change is expected to affect human health directly—from heat waves, floods, and storms—and indirectly—by increasing smog and ozone in cities, contributing to the spread of infectious diseases, and reducing the availability and quality of food and water. The USGCRP report says that children, the elderly, and the poor are at the greatest risk of negative health impacts in the United States.

The U.S. Centers for Disease Control and Prevention have identified a number of health effects associated with climate change, including an increase in heat-related illnesses and deaths from more frequent heat waves, a rise in asthma and other respiratory illnesses due to increased air pollution, higher rates of food- and water-related diseases, and an increase in the direct and indirect impacts of extreme weather events, like hurricanes.

Threats to Ecosystems

Climate change is threatening ecosystems around the world, affecting plants and animals on land, in oceans, and in freshwater lakes and rivers. Some ecosystems are especially at risk, including the Arctic and sub-Arctic because they are sensitive to temperature and likely to experience the greatest amount of warming; coral reefs because they are sensitive to high water temperatures and ocean acidity, both of which are rising with atmospheric CO_2 levels; and tropical rainforests because they are sensitive to small changes in temperature and precipitation.

Clear evidence exists that the recent warming trend is already affecting ecosystems. Entire ecosystems are shifting toward the poles and to higher altitudes. This poses unique challenges to species that already live at the poles, like polar bears, as well as mountain-dwelling species already living at high altitudes. Spring events, like the budding of leaves and migration of birds, are occurring earlier in the year. Different species are responding at different rates and in different ways, which has caused some species to get out of sync with their food sources. The risks to species increase with increasing temperatures; scientists say that an additional 2°F of warming will increase the risk of extinction for up to 30 percent of species.

Shrinking Arctic Sea Ice

Arctic sea ice has seen dramatic declines in recent years. In 2007, Arctic sea ice shrank to its smallest summertime extent ever observed, opening the Northwest Passage for the first time in human memory. This new sea ice minimum came only a few months after a study reported that since the 1950s, summer sea ice extents have declined three times faster than projected by climate models. In the summer of 2010, Arctic sea ice set a new kind of record: It decreased to the lowest volume ever observed. While the extent (the area of the Arctic Ocean covered by ice) in 2010 was slightly higher than in 2007, the ice was considerably thinner in 2010, making the volume lower than in 2007. Scientists are concerned that this historically low volume of ice could be more susceptible to melting in the future, causing sea ice loss to accelerate.

The importance of sea ice decline comes from the role it plays in both the climate system and large Arctic ecosystems. Snow and ice reflect sunlight very effectively, while open water tends to absorb it. As sea ice melts, the earth's surface will reflect less light and absorb more. Consequently, the disappearance of Arctic ice will actually intensify climate change.

Moreover, as the edge of the sea ice retreats farther from land during the summer, many marine animals that depend on the sea ice, including seals, polar bears, and fish, will lose access to their feeding grounds for longer periods (Figure 13.17). Eventually, this shift will deprive these organisms of their food sources and their populations will not be sustained.

If warming continues, scientists are sure that the Arctic Ocean will become largely free of ice during the summer. Depending in part on the rate of future greenhouse gas emissions, the latest model projections indicate that the opening of the Arctic is likely to occur sometime between the 2030s and 2080s. The opening of the Arctic has enormous implications, ranging from global climate disruption to national security issues to dramatic ecological

FIGURE 13.17 **Polar bears depend on the Arctic ice sheet to hunt prey from the ocean.** They are marine animals, but they can only swim so far between ice patches before they drown.

Shutterstock/Vladimir Melnik

shifts. The Arctic may seem far removed from our daily lives, but changes there are likely to have serious global implications.

Rising Sea Level

Among the most serious and potentially catastrophic effects of climate change is sea level rise, which is caused by a combination of the "thermal expansion" of ocean water as it warms and the melting of land-based ice. To date, most climate-related sea level rise can be attributed to thermal expansion. Going forward, however, the largest potential source of sea level rise comes from melting land-based ice, which adds water to the oceans. **By the end of the century, if nothing is done to rein in GHG emissions, global sea level could be three to six feet higher than it is today**, depending on how much land-based ice melts. Moreover, if one of the polar ice sheets on Greenland or West Antarctica becomes unstable because of too much warming, sea level is likely to continue to rise for more than a thousand years and could rise **by 20 feet or more, which would permanently flood virtually all of America's major coastal cities**.

FIGURE 13.18 **The island nation of Kiribati is threatened by rising sea levels, because most of the surface area of the islands in this chain are less than 10 feet above sea level.**

Even small amounts of sea level rise will have severe impacts in many low-lying coastal communities throughout the world, especially when storm surges are added on top of sea level rise. High population densities and low elevations make some regions especially vulnerable (Figure 13.18), including Bangladesh and the Nile River Delta in Egypt. In the United States, about half of the population lives near the coast. The most vulnerable areas are the Mid-Atlantic and Gulf Coasts, especially the Mississippi Delta. Also at risk are low-lying areas and bays, such as North Carolina's Outer Banks, much of the Florida Coast, and California's San Francisco Bay and Sacramento/San Joaquin Delta.

Loss of Glaciers, Ice Sheets, and Snow Pack

Land-based snow and ice cover are declining because of climate change and contributing to sea level rise. Mountain glaciers at all latitudes are in retreat, from the Himalayas in Central Asia to the Andes in tropical South America to the Rockies and Sierras in the western United States. As a consequence of warming, many mountain glaciers will be gone by mid-century; Glacier National Park, for example, will likely lose its glaciers by 2030.

The polar ice sheets on Greenland and Antarctica have both experienced net losses of ice in recent years. Melting polar ice sheets add billions of tons of water to the oceans each year. Recent peer-reviewed research found that the Greenland Ice Sheet is losing ice twice as fast as scientists had previously estimated and ice loss has accelerated on both Greenland and Antarctica over the past decades.

Antarctica is losing ice to the melting and slipping of glacier ice into the ocean at a rate enhanced by climate change. Scientists who study the ice sheet fear that the loss of ice could be accelerated by rising sea levels and the warming of ocean water around the fringe of the ice sheet, which rests on the seabed around the coast of West Antarctica. Beyond some threshold amount of warming, the ice sheet could become unstable and ongoing rapid sea level rise could then be unstoppable. Not knowing exactly what level of warming would destabilize this ice sheet calls for caution in how much more warming we allow.

What Can Be Done

The GHGs that are already in the atmosphere because of human activity will continue to warm the planet for decades to come. In other words, some level of continued climate change is inevitable, which means humanity is going to have to take action to adapt to a warming world. However, it is still possible—and necessary—to reduce the magnitude of climate change. A growing body of scientific research has clarified that climate change is already underway and some dangerous impacts have occurred. Avoiding much more severe impacts in the future requires large reductions in human-induced CO_2 emissions in the coming decades. Consequently, many governments have committed to reduce their countries' emissions by between 50 and 85 percent below 2000 levels by 2050. Global emissions reductions on this scale will reduce the costs of damages and of adaptation, and will dramatically reduce the probability of catastrophic outcomes.

Adapted from Global Warming and Global Climate Change, Pew Center on Global Climate Change. 2011. Climate Change 101: Science and Impacts. Available online at: http://www.pewclimate.org/climate-change-101/science-impacts. Used by permission of the Pew Center on Global Climate Change, www.pewclimate.org.

CHAPTER 14

Nuclear Energy

14.1 Harnessing Nuclear Energy to Generate Electricity

Atoms are tiny particles that make up every object in the universe. Nuclear energy is energy in the nucleus (core) of an atom. There is enormous energy present in the bonds that hold the nucleus together. Energy is released when those bonds are broken.

Nuclear energy can be used to make electricity, but it must first be released. Nuclear energy can be released from atoms through nuclear fusion and nuclear fission.

In nuclear fission, atoms are split apart to form smaller atoms, releasing energy. Nuclear power plants use this energy to produce electricity. In nuclear fusion, energy is released when atoms are combined or fused together to form a larger atom. This is how the sun produces energy (Figure 14.1). Nuclear fusion is the subject of ongoing research, but it is not yet clear whether or not it will be a commercially viable technology for electricity generation.

Trif/Shutterstock

FIGURE 14.1 The Sun gives off energy from a reaction called nuclear fusion.

Sections 14.1, 14.2, and 14.3 courtesy of the United States Energy Information Administration
Section 14.4 courtesy of the United States Nuclear Regulatory Commission

Most power plants, including nuclear plants, use heat to produce electricity. They rely on steam from heated water to spin large turbines, which generate electricity. Instead of burning fossil fuels to produce the steam, nuclear plants use heat given off during fission.

Uranium is the fuel most widely used by nuclear plants for nuclear fission. Uranium is a nonrenewable energy source, but it is a common metal found in rocks worldwide. Nuclear power plants use a certain kind of uranium, referred to as U-235, for fuel because its atoms are easily split apart. Although uranium is about 100 times more common than silver, U-235 is relatively rare. Once uranium is mined, the U-235 must be extracted and processed before it can be used as a fuel. The uranium fuel is formed into ceramic pellets. The pellets are about the size of your fingertip, but each one produces roughly the same amount of energy as 150 gallons of oil. These energy-rich pellets are stacked end-to-end in 12-foot metal fuel rods. A bundle of fuel rods, sometimes hundreds, is called a fuel assembly. A reactor core contains many fuel assemblies.

During nuclear fission, a small particle called a neutron hits the uranium atom and splits it, releasing a great amount of energy in the form of heat and radiation. More neutrons are also released when the uranium atom splits. These neutrons go on to bombard other uranium atoms, and the process repeats itself over and over again (Figure 14.2). This is called a chain reaction.

Fission takes place inside the reactor of a nuclear power plant. At the center of the reactor is the core, which contains the uranium fuel.

The heat given off during fission in the reactor core is used to boil water into steam, which turns the turbine blades. As they turn, they drive generators that make electricity. Afterward, the steam is cooled back into water in a separate structure at the power plant called a cooling tower. The water can be used again and again.

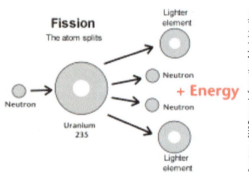

FIGURE 14.2 Fission of U-235.

Did You Know

On December 2, 1942, under the bleachers of the football stadium at the University of Chicago, Dr. Enrico Fermi initiated the first controlled nuclear chain reaction. The experiment, conducted as part of the wartime atomic bomb program, also led to peaceful uses of the atom, including construction of the first U.S. nuclear power plant at Shippingport, Pennsylvania, in 1957.

Nuclear Power Plants Generate about One-Fifth of U.S. Electricity

The United States has 65 nuclear power plants with 104 nuclear reactors. Thirty-six of the plants have two or more reactors. Nuclear power plants are located in 31 different states but most are located east of the Mississippi. Nuclear power has generated about one-fifth of U.S. electricity each year since 1990. Nuclear power provides about as much electricity as they use in California, Texas, and New York, the three states with the most people.

Nuclear reactors look like large concrete domes from the outside. Not all nuclear power plants have cooling towers (Figure 14.3).

Recent Nuclear Construction Activity

In February 2012, the U.S. Nuclear Regulatory Commission voted to approve Southern Company's application to build and operate two new nuclear reactors, Units 3 and 4, at its Vogtle plant. The Vogtle units are the first reactors to receive construction approval in over 30 years, and are expected to be operational in 2016 and 2017.

marlee/Shutterstock

FIGURE 14.3   Large concrete domes obscure the nuclear reactors inside a nuclear power plant.

The last new reactor to enter commercial service was the Tennessee Valley Authority's (TVA) Watts Bar 1 in Tennessee in 1996. In 2002, the TVA returned Browns Ferry Unit 1 to service; the unit had been shut down since 1985. In 2007, construction resumed on a partially built reactor, Watts Bar 2, which is slated for initial operation in 2013. Construction on two other reactors, Bellefonte 1 and 2 in Alabama, remains suspended, but TVA has left open the possibility that the reactors eventually might be completed.

What Is the Status of the U.S. Nuclear Industry?

Did You Know
The Grand Coulee Dam in the state of Washington has the most capacity of any electric power plant in the United States, at 7,079 net megawatts. The Palo Verde nuclear plant in Arizona ranks second with a capacity of 3,937 net megawatts. But nuclear plants are able to use more of their capacity than hydropower facilities. In 2012, Grand Coulee generated over 26,461 gigawatthours of electricity, while Palo Verde generated more than 31,934 gigawatthours.

There are currently 100 operating commercial nuclear reactors at 62 nuclear power plants in the United States. The average age of U.S. reactors is about 33 years. The oldest operating reactors, Nine Mile Point Unit 1 and Oyster Creek, began commercial operation in December 1969. Thirty-four reactors began commercial operation between 1985 and 1996. The last reactor to enter service was the Watts Bar Unit 1 in Tennessee in 1996.

Since 1990, the share of total annual U.S. electricity generation provided by nuclear power has averaged about 20%. Nuclear generation generally increased through plant modifications to increase capacity (known as uprates) and has also increased by shortening the length of time a reactor is offline for refueling. The U.S. Energy Information Administration expects nuclear power generation to grow, although at a rate about half that of total electricity generation.

Most of the commercial reactors in the United States are located east of the Mississippi River. Illinois has the most reactors (11) and the most nuclear capacity. The largest reactor in the United States, with a capacity above 1,350 net megawatts, is the Grand Gulf Nuclear Station, located in Port Gibson, Mississippi. The smallest reactor, with a capacity of 478 net megawatts, is at Fort Calhoun, Nebraska.

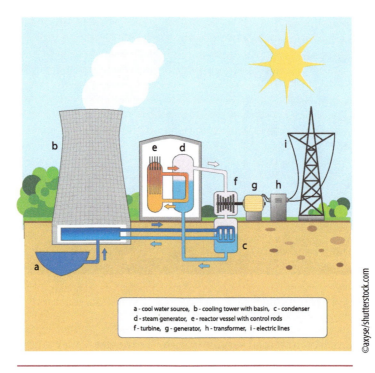

a - cool water source, b - cooling tower with basin, c - condenser
d - steam generator, e - reactor vessel with control rods
f - turbine, g - generator, h - transformer, i - electric lines

©axyse/shutterstock.com

FIGURE 14.4   A simplied model of a Nuclear Reactor.

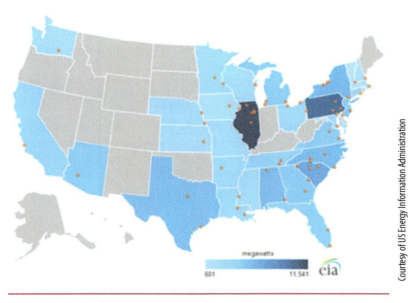

Courtesy of US Energy Information Administration

FIGURE 14.5   Nuclear power generating capacity across the U.S.

Four reactors were permanently shut down in 2013:

1. Crystal River Unit 3, Florida
2. Kewaunee Power Station, Wisconsin
3. San Onofre Nuclear Generating Station Units 2 and 3, California

Many Plants Have More than One Reactor

The term power plant refers to an entire facility. A plant may contain nuclear as well as non-nuclear electric generating units. Each nuclear reactor located at a commercial nuclear plant is unique with its own personnel and equipment. The reactor provides heat to make steam, which drives a turbine and in turn drives the generator that produces electricity.

Thirty-five U.S. nuclear plants have at least two reactors. Although some foreign nuclear power plants have as many as eight reactors, only three U.S. plants have more than two operational reactors: Palo Verde Nuclear Generating Station in Arizona, Browns Ferry Nuclear Power Plant in Alabama, and Oconee Nuclear Station in South Carolina.

Nuclear Plants Are Generally Used More Intensively than Other Plants

For cost and technical reasons, nuclear power plants are generally used more intensively than coal or natural gas units. In 2012, the nuclear share of electricity generating capacity was 10%, while nuclear's share of national power output was 19%.

As of December 2013, the NRC has applications for a total of 22 new reactors, although the NRC review for eight of these reactors has been suspended or deferred. It is unknown how many of the proposed reactors will be built. The NRC application review process is a detailed review that takes 30 to 60 months. Under current licensing regulations, a utility that seeks to build a new reactor can use off-the-shelf reactor designs that have previously been approved and certified by the NRC. Issuance of a design certification by the NRC is independent of applications to construct or operate a new nuclear power plant. When the utility uses an NRC-certified reactor design, safety issues related to the design have been resolved, and the focus of the NRC's review is the quality of construction. Construction may take around six years for each reactor. EIA projects that the nuclear industry will add approximately 10.4 gigawatts (10,400 megawatts) of new capacity between 2013 and 2040, 9.7 gigawatts are expected from new reactors and 0.7 gigawatts are expected from uprates of existing plants.

What Are Small Modular Reactors?

Small Modular Reactors (SMRs) are about one-third the size of the reactors currently under construction. SMRs would have simple compact designs that could be assembled in a factory and transported by train or truck directly to the plant site, potentially reducing the time it takes to construct a new nuclear power plant. The U.S. Department of Energy has awarded $452 million to support the design, certification and commercialization of SMRs in the United States. SMRs may help meet some of the nation's future electricity demand and could be deployed within the next 10 to 15 years.

14.2 Where Our Uranium Comes From

Economically recoverable uranium deposits have been discovered primarily in the western United States, Australia, Canada, Central Asia, Africa, and South America. Once uranium is mined, the U-235 must be extracted and processed before it can be used as a fuel. Mined uranium ore typically yields one to four pounds of uranium concentrate (U_3O_8 or yellowcake) per ton, or 0.05% to 0.20% U_3O_8. The Energy Explained page about the nuclear fuel cycle describes uranium processing in more detail.

Most of Our Uranium Is Imported

Owners and operators of U.S. nuclear power reactors purchased the equivalent of 57 million pounds of uranium in 2013. Uranium delivered to U.S. reactors in 2013 came from six continents:

- 17% of delivered uranium came from the United States.
- 83% of delivered uranium was from other countries:
- 35% of delivered uranium came from Kazakhstan, Russia, and Uzbekistan.
- 19% of delivered uranium came from Australia.
- 15% of delivered uranium came from Malawi, Nambia, Niger, and South Africa.
- 14% of delivered uranium came from Canada.
- 1% of delivered uranium came from Brazil, China, Czech Republic, Germany, Hungary, and Portugal.

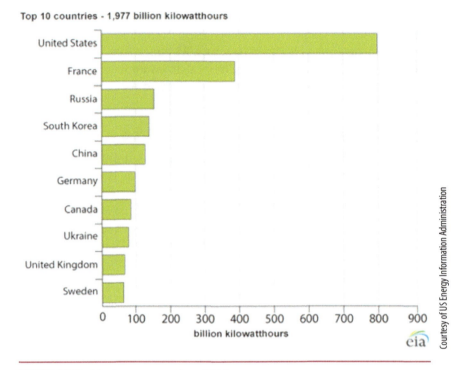

FIGURE 14.6 Nuclear generation, 2012.

14.3 Nuclear Power Generation Around the World

The United States has the most nuclear capacity and generation among the 30 countries in the world that have commercial nuclear power plants. France, the country with the second highest nuclear capacity, relies on nuclear power for nearly 75% of its electricity. Fourteen other countries generate over 20% of their electricity from nuclear power (Figure 14.6).

Which countries have the most nuclear power generation?

14.4 Nuclear Power and the Environment

Nuclear Power Plants Produce No Carbon Dioxide

Unlike fossil fuel-fired power plants, nuclear reactors do not produce air pollution or carbon dioxide while operating. However, the processes for mining and refining uranium ore and making reactor fuel require large amounts of energy. Nuclear power plants have large amounts of metal and concrete, which also require large amounts of energy to manufacture. If fossil fuels are used to make the electricity and manufacture the power plant materials, then the emissions from burning those fuels could be associated with the electricity that nuclear power plants generate.

Nuclear Energy and Radioactive Waste

The main environmental concerns for nuclear power are radioactive wastes such as uranium mill tailings, spent (used) reactor fuel, and other radioactive wastes. These materials can remain radioactive and dangerous to human health for thousands of years. They are subject to special regulations that govern their handling, transportation, storage, and disposal to protect human health and the environment.

Background

Radioactive (or nuclear) waste is a byproduct from nuclear reactors, fuel processing plants, and institutions such as hospitals and research facilities. It also results from the decommissioning of nuclear reactors and other nuclear facilities that are permanently shut down. The Nuclear Regulatory Commission separates wastes into two broad classifications: high-level or low-level waste. High-level radioactive waste results primarily from the fuel used by reactors to produce electricity. Low-level radioactive waste results from reactor operations and from medical, academic, industrial, and other commercial uses. These are described in more detail below.

The NRC has regulatory authority over storage and disposal of all commercially generated wastes in the United States and those high-level wastes generated by the Department of Energy that are subject to longterm storage and that are not used for, or part of, research and development activities. Regulations require conformance with minimum acceptable performance criteria for waste management activities, while providing for flexibility in technological approach. These criteria and guidelines are designed to ensure adequate protection of the public health and safety and the environment.

High-Level Waste

High-level radioactive waste is uranium fuel that has been used in a nuclear power reactor and is "spent" or is no longer efficient in generating power to the reactor to produce electricity. Spent fuel is thermally hot as well as being highly radioactive, requiring remote handling and shielding. The basic fuel of a nuclear power reactor contains uranium 235, which is in ceramic pellets inside of metal rods. Before these fuel rods are used, they are only slightly radioactive and may be handled without special shielding.

The splitting of relatively heavy uranium atoms during reactor operation creates radioactive isotopes of several lighter elements, such as cesium-137 and strontium-90, called "fission products," that account for most of the heat and penetrating radiation in high-level waste. Some uranium atoms also capture neutrons from fissioning uranium atoms nearby to form heavier elements like plutonium. These heavier-thanuranium, or "transuranic," elements do not produce nearly the amount of heat or penetrating radiation that fission products do, but they take much longer to decay. Transuranic wastes, also called "TRU," therefore account for most of the radioactive hazard remaining in high-level waste after a thousand years.

Radioactive isotopes will eventually decay, or disintegrate, to harmless materials. However, while they are decaying, they emit radiation. Some isotopes decay in hours or even minutes, but others decay very slowly. Strontium-90 and cesium-137 have half-lives of about 30 years (that means that half the radioactivity of a given quantity of strontium-90, for example, will decay in 30 years). Plutonium-239 has a half-life of 24,000 years.

High-level wastes are hazardous to humans and other life forms because of their high radiation levels that are capable of producing fatal doses during short periods of direct exposure. For example, ten years after removal from a reactor, the surface dose rate for a typical spent fuel assembly exceeds 10,000 rem/hour, whereas a fatal whole-body dose for humans is about 500 rem (if received all at one time). Furthermore, if constituents of these high-level wastes were to get into ground water or rivers, they could enter into food chains. Although the dose produced through this indirect exposure is much smaller than a direct exposure dose, there is a greater potential for a larger population to be exposed.

Reprocessing separates residual uranium and unfissioned plutonium from the fission products. The uranium and plutonium can be used again as fuel. Most of the high-level waste (other than spent fuel) generated over the last 35 years has come from reprocessing of fuel from government-owned plutonium production reactors and from naval, research and test reactors. A small amount of liquid high-level waste was generated from the reprocessing of commercial power reactor fuel in the 1960's and early 1970's. There is no commercial reprocessing of nuclear power fuel in the United States at present; almost all existing commercial high-level waste is in the form of unreprocessed spent fuel.

Storage and Disposal

At this time there are no facilities for permanent disposal of high-level radioactive waste. Since the only way radioactive wastes finally become harmless is through decay, which for some isotopes contained in high-level wastes can take hundreds of thousands of years, the wastes must be stored in a way that provides adequate protection for very long times.

The spent fuel rods from nuclear power plants must be handled and stored with the same care as separated high-level waste, since they contain the highly-radioactive fission products plus uranium and plutonium. Spent fuel is currently being stored in large water-cooled pools and dry storage casks at nuclear power plants. Some is also stored at facilities at West Valley, New York, Morris, Illinois, and Idaho National Engineering and Environmental Laboratory.

Existing high-level wastes from reprocessing are presently stored at West Valley, New York; Hanford, Washington; Idaho Falls, Idaho; and Savannah River, South Carolina. Liquid high-level wastes are stored in large underground tanks of either stainless steel or carbon steel, depending on whether they are acid or alkaline. Some of the liquid waste has been solidified into glass, ceramic slag, salt cake and sludge.

In 1982, the Congress enacted the Nuclear Waste Policy Act (NWPA) and on January 7, 1983, the President signed it into law. This legislation defined the Federal Government's responsibility to provide permanent disposal in a deep geologic repository for spent fuel and high-level radioactive waste from commercial and defense activities. Under amended provisions (1987) of this Act, the Department of Energy (DOE) has the responsibility to locate, build, and operate a repository for such wastes. The NRC has the responsibility to establish regulations governing the construction, operation, and closure of the repository, consistent with environmental standards established by the U.S. Environmental Protection Agency.

The 1987 amendments required DOE to evaluate only the suitability of the site at Yucca Mountain, Nevada, for a geologic disposal facility. In addition, the amendments outlined a detailed approach for the disposal of high-level radioactive waste involving review by the President, Congress, State and Tribal governments, NRC and other Federal agencies.

In February 2002, after many years of studying the suitability of the site, DOE recommended to the President that the Yucca Mountain site be developed as a long-term geologic repository for high-level waste. In April 2002, the Governor of Nevada notified Congress of his State's objection to the proposed repository. Subsequently, Congress voted to override the objection of the state.

DOE submitted a license application to the NRC for construction authorization for a repository at Yucca Mountain in June 2008. The NRC will issue a license only if DOE can demonstrate that it can construct and operate the repository safely and comply with NRC regulations.

Spent Reactor Fuel Storage and Power Plant Decommissioning

When a nuclear power plant stops operating, the facility must be decommissioned. This involves safely removing the plant from service and reducing radioactivity to a level that permits other uses of the property. The U.S. Nuclear Regulatory Commission, for example, has strict rules governing nuclear power plant decommissioning that involve cleanup of radioactively contaminated plant systems and structures, and removal of the radioactive fuel.

Nuclear Reactors and Power Plants Have Complex Safety and Security Features

An uncontrolled nuclear reaction in a nuclear reactor can potentially result in widespread contamination of air and water with radioactivity for hundreds of miles around a reactor. The risk of this happening at nuclear power plants in the United States is considered to be very small due to the diverse and redundant barriers and numerous safety systems at nuclear power plants, the training and skills of the reactor operators, testing and maintenance activities, and the regulatory requirements and oversight of the Nuclear Regulatory Commission. A large area surrounding nuclear power plants is restricted and guarded by armed security teams. U.S. reactors have containment vessels that are designed to withstand extreme weather events and earthquakes.

Responsibilities of U.S. Government Agencies: Radioactive Waste Materials

The NRC is responsible for licensing and regulating the receipt and possession of high-level waste, including spent fuel as well as reprocessing waste, at privately-owned facilities and at certain facilities of the DOE. The DOE facilities which are or will be subject to NRC regulation are defined by law to include: (1) facilities used primarily for the receipt and storage of high-level waste resulting from activities licensed under the Atomic Energy Act and (2) facilities other than Research and Development facilities authorized for the express purpose of subsequent long-term storage of DOE-generated waste. Facilities for permanent disposal will require a license from NRC under these provisions.

The responsibilities of other government agencies in the management of high-level waste include the following.

The Department of Energy (DOE) plans and carries out programs for safe handling of DOE-generated radioactive wastes, develops waste disposal technologies, and will design, construct and operate disposal facilities for DOE-generated and commercial high-level wastes. DOE has completed solidifying the liquid wastes that are currently in storage at West Valley. The Nuclear Waste Policy Act of 1982 sets specific roles and schedules for the DOE to follow in developing HLW repositories. (The repositories will be licensed by the NRC.)

The Environmental Protection Agency (EPA) develops environmental standards and Federal radiation protection guidance for offsite radiation due to the disposal of spent nuclear fuel and high-level and transuranic radioactive wastes. The standards limit the amount of radioactivity entering the biosphere outside the boundaries of the facility and also limit the radiation exposure to the public from management of spent fuel and waste prior to disposal. The guidance establishes criteria to be followed when these wastes are disposed of.

The Department of Transportation (DOT) regulates both the packaging and carriage of all hazardous materials including high-level nuclear waste. Packaging must meet NRC regulations, which are compatible with and generally derived from internationally developed standards, and the package design must be reviewed and certified by NRC. DOT prescribes limits for external radiation levels and contamination, and controls the mechanical condition of carrier equipment and qualifications of carrier personnel.

The Department of the Interior (DOI), through the U.S. Geological Survey, conducts laboratory and field geologic investigations in support of DOE's waste disposal programs, and collaborates with DOE on the earth science technical activities. The Bureau of Land Management, within DOI, manages certain public lands. DOI may withdraw such public lands for the limited exclusive use of DOE in support of radioactive waste disposal actions.

Low-Level Waste

Low-level wastes, which are generally defined as radioactive wastes other than high-level and wastes from uranium recovery operations, are commonly disposed of in near-surface facilities rather than in a geologic repository that is required for high-level wastes. There is no intent to recover the wastes once they are disposed of.

Low-level waste includes items that have become contaminated with radioactive material or have become radioactive through exposure to neutron radiation. This waste typically consists of contaminated protective shoe covers and clothing, wiping rags, mops, filters, reactor water treatment residues, equipments and tools, luminous dials, medical tubes, swabs, injection needles, syringes, and laboratory animal carcasses and tissues. The radioactivity can range from just above background levels found in nature to much higher levels in certain cases such as parts from inside the reactor vessel in a nuclear power plant.

Low-level waste is typically stored on-site by licensees, either until it has decayed away and can be disposed of as ordinary trash, or until amounts are large enough for shipment to a low-level waste disposal site in containers approved by the Department of Transportation.

There have been eight operating commercial facilities in the United States licensed to dispose of low-level radioactive wastes. They are located at (1) West Valley, New York; (2) Maxey Flats near Morehead, Kentucky; (3) Sheffield, Illinois; (4) Beatty, Nevada; (5) Hanford, Washington; (6) Clive, Utah; (7) Barnwell, South Carolina;

and (8) Andrews, Texas. At the present time, only the latter four sites are receiving waste for disposal; they are regulated by the states. The West Valley, Maxey Flats, Sheffield and Beatty sites have permanently stopped receiving wastes. Burial of transuranic waste is limited at all of the sites. Transuranic waste includes material contaminated with radioactive elements (e.g., neptunium, americium, plutonium) that are artificially made and is produced primarily from reprocessing spent fuel and from use of plutonium in fabrication of nuclear weapons.

In 2000, low-level waste disposal facilities received about 3.3 million cubic feet of commercially generated radioactive waste. Of this, 8.2% came from nuclear reactors, 83.8% from industrial users, 7.6% from government sources (other than nuclear weapons sites), 0.2% from academic users, and the rest was undefined.

Mill Tailings

Another type of radioactive waste consists of tailings generated during the milling of certain ores to extract uranium or thorium. These wastes have relatively low concentrations of radioactive materials with long half-lives. Tailings contain radium (which, through radioactive decay, becomes radon), thorium, and small residual amounts of uranium that were not extracted during the milling process.

Doses in Our Daily Lives

On average, Americans receive a radiation dose of about 0.62 rem (620 millirem) each year. Half of this dose comes from natural background radiation. Most of this background exposure comes from radon in the air, with smaller amounts from cosmic rays and the Earth itself. (The chart to the right shows these radiation doses in perspective.) The other half (0.31 rem or 310 mrem) comes from man-made sources of radiation, including medical, commercial, and industrial sources. In general, a yearly dose of 620 millirem from all radiation sources has not been shown to cause humans any harm.

Doses from Medical Procedures

Medical procedures account for nearly all (96%) human exposure to man-made radiation. For example, a chest x-ray typically gives a dose of about 0.01 rem (10 millirem) and a full-body CT gives a dose of 1 rem (1,000 mrem), as shown in the table to the left.

Among these medical procedures, x-rays, mammography, and CT use radiation or perform functions similar to those of radioisotopes. However, they do not involve radioactive material and, hence, are not regulated by the U.S. Nuclear Regulatory Commission (NRC). Instead, most of these procedures are regulated by State health agencies. In fact, among these procedures, the NRC and its Agreement States only license and regulate the possession and use of radioactive materials for nuclear medicine.

Medical Procedure Doses	
Procedure	Dose (mrem)
X-Rays-single exposure	
Pelvis	70
Abdomen	60
Chest	10
Dental	1.5
Hand/Foot	0.5
Mammogram (2 views)	72
Nuclear Medicine	400
CT	
Full body	1,000
Chest	700
Head	200

Radioactivity in Food

All organic matter (both plant and animal) contains some small amount of radiation from radioactive potassium-40 (^{40}K), radium-226 (^{226}Ra), and other isotopes. In addition, all water on Earth contains small amounts of dissolved uranium and thorium. As a result, the average person receives an average internal dose of about 30 millirem of these materials per year from the food and water that we eat and drink, as illustrated by the following table. (Amounts are shown in picocuries per kilogram.)

Natural Radioactivity in Food		
Food	^{40}K (pCi/kg)	^{226}Ra (pCi/kg)
Bananas	3,520	1
Carrots	3,400	0.6–2
White Potatoes	3,400	1–2.5
Lima Beans (raw)	4,640	2–5
Red Meat	3,000	0.5
Brazil Nuts	5,600	1,000–7,000
Beer	390	—
Drinking Water	—	0–0.17

14.5 Case Study: Chernobyl Nuclear Power Plant Accident

Background

Ionizing radiation sign warns of radiation hazards near the damaged Chernobyl nuclear power plant.

On April 26, 1986, a sudden surge of power during a reactor systems test destroyed Unit 4 of the nuclear power station at Chernobyl, Ukraine, in the former Soviet Union. The accident and the fire that followed released massive amounts of radioactive material into the environment.

Emergency crews responding to the accident used helicopters to pour sand and boron on the reactor debris. The sand was to stop the fire and additional releases of radioactive material; the boron was to prevent additional nuclear reactions. A few weeks after the accident, the crews completely covered the damaged unit in a temporary concrete structure, called the "sarcophagus," to limit further release of radioactive material. The Soviet government also cut down and buried about a square mile of pine forest near the plant to reduce radioactive contamination at and near the site. Chernobyl's three other reactors were subsequently restarted but all eventually shut down for good, with the last reactor closing in 1999. The Soviet nuclear power authorities presented their initial accident report to an International Atomic Energy Agency meeting in Vienna, Austria, in August 1986.

After the accident, officials closed off the area within 30 kilometers (18 miles) of the plant, except for persons with official business at the plant and those people evaluating and dealing with the consequences of the accident and operating the undamaged reactors. The Soviet (and later on, Russian) government evacuated about 115,000 people from the most heavily contaminated areas in 1986, and another 220,000 people in subsequent years (Source: UNSCEAR Report at http://www.unscear.org/docs/reports/2008/11-80076_Report_2008_Annex_D.pdf, 2008, pg. 53).

Health Effects from the Accident

The Chernobyl accident's severe radiation effects killed 28 of the site's 600 workers in the first four months after the event. Another 106 workers received high enough doses to cause acute radiation sickness. Two workers died within hours of the reactor explosion from non-radiological causes. Another 200,000 cleanup workers in 1986 and 1987 received doses of between 1 and 100 rem (The average annual radiation dose for a U.S. citizen is about .6 rem). Chernobyl cleanup activities eventually required about 600,000 workers, although only a small fraction of these workers were exposed to elevated levels of radiation. Government agencies continue to monitor cleanup and recovery workers' health. (UNSCEAR 2008, pg. 47, 58, 107, and 119)

The Chernobyl accident contaminated wide areas of Belarus, the Russian Federation, and Ukraine inhabited by millions of residents. Agencies such as the World Health Organization have been concerned about radiation exposure to people evacuated from these areas. The majority of the five million residents living in contaminated areas, however, received very small radiation doses comparable to natural background levels (0.1 rem per year). (UNSCEAR 2008, pg. 124–25) Today the available evidence does not strongly connect the accident to radiation-induced increases of leukemia or solid cancer, other than thyroid cancer. Many children and adolescents in the area in 1986 drank milk contaminated with radioactive iodine, which delivered substantial doses to their thyroid glands. To date, about 6,000 thyroid cancer cases have been detected among these children. Ninety-nine percent of these children were successfully treated; 15 children and adolescents in the three countries died from thyroid cancer by 2005. The available evidence does not show any effect on the number of adverse pregnancy outcomes, delivery complications, stillbirths or overall health of children among the families living in the most contaminated areas. (UNSCEAR 2008, pg. 65)

Experts conclude some cancer deaths may eventually be attributed to Chernobyl over the lifetime of the emergency workers, evacuees and residents living in the most contaminated areas. These health effects are far lower than initial speculations of tens of thousands of radiation-related deaths.

US Reactors and NRC's Response

The NRC continues to conclude that many factors protect U.S. reactors against the combination of lapses that led to the accident at Chernobyl. Differences in plant design, broader safe shutdown capabilities and strong structures to hold in radioactive materials all help ensure U.S. reactors can keep the public safe. When the NRC reviews new information it takes into account possible major accidents; these reviews consider whether safety requirements should be enhanced to ensure ongoing protection of the public and the environment.

The NRC's post-Chernobyl assessment emphasized the importance of several concepts, including:

- designing reactor systems properly on the drawing board and implementing them correctly during construction and maintenance;
- maintaining proper procedures and controls for normal operations and emergencies;
- having competent and motivated plant management and operating staff; and
- ensuring the availability of backup safety systems to deal with potential accidents.

The post-Chernobyl assessment also examined whether changes were needed to NRC regulations or guidance on accidents involving control of the chain reaction, accidents when the reactor is at low or zero power, operator training, and emergency planning.

Discussion

The Chernobyl reactors, called RBMKs, were high-powered reactors that used graphite to help maintain the chain reaction and cooled the reactor cores with water. When the accident occurred the Soviet Union was using 17 RBMKs and Lithuania was using two. Since the accident, the other three Chernobyl reactors, an additional Russian RMBK and both Lithuanian RBMKs have permanently shut down. Chernobyl's Unit 2 was shut down in 1991 after a serious turbine building fire; Unit 1 was closed in November 1996; and Unit 3 was closed in December 1999, as promised by Ukrainian President Leonid Kuchma. In Lithuania, Ignalina Unit 1 was shut down in December 2004 and Unit 2 in 2009 as a condition of the country joining the European Union.

Closing Chernobyl's reactors required a combined effort from the world's seven largest economies (the G-7), the European Commission and Ukraine. This effort supported such things as short-term safety upgrades at Chernobyl Unit 3, decommissioning the entire Chernobyl site, developing ways to address shutdown impacts on workers and their families, and identifying investments needed to meet Ukraine's future electrical power needs.

On the accident's 10th anniversary, the Ukraine formally established the Chernobyl Center for Nuclear Safety, Radioactive Waste and Radio-ecology in the town of Slavutych. The center provides technical support to Ukraine's nuclear power industry, the academic community and nuclear regulators.

Sarcophagus

The Soviet authorities started the concrete sarcophagus to cover the destroyed Chernobyl reactor in May 1986 and completed the extremely challenging job six months later. Officials considered the sarcophagus a temporary fix to filter radiation out of the gases from the destroyed reactor before the gas was released to the environment. After several years, experts became concerned that the high radiation levels could affect the stability of the sarcophagus.

In 1997, the G-7, the European Commission and Ukraine agreed to jointly fund the Chernobyl Shelter Implementation Plan to help Ukraine transform the existing sarcophagus into a stable and environmentally safe system. The European Bank for Reconstruction and Development manages funding for the plan, which will protect workers, the nearby population and the environment for decades from the very large amounts of radioactive material still in the sarcophagus. The existing sarcophagus was stabilized before work began in late 2006 to replace it with a new safe shelter. The new confinement design includes an arch-shaped steel structure, which will slide across the existing sarcophagus via rails. This new structure is designed to last at least 100 years.

CHAPTER 15

Renewable Energy

Introduction

Unlike fossil fuels, which are exhaustible, renewable energy sources regenerate and can be sustained indefinitely.

The five renewable sources used most often are: (Figure 15.1)

- Biomass—including:
  - ❏ wood and wood waste,
  - ❏ municipal solid waste,
  - ❏ landfill gas and biogas,
  - ❏ ethanol,
  - ❏ biodiesel
- Water (hydropower)
- Geothermal
- Wind
- Solar

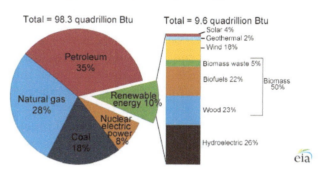

U.S. energy consumption by energy source, 2014

FIGURE 15.1 **U.S. energy consumption by energy source, 2012.**

Source: U.S. Energy Information Administration, *Monthly Energy Review*, Table 1.3 and 10.1 (March 2015), preliminary data.
Note: Sum of components may not equal 100% as a result of independent rounding.

Many paper mills use wood waste to produce steam and electricity.

The use of renewable energy is not new. More than 150 years ago, wood, which is one form of biomass, supplied up to 90% of our energy needs. As the use of coal, petroleum, and natural gas expanded, the United States became less reliant on wood as an energy source. Today, we are looking again at renewable sources to find new ways to use them to help meet our energy needs.

In 2012, consumption of renewable sources in the United States totaled about 9 quadrillion Btu—1 quadrillion is the number 1 followed by 15 zeros—or about 9% of **all** energy used nationally. About 12% of U.S. **electricity** was generated from renewable sources in 2012.

Over half of renewable energy goes to producing electricity. The next largest use of renewable energy is biomass (wood and waste) for the production of heat and steam for industrial purposes and for space heating, mostly in homes. Biomass also includes biofuels, such as ethanol and biodiesel, used for transportation.

Renewable energy plays an important role in the supply of energy. When renewable energy sources are used, the demand for fossil fuels is reduced. Unlike fossil fuels, non-biomass renewable sources of energy (hydropower, geothermal, wind, and solar) do not directly emit greenhouse gases.

Why Don't We Use More Renewable Energy?

In the past, renewable energy has generally been more expensive to produce and use than fossil fuels. Renewable resources are often located in remote areas, and it is expensive to build power lines to the cities where the

259

electricity they produce is needed. The use of renewable sources is also limited by the fact that they are not always available—cloudy days reduce solar power; calm days reduce wind power; and droughts reduce the water available for hydropower.

The production and use of renewable fuels has grown more quickly in recent years as a result of higher prices for oil and natural gas, and a number of state and federal government incentives for renewable energy. The use of renewable fuels is expected to continue to grow over the next 30 years, although EIA projects that we will still rely on non-renewable fuels to meet most of our energy needs.

15.1 Hydropower

Energy from Moving Water

Solar energy heats water on the surface, causing it to evaporate. This water vapor condenses into clouds and falls back onto the surface as precipitation. The water flows through rivers back into the oceans, where it can evaporate and begin the cycle over again (Figure 15.2).

The water flows from behind the dam through penstocks, turns the turbines, and causes the generators to generate electricity. The electricity is carried to users by a transmission line. Other water flows from behind the dam over spillways and into the river below (Figure 15.3).

Hydropower is the largest renewable energy source for electricity generation in the United States. In 2013, hydropower accounted for about 6% of total U.S. electricity generation and 52% of generation from all renewables.

Because the source of hydroelectric power is water, hydroelectric power plants are usually located on or near a water source.

The amount of available energy in moving water is determined by its flow or fall. Swiftly flowing water in a big river, like the Columbia River that forms the border between Oregon and Washington, carries a great deal of energy in its flow. Water descending rapidly from a very high point, like Niagara Falls in New York, also has substantial energy in its flow.

FIGURE 15.2   The Water Cycle.

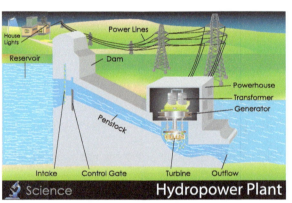

FIGURE 15.3   Basics of a hydroelectric plant.

In both instances, the water flows through a pipe, or penstock, then pushes against and turns blades in a turbine to spin a generator to produce electricity. In a run-of-the-river system, the force of the current applies the needed pressure, while in a storage system, water is accumulated in reservoirs created by dams, then released as needed to generate electricity.

History of Hydropower

Hydropower is one of the oldest sources of energy. It was used thousands of years ago to turn paddle wheels to help grind grain. Our nation's first industrial use of hydropower to generate electricity occurred in 1880, when 16 brush-arc lamps were powered using a water turbine at the Wolverine Chair Factory in Grand Rapids, Michigan.

The first U.S. hydroelectric power plant opened on the Fox River near Appleton, Wisconsin, on September 30, 1882.

Hydropower Use Around the World

According to the United Nations World Water Assessment Programme (UN WWAP), approximately 20% of the world's electricity comes from hydropower. The U.S., Europe and China lead the world in the amount of electricity generated by this source of power. China has the largest number of dams of any country in the world, with 86,000 and counting. New hydropower projects are limited by 3 primary factors. The first factor relates to geology, as large dams must be built along rivers flowing through canyons of non-fractured rocks so that the canyon walls can help hold back and funnel the water. Additionally, dams are expensive to build, and this can be a limiting factor for many developing countries. Many countries in Africa have yet to tap potential areas for hydropower due to the expense in building large dams. Finally, social and environmental factors may prevent the development of hydropower in certain regions. For example, the Three Gorges Dam in China (which is 5 times larger than the Hoover Dam in the U.S.) caused more than 1 million people to be relocated as the rivers behind the dam were flooded with water. Additionally, numerous architectural and cultural sites disappeared as a result of the new dammed up lakes.

FIGURE 15.4 **Fish ladders provide a way for migrating fish to bypass dams and get upstream to spawning grounds.**

Shutterstock/Carsten Medom Madsen

Hydropower and the Environment

Most dams in the United States were built mainly to control floods and to help supply water for cities and irrigation. A small number of dams were built specifically for hydropower generation. While hydropower generators do not directly produce emissions of air pollutants, hydropower dams, reservoirs, and the operation of hydropower electric generators can affect the environment.

A dam that creates a reservoir (or a dam that diverts water to a run-of-river hydropower plant) may obstruct fish migration. A reservoir and dam can also change natural water temperatures, water chemistry, river flow characteristics, and silt loads. All of these changes can affect the ecology and the physical characteristics of the river. These changes may have negative impacts on native plants and on animals in and around the river. Reservoirs occasionally cover important natural areas, agricultural land, or archeological sites. A reservoir and operation of the dam may also result in the relocation of people. The physical impacts of a dam and reservoir, the operation of the dam, and the use of the water can change the environment over a much larger area than the area covered by a reservoir.

While no new hydropower dams have been built recently in the United States, they are being built in other countries like China. Manufacturing the concrete and steel used to construct these dams requires equipment that may produce emissions. If fossil fuels are used as the energy source to make these materials, then the emissions from the equipment could be associated with the electricity that hydropower facilities generate. However, given the long operating lifetime of a hydropower plant (50 years to 100 years) these emissions are offset.

Greenhouse gases, carbon dioxide, and methane may also form in reservoirs and be emitted into the atmosphere. The exact amount of greenhouse gases produced from hydropower plant reservoirs is uncertain. The emissions from reservoirs in tropical and temperate regions, including the United States, may be equal to or greater than the greenhouse effect of the carbon dioxide emissions from an equivalent amount of electricity generated with fossil fuels. Scientists at Brazil's National Institute for Space Research designed a system to capture methane in a reservoir and burn it to produce electricity.

Hydro turbines kill and injure some of the fish that pass through the turbine. The U.S. Department of Energy has sponsored research and development of turbines that could reduce fish deaths to less than 2%, in comparison to fish kills of 5% to 10% for the best existing turbines.

In the Columbia River, along the border of Oregon and Washington, salmon must swim upstream to their spawning grounds to reproduce, but the series of dams along the river gets in their way. Different approaches to fixing this problem have been used, including the construction of *fish ladders* that help salmon move through the dam to the spawning grounds upstream. Fish ladders are used at many dams in which fish migration is an issue.

15.2 Biomass

Biomass is organic material that comes from plants and animals. Biomass contains stored energy from the sun. Plants absorb the sun's energy in a process called photosynthesis (Figure 15.5). The chemical energy in plants is passed to animals and people when they consume plants and plant products.

Biomass is a renewable energy source. Some examples of biomass fuels are wood, crops, animal manure, and human sewage. The chemical energy in biomass is released as heat when it is burned. The wood you burn in a fireplace is a biomass fuel. Wood and waste materials made from wood and garbage are burned to produce steam for making electricity or heat for industries.

Burning biomass is not the only way to release its energy. Biomass can be converted to other useable forms of energy like methane gas, or transportation fuels like ethanol and biodiesel. Methane gas is the main ingredient of natural gas. Garbage, and agricultural and human waste, release methane gas—also called landfill gas or biogas. Crops like corn and sugar cane can be fermented to produce ethanol. Biodiesel, another transportation fuel, can be produced from vegetable oils and animal fats.

Biomass fuels provided about 5% of the energy used in the United States in 2013. Of the 5%, about 45% was from wood and wood-derived biomass, 44% was from biofuels (mainly ethanol), and about 11% was from municipal waste. Researchers are trying to develop ways to use more biomass for fuel.

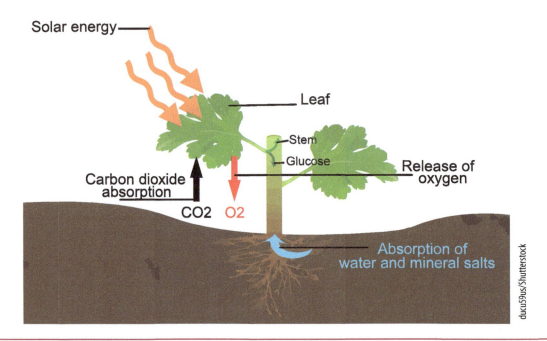

Solar energy

Leaf

Stem

Glucose

Release of oxygen

Carbon dioxide absorption

CO2   O2

Absorption of water and mineral salts

ducu59us/Shutterstock

FIGURE 15.5   Photosynthesis.

Wood and Wood Waste

Wood has been used as a fuel source for thousands of years. Wood was the main source of energy in the United States and the rest of the world until the mid-1800s. Wood continues to be a major source of energy for cooking and heating in much of the developing world. In recent years, using wood as a main heating source has also become more popular in the United States, primarily in the Northeast. From 2005 to 2012, all of the states in the New England and the Middle Atlantic Census divisions experienced at least a 50% increase in the number of homes that use wood as the primary heating source according to the March 17, 2014 Today in Energy article. In the United States, wood and wood waste (bark, sawdust, wood chips, wood scrap, and paper mill residues) provide about 2% of the total amount of energy we use today (Figure 15.7).

About 80% of the wood and wood waste fuel used in the United States is consumed by industry, electric power producers, and commercial businesses. The rest of the wood and wood waste fuel consumed in the United States is used in homes for heating and cooking. Many manufacturing plants in the wood and paper products industry use wood waste to produce their own steam and electricity. This saves money because it reduces the amount of fuel and electricity that must be purchased.

Energy from Garbage

Garbage, often called municipal solid waste (MSW), is used to produce energy in waste-to-energy plants and in landfills in the United States. MSW contains biomass (or biogenic) materials like paper, cardboard, food scraps, grass clippings, leaves, wood, and leather products, and other non-biomass combustible materials, mainly plastics and other synthetic materials made from petroleum.

In 1960, the average American threw away 2.7 pounds of trash a day. Today, each American throws away about 4.4 pounds of trash every day. Of that, about 1.5 pounds are recycled or composted. What do we do with the rest? One option is to burn it. (Burning is sometimes called combustion.) About 85% of our household trash is material that will burn, and most of that is biogenic, or material that is made from biomass (plant or animal products). About 71% of MSW (by weight) is biogenic (Figures 15.8 a and b).

MSW is burned in special waste-to-energy plants that use its heat energy to make steam to heat buildings or to generate electricity. There are about 86 waste-to-energy plants in the United States that generate electricity or produce steam. In 2011, waste-to-energy plants generated 14 billion kilowatt hours of electricity, about the same amount used by 1.3 million U.S. households. The biogenic material in MSW contributed about 51% of the energy of the MSW that was burned in waste-to-energy facilities that generated electricity. Many large

FIGURE 15.6   **Wood is still a viable source of energy in the U.S. and around the world.**

FIGURE 15.7   **Wood chips and saw dust can be burned as a source of energy.**

M. Niebuhr/Shutterstock

Konstantin Romanov/Shutterstock

landfills also generate electricity with the methane gas that is produced as biomass decomposes in the landfills.

Providing electricity is not the major advantage of waste-to-energy plants. It actually costs more to generate electricity at a waste-to-energy plant than it does at a coal, nuclear, or hydropower plant. The major advantage of burning waste is that it reduces the amount of material that we bury in landfills. Waste-to-energy plants burned about 29 million tons of MSW in 2011. Burning MSW reduces the volume of waste by about 87%.

Landfills can be a source of energy. Anaerobic bacteria that live in landfills decompose organic waste to produce a gas called biogas that contains methane.

Methane is the same energy-rich gas that is in natural gas, which is the fuel used for heating, cooking, and producing electricity. Methane is colorless and odorless, and a very strong greenhouse gas. Natural gas utilities add an odorant (bad smell) so people can detect natural gas leaks from pipelines. Landfill biogas can also be dangerous to people or the environment. New rules require landfills to collect methane gas for safety and pollution control (Figure 15.9).

Some landfills simply burn the methane gas in a controlled way to get rid of it. But the methane can also be used as an energy source. Landfills can collect the methane gas, treat it, and then sell it as a commercial fuel. It can then be burned to generate steam and electricity. As of July 2013, there were 621 operational landfill gas energy projects in the United States. California had the most landfill gas energy projects in operation (77), followed by Pennsylvania (44), and Michigan (41).

Biomass and the Environment

Using biomass for energy can have both positive and negative impacts on the environment. Using biomass for energy provides an alternative to using fossil fuels like coal, petroleum, or natural gas. Burning fossil fuels and burning biomass releases carbon dioxide (CO_2), a greenhouse gas, but when the plants that are the source of biomass are grown, a nearly equivalent amount of CO_2 is captured through photosynthesis.

Using Animal Waste

Some farmers produce biogas in large tanks called "digesters" where they put manure and bedding material from their barns. Some cover their manure ponds (also called lagoons) to capture biogas. Biogas digesters and manure ponds contain the same anaerobic bacteria in landfills. The biogas can be used to generate electricity or heat for use on the farm, or to sell electricity to an electric utility.

Burning Wood

Using wood, and charcoal made from wood, for heating and cooking can replace fossil fuels and may result in lower CO_2 emissions overall. Wood may be harvested from forests or woodlots that have to be thinned, or it may come from urban trees that fall down or that have to be cut down. Wood smoke contains harmful pollutants like carbon monoxide and particulate matter. Burning wood in an open fireplace for heating is an

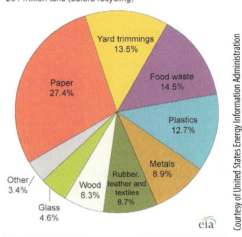

Total MSW generation (by material), 2012
251 million tons (before recycling)

FIGURE 15.8a **Total MSW generation (by materail), 2012.**

Source: U.S. Environmental Protection Agency, *Municipal Solid Waste in the United States: 2012 Facts and Figures* (February 2014)

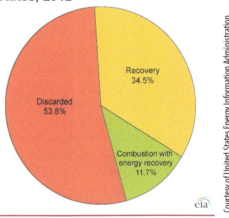

Management of MSW in the United States, 2012

FIGURE 15.8b **Management of MSW in the United States, 2012.**

Source: U.S. Environmental Protection Agency, *Municipal Solid Waste in the United States: 2012 Facts and Figures* (February 2014)

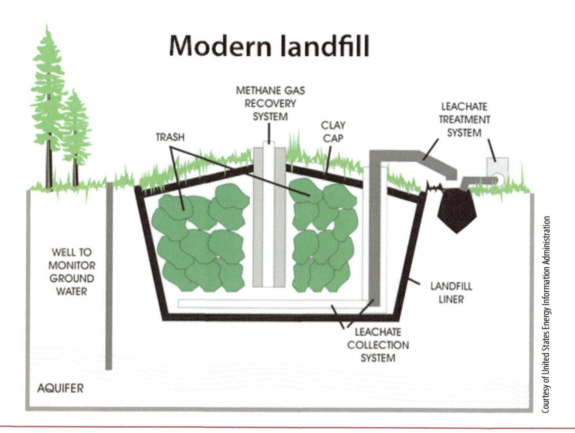

Modern landfill

METHANE GAS
RECOVERY
SYSTEM

LEACHATE
TREATMENT
SYSTEM

CLAY
CAP

TRASH

WELL TO
MONITOR
GROUND
WATER

LANDFILL
LINER

LEACHATE
COLLECTION
SYSTEM

AQUIFER

Courtesy of United States Energy Information Administration

FIGURE 15.9   **A diagram showing how a modern landfill works.**

Source: The National Energy Education Project (public domain)

inefficient way to produce heat, but it can also produce air pollution. Modern wood burning stoves and fire-place inserts are designed to reduce the amount of particulates emitted by the appliance. Wood and charcoal are major cooking and heating fuels in poor countries, and the wood may be harvested faster than trees can grow. This results in deforestation. Planting fast-growing trees for fuel and using fuel-efficient cooking stoves can help slow deforestation and improve the environment.

Burning Waste

Producing energy by burning municipal solid waste (MSW, or garbage) and by burning wood waste in facilities like waste-to-energy plants means that less waste must be buried in landfills. Waste-to-energy plants produce air pollution when MSW is burned to produce steam or electricity. Burning garbage also releases the chemicals and substances in the waste being burned. Some of these chemicals can be hazardous to people and the environment if they are not properly controlled.

The U.S. Environmental Protection Agency (EPA) applies strict environmental rules to waste-to-energy plants, and requires that waste-to-energy plants use air pollution control devices, such as scrubbers, fabric filters, and electrostatic precipitators to capture air pollutants. Scrubbers clean emissions from these facilities by spraying a liquid into the chemical gas to neutralize the acids present in the stream of emissions. Fabric filters and electrostatic precipitators also remove particles from the combustion gases. The particles—called fly ash—are then mixed with the ash that is removed from the bottom of the waste-to-energy plant's furnace. A waste-to-energy furnace burns at such high temperatures (1,800 to 2,000°F) that many complex chemicals break down into simpler, less harmful compounds.

Ash can contain high concentrations of various metals that were present in the original waste. Textile dyes, printing inks, and ceramics, for example, may contain lead and cadmium. Separating waste before combustion can solve part of the problem. Because batteries are the largest source of lead and cadmium in municipal waste, they should not be included in regular trash. Florescent light bulbs should also not be included in regular trash because they contain small amounts of mercury. The EPA tests ash from waste-to-energy plants to make sure that it is not hazardous. The test looks for chemicals and metals that would contaminate ground water. Ash that is considered safe is used in municipal solid waste landfills as a cover layer. It is also used to build roads and make cement blocks.

Collecting Landfill Gas or Biogas

As previously mentioned, biogas is composed mainly of methane and CO_2 that forms as a result of biological processes in sewage treatment plants, waste landfills, and livestock manure management systems. The electricity generated from biogas is considered renewable and is used in many states to meet state Renewable Portfolio Standards (RPS). This electricity may replace electricity produced by burning fossil fuels and could result in a net reduction in CO_2 emissions.

15.3 Biofuels: Ethanol and Biodiesel

"Biofuels" are transportation fuels like ethanol and biodiesel that are made from biomass materials. These fuels are usually blended with the petroleum fuels—gasoline and diesel fuel, but they can also be used on their own. Using ethanol or biodiesel means we don't burn quite as much fossil fuel. Ethanol and biodiesel are usually more expensive than the fossil fuels that they replace, but they are also cleaner-burning fuels, producing fewer air pollutants.

What is Ethanol?

Ethanol is a clear, colorless alcohol fuel made from the sugars found in grains (Figure 15.10), such as:

- Corn
- Sorghum
- Barley

Other sources of sugars to produce ethanol include:

- Potato skins
- Rice
- Sugar cane
- Sugar beets
- Yard clippings
- Bark
- Switchgrass

Ethanol is a renewable fuel because it is made from plants. There are several ways to make ethanol from different plant sources. Most of the ethanol used in the United States today is distilled from corn. Scientists are working on cheaper ways to make ethanol by using all parts of plants and trees rather than just the grain. Farmers are experimenting with "woody crops," mostly small poplar trees and switchgrass, to see if they can be grown cheaply and abundantly.

The most common ethanol production processes today use yeast to ferment the sugars and starch in corn. Corn is the main ingredient for fuel ethanol in the United States because of its

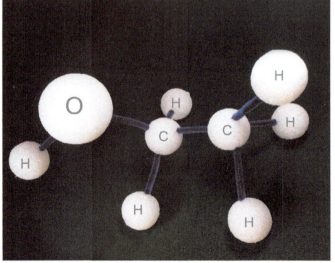

FIGURE 15.10 **A ball and stick molecular model of ethanol.**

abundance and low price. Most ethanol is produced in corn-growing states in the Midwest. The starch in the corn is fermented into sugar, which is then fermented into alcohol.

Sugar cane and sugar beets are the most common ingredients used to make ethanol in other parts of the world. Since alcohol is created by fermenting sugar, sugar crops are the easiest ingredients to convert into alcohol. Brazil, the world's second-largest fuel ethanol producer, makes most of its ethanol from sugar cane. Most of the cars in Brazil are capable of running on pure ethanol or on a blend of gasoline and ethanol.

Switchgrass can yield almost twice as much ethanol as corn, estimates geneticist Ken Vogel, who is conducting breeding and genetics research on switchgrass to improve its biomass yield and its ability to recycle carbon as a renewable energy crop (Figure 15.11).

FIGURE 15.11   Panicum virgatum (switchgrass) being grown.

Nearly all gasoline sold now in the U.S. contains some ethanol. About 99% of the fuel ethanol consumed in the U.S. is added to gasoline in mixtures of up to 10% ethanol and 90% gasoline. Any gasoline powered engine in the U.S. can use E10 (gasoline with 10% ethanol), but only specific types of vehicles can use mixtures with greater than 10% ethanol. The U.S. Environmental Protection Agency ruled in October 2010, that cars and light trucks of model year 2007 and newer can use E15. A flex-fuel vehicle (FFV) is necessary to mixtures with higher amounts of ethanol. E85, a fuel that is 85% ethanol and 15% gasoline, is mainly sold in the Midwest. In 2012, there were about 110 million vehicles in the U.S. capable of running on E85, but only about 10% of them actually used E85.

Ethanol can also be produced by breaking down cellulose in woody fibers. Cellulosic ethanol is considered an advanced biofuel and involves a more complicated production process than the process used to make conventional ethanol.

Trees and grasses are potential feedstocks (the raw material needed to make a product) for cellulosic ethanol production. Trees and grasses require less energy, fertilizers, and water than grains and can also be grown on lands that are not suitable for growing food. Scientists have developed fast-growing trees that grow to size in 10 years. Many grasses can produce two harvests a year for many years without annual replanting.

History of Ethanol

In the 1850s, ethanol was a major lighting fuel. During the Civil War, a liquor tax was placed on ethanol to raise money for the war. The tax increased the price of ethanol so much that it could no longer compete with other fuels like kerosene. Ethanol production declined sharply because of this tax, and production levels did not begin to recover until the tax was repealed in 1906.

In 1908, Henry Ford designed his Model T, a very early automobile (Figure 15.13), to run on a mixture of gasoline and alcohol. Ford called this mixture the fuel of the future. In 1919, when Prohibition

FIGURE 15.12   Biodiesel can be made from recycled restaurant grease.

Julie Clopper/Shutterstock

FIGURE 15.13 **The Model T Car was designed to run on a blend of ethanol and gasoline.**

began, ethanol was banned because it was considered an alcoholic beverage. It could only be sold when mixed with petroleum. Ethanol was used as a fuel again after Prohibition ended in 1933.

Ethanol use increased temporarily during World War II when oil and other resources were scarce. In the 1970s, interest in ethanol as a transportation fuel was revived as oil embargoes, rising oil prices, and growing dependence on imported oil increased interest in alternative fuels. Since that time, ethanol use and production has been encouraged by tax benefits and by environmental requirements for cleaner-burning fuels.

In 2005, Congress enacted a renewable fuel standard (RFS) that set minimum requirements for the use of renewable fuels, including ethanol. In 2007, the RFS renewable fuel use targets were set to rise steadily to a level of 36 billion gallons by 2022. In 2013, about 13 billion gallons of ethanol were added to the gasoline consumed in the United States.

Ethanol as a Transportation Fuel

Joshua Rainey Photography/Shutterstock

FIGURE 15.14 **Gasoline pump.**

Ethanol and the Environment

Unlike gasoline, pure ethanol is nontoxic and biodegradable, and it quickly breaks down into harmless substances if spilled. Chemical denaturants are added to fuel ethanol (about 2% by volume), and many of the denaturants used are toxic. Similar to gasoline, ethanol is a highly flammable liquid and must be transported carefully.

Ethanol and ethanol-gasoline mixtures burn cleaner and have higher octane than pure gasoline, but they also have higher evaporative emissions from fuel tanks and dispensing equipment. These evaporative emissions contribute to the formation of harmful, ground-level ozone and smog. Gasoline requires extra processing to reduce evaporative emissions before it is blended with ethanol.

Ethanol can be considered carbon-neutral because the plants used to make fuel ethanol (such as corn and sugarcane) absorb carbon dioxide (CO_2) as they grow and may offset the CO_2 produced when ethanol is made and burned. In the United States, coal and natural gas are used as the heat sources in the fermentation process to make fuel ethanol. The impact of greater ethanol use on net CO_2 emissions depends on how ethanol is made and depends on whether or not indirect impacts on land use are included in the calculations. Growing plants for fuel is a controversial topic because some believe the land, fertilizers, and energy used to grow biofuel crops should be used to grow food crops instead.

The U.S. government is supporting efforts to produce ethanol with methods that use less energy than conventional fermentation, and that use "cellulosic" biomass, which requires less cultivation, fertilizer, and pesticides than corn and sugar cane. Cellulosic ethanol feedstock includes native prairie grasses, fast growing trees, sawdust, and even waste paper.

What Is Biodiesel?

Biodiesel is a renewable fuel that can be used instead of diesel fuel, which is made from petroleum. Biodiesel can be made from vegetable oils, animal fats, or greases. Today, most biodiesel is made from soybean oil. About half of biodiesel producers are able to make biodiesel from used oils or fats, including recycled restaurant grease.

Biodiesel is most often blended with petroleum diesel in ratios of 2% (B2), 5% (B5), or 20% (B20). Biodiesel can also be used as pure biodiesel (B100). Biodiesel fuels can be used in regular diesel vehicles without making any changes to the engines. It can also be stored and transported using diesel tanks and equipment.

The practice of fueling engines with biodiesel is becoming more common, but it isn't a new idea. Before petroleum diesel fuel became popular, Rudolf Diesel, the inventor of the diesel engine in 1897, experimented with using vegetable oil (biodiesel) as fuel.

Most trucks, buses, and tractors in the United states use diesel fuel. Diesel is a nonrenewable fuel made from petroleum. Using biodiesel means that we use a little bit less petroleum. Biodiesel results in less pollution than petroleum diesel. Any vehicle that operates on diesel fuel can switch to biodiesel without changes to its engine.

Biodiesel has chemical characteristics much like petroleum-based diesel, so it can be used as a direct substitute for diesel fuel. Biodiesel can also be blended with petroleum diesel in any percentage without suffering any significant loss of fuel economy.

Low-level biodiesel blends like B2 through B5 are popular fuels in the trucking industry because biodiesel has excellent lubricating properties, and therefore usage of the blends can be beneficial for engine performance. Pure biodiesel (often called B100) and biodiesel blends are sensitive to cold weather and may require a special type of anti-freeze, just like petroleum-based diesel fuel does. Biodiesel acts like a detergent additive, loosening, and dissolving sediments in storage tanks. Because biodiesel is a solvent, B100 may cause rubber and other components to fail in older vehicles. This problem does not occur with biodiesel blends.

Biodiesel Use Is Increasing

Due to environmental benefits, ease of use, and available subsidies, biodiesel use in the U.S. grew from about 10 million gallons in 2001 to 358 million gallons in 2007. In 2011, substantial quantites of biodiesel were needed to meet the Renewable Fuels Standard. Biodiesel consumption was about 870 million gallons in 2012.

Fleet vehicles such as school and transit buses, snowplows, garbage trucks, mail trucks, and military vehicles often use B20, and many fleets are required by federal or state mandates to use biodiesel blends. There are public fueling stations that offer biodiesel blends for sale to the public in nearly every state.

In 2011, 6.35 billion gallons of biodiesel were consumed in about 64 countries; 54% was consumed in five countries (Figure 15.15).

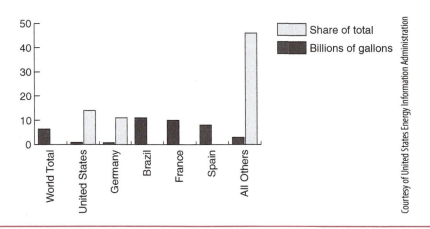

FIGURE 15.15 **World Biodiesel Consumption, 2011.**

Biodiesal and the Environment

Biodiesel is nontoxic and biodegradable. Compared to diesel fuel made from petroleum, biodiesel produces fewer air pollutants like particulates, carbon monoxide, sulfur dioxide, hydrocarbons, and air toxics. However, it does slightly increase emissions of nitrogen oxides. Biodiesel does not smell as bad as regular diesel fuel when it burns; sometimes biodiesel exhaust smells like french fries!

Using a gallon of biodiesel produced in the United States avoids the CO_2 emissions that result from burning about a gallon of petroleum diesel. Biodiesel may be considered to be carbon-neutral when the plants that are used to make biodiesel, such as soy beans and palm oil trees, absorb CO_2 as they grow and offset the CO_2 produced while making and using biodiesel. Most of the biodiesel used in the United States is made from soybean oil that is a by product of processing soybeans for animal feed and numerous other food and non-food products, and from waste animal fat and grease (Figure 15.16).

However, in some parts of the world, large areas of natural vegetation and forests have been cleared and burned to grow soybeans and palm oil trees to make biodiesel. The negative environmental impacts of these activities may be greater than any potential benefits from using the biodiesel produced from the plants grown to make biodiesel.

15.4 Wind

Wind is caused by the uneven heating of the earth's surface by the sun. Because the earth's surface is made of different types of land and water, it absorbs the sun's heat at different rates. One example of this uneven heating can be found in the daily wind cycle.

During the day, the air above the land heats up faster than the air over water (Figure 15.17). The warm air over the land expands and rises, and the heavier, cooler air rushes in to take its place, creating wind. At night,

Kenneth Summers/Shutterstock

FIGURE 15.16 **Biodiesel Burns Much Cleaner than Diesel and can be made from food waste and food by-products.**

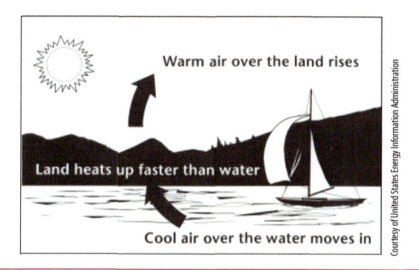

Courtesy of United States Energy Information Administration

FIGURE 15.17 **Energy from Moving Air.**

Source: National Energy Education Development Project (public domain)

the winds are reversed because the air cools more rapidly over land than over water (Figure 15.17). In the same way, the atmospheric winds that circle the earth are created because the land near the earth's equator is heated more by the sun than the land near the North Pole and the South Pole.

Wind Energy for Electricity Generation

Today, wind energy is mainly used to generate electricity, although water pumping windmills were once used throughout the United States. Wind turbines use blades to collect the wind's kinetic energy. Wind flows over the blades creating lift (similar to the effect on airplane wings), which causes the blades to turn. The blades are connected to a drive shaft that turns an electric generator, which produces electricity. In 2013, wind turbines in the United States generated about 4% of total U.S. electricity generation. Although this is a small share of the country's total electricity production, it was equal to the electricity use of about 15 million households in 2012.

The amount of electricity generated from wind has grown significantly in recent years. Electricity generation from wind in the United States increased from about 6 billion kilowatthours (kWh) in 2000, to about 168 billion kWh in 2013. In fact, in 2008, 2009, and 2012, wind was the fastest growing source of electricity production in the United States.

New technologies have decreased the cost of producing electricity from wind, and growth in wind power has been encouraged by state renewable energy portfolio standards and goals, tax breaks, and green pricing programs.

Wind Power Plants

Operating a wind power plant is not as simple as just building a windmill in a windy place. Wind plant owners must carefully plan where to locate their machines. It is important to consider how fast and how much the wind blows at the site. As a rule, wind speed increases with altitude and over open areas that have no windbreaks. Good sites for wind plants are the tops of smooth, rounded hills, open plains or shorelines, and mountain gaps that produce wind funneling.

Wind speed varies throughout the United States (Figure 15.18). It also varies from season to season. In Tehachapi, California, the wind blows more from April through October than it does in the winter. This is because of the extreme heating of the Mojave Desert during the summer months. The hot air over the desert rises, and the cooler, denser air above the Pacific Ocean rushes through the Tehachapi mountain pass to take its place. In a state like Montana, on the other hand, the wind blows more during the winter.

Fortunately, these seasonal variations are a good match for the electricity demands of the regions. In California, people use more electricity during the summer for air conditioners. In Montana, people use more electricity during the winter.

Large wind turbines (sometimes called wind machines) generated electricity in 36 different states in 2011 (Figure 15.19). The top five states with the largest generation of electricity from wind were Texas, Iowa, California, Minnesota, and Illinois (Figure 15.20).

Wind power plants, or wind farms, as they are sometimes called, are clusters of wind machines used to produce electricity. A wind farm usually has dozens of wind machines scattered over a large area. The world's largest wind farm, the Horse Hollow Wind Energy Center in Texas, has 421 wind turbines that generate enough electricity to power 220,000 homes per year.

Many wind plants are not owned by public utility companies. Instead, they are owned and operated by business people who sell the electricity produced on the wind farm to electric utilities. These private companies are known as Independent Power Producers.

International Wind Power

In 2009, most of the wind power plants in the world were located in Europe and in the United States where government programs have helped support wind power development. The United States ranked first in the world in wind power generation, followed by Germany, Spain, China, and India. Denmark ranked ninth in the world in wind power generation, but generated about 19% of its electricity from wind, the largest share of any country.

Offshore Wind Power

Conditions are well suited along much of the coasts of the United States to use wind energy. However, there are people who oppose putting turbines just offshore, near the coastlines, because they think the wind turbines will spoil the view of the ocean. There is a plan to build an offshore wind plant off the coast of Cape Cod, Massachusetts.

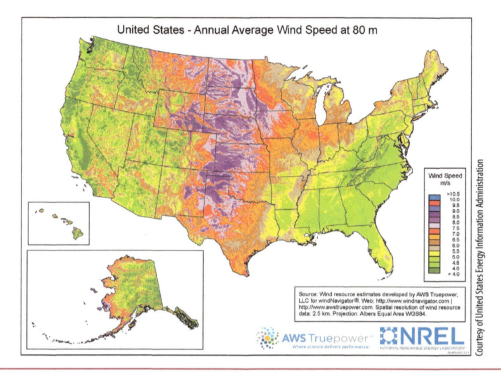

FIGURE 15.18   Map of U.S. Wind Resources. Areas in green (Western and Eastern states) have lower wind potential, but the middle 'Plains' states have higher wind speeds as shown by the pink, purple and orange shaded regions on this map.

Source: National Renewable Energy Laboratory, U.S. Department of Energy (Public Domain)

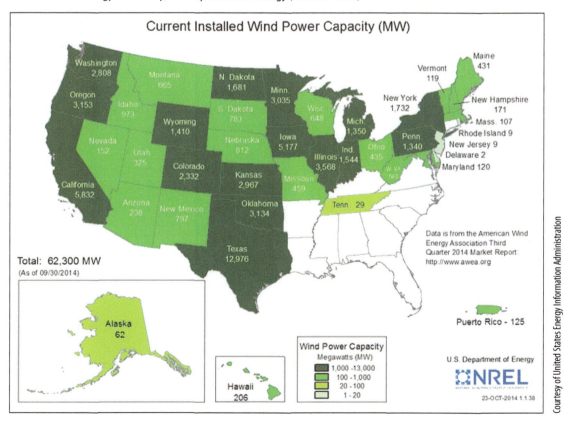

FIGURE 15.19   Map of wind production in U.S.

Source: National Renewable Energy Laboratory, U.S. Department of Energy (public domain)

Top wind power producing states, 2013

FIGURE 15.20 **Top 5 wind producing states in 2012 from most to least include; Texas (#1), Iowa, California, Oklahoma, and Illinois (#5).**

Source: U.S. Energy Information Administration, *Electric Power Monthly*, Table 1.17.B (February 2014)

Wind is a renewable energy source that does not pollute, so some people see it as a good alternative to fossil fuels.

Types of Wind Machines (Turbines)

There are two types of wind machines (turbines) used today, based on the direction of the rotating shaft (axis): horizontal-axis wind machines and vertical-axis wind machines. The size of wind machines varies widely. Small turbines used to power a single home or business may have a capacity of less than 100 kilowatts. Some large commercial-sized turbines may have a capacity of 5 million watts, or 5 megawatts. Larger turbines are often grouped together into wind farms that provide power to the electrical grid.

Most wind machines being used today are the horizontal-axis type (Figure 15.22). Horizontal-axis wind machines have blades like airplane propellers. A typical horizontal wind machine stands as tall as a 20-story building and has three blades that span 200 feet across. The largest wind machines in the world have blades longer than a football field. Wind machines stand tall and wide to capture more wind.

Vertical-axis wind machines have blades that go from top to bottom. The most common type—the Darrieus wind turbine, named after the French engineer Georges Darrieus who patented the design in 1931—looks like a giant, two-bladed egg beater. This type of vertical wind machine typically stands 100 feet tall and 50 feet wide. Vertical-axis wind machines make up only a very small share of the wind machines used today.

The History of Wind Energy

Since early recorded history, people have been harnessing the energy of the wind. Wind energy propelled boats along the Nile River as early as 5000 B.C. By 200 B.C., simple windmills in China were pumping water, while vertical-axis windmills with woven reed sails were grinding grain in Persia and the Middle East.

New ways of using the energy of the wind eventually spread around the world. By the 11th century, people in the Middle East were using windmills extensively for food production; returning merchants and crusaders carried

FIGURE 15.21 **Wind Turbines in the Ocean.**

this idea back to Europe. The Dutch refined the windmill and adapted it for draining lakes and marshes in the Rhine River Delta (Figure 15.23). When settlers took this technology to the New World in the late 19th century, they began using windmills to pump water for farms and ranches, and later, to generate electricity for homes and industry.

American colonists used windmills to grind wheat and corn, to pump water, and to cut wood at sawmills. As late as the 1920s, Americans used small windmills to generate electricity in rural areas without electric service. When power lines began to transport electricity to rural areas in the 1930s, local windmills were used less and less, though they can still be seen on some Western ranches.

The oil shortages of the 1970s changed the energy picture for the country and the world. It created an interest in alternative energy sources, paving the way for the re-entry of the windmill to generate electricity. In the early 1980s, wind energy really took off in California, partly because of state policies that encouraged renewable energy sources.

In the 1990s, the push came from a renewed concern for the environment in response to scientific studies indicating potential changes to the global climate if the use of fossil fuels continues to increase. Wind energy is an economical power resource in many areas of the country.

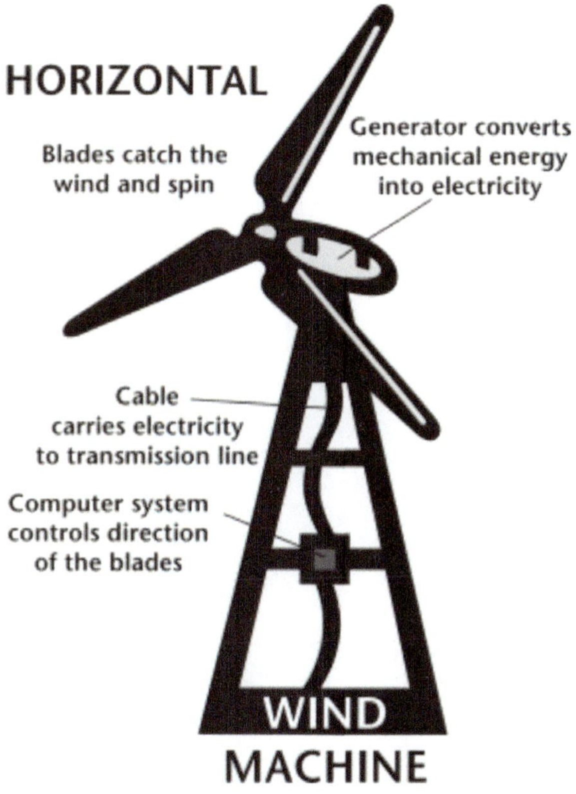

Growing concern about emissions from fossil fuel generation, increased government support, and higher costs for fossil fuels (especially natural gas and coal) have helped wind power capacity in the United States grow substantially over the past 10 years.

HORIZONTAL

Blades catch the wind and spin

Generator converts mechanical energy into electricity

Cable carries electricity to transmission line

Computer system controls direction of the blades

WIND MACHINE

Courtesy of United States Energy Information Administration

FIGURE 15.22

Wind Power and the Environment

Wind is a clean source of renewable energy, and overall, the use of wind for energy has fewer environmental impacts than using many other energy sources. Wind turbines (often called windmills) do not release emissions that pollute the air or water (with rare exceptions), and they do not require water for cooling. They may also reduce the amount of electricity generated from fossil fuels and therefore reduce the amount of air pollution, carbon dioxide emissions, and water use of fossil fuel power plants.

A wind turbine has a small physical footprint relative to the amount of electricity it can produce. Many wind projects, sometimes called wind farms, are located on farm, grazing, and forest land. The extra income from the turbines may allow farmers and ranchers to stay in business and keep their property from being developed for other uses (Figure 15.24). For example, wind power projects have been proposed as alternatives to mountain top removal coal mining projects in the Appalachian mountains of the United States. Offshore wind turbines on lakes or the ocean may have smaller environmental impacts than turbines on land.

Symbiot/Shutterstock

FIGURE 15.23 Traditional Dutch-type windmill.

Arina P. Habich/Shutterstock

FIGURE 15.24 **Wind turbines In Iowa.**

Modern wind turbines are very large machines, and some people do not like their visual impact on the landscape. A few wind turbines have caught on fire, and some have leaked lubricating fluids, though this is relatively rare. Some people do not like the sound that wind turbine blades make. Some types of wind turbines and wind projects cause bird and bat deaths. These deaths may contribute to declines in species that are also being affected by other human-related impacts. Many more birds are killed from collisions with vehicles and buildings, by house cats and hunters, and by pesticides as compared to wind turbines. Furthermore, their natural habitats may be altered or destroyed by human development and by the changes in the climate that most scientists believe are caused by greenhouse gases emissions from human activities (which wind energy use can help reduce). The wind energy industry and the U.S. government are researching ways to reduce the impact of wind turbines on birds and bats.

Finally, most wind power projects on land also require service roads that add to their physical impact on the environment. Making the metals and other materials in wind turbines and the concrete for their foundations requires the use of energy, which may be from fossil fuels. Some studies have shown that wind turbines produce much more clean electricity over their operating life than the equivalent amount of energy used to make and install them.

Wind turbines do have negative impacts on the environment, but the negative impacts have to be balanced with our need for electricity and the overall lower environmental impact of using wind for energy relative to other sources of energy to make electricity.

15.5 Geothermal

The word geothermal comes from the Greek words geo (earth) and therme (heat). Geothermal energy is heat from within the earth. We can recover this heat as steam or as hot water and use it to heat buildings or to generate electricity. Geothermal energy is a renewable energy source because the heat is continuously produced inside the earth.

Ancient Romans, Chinese, and Native American cultures used hot mineral springs for bathing, cooking, and heating. Today, many hot springs are still used for bathing, and many people believe the hot, mineral-rich waters have natural healing powers.

Temperatures hotter than the sun's surface are continuously produced inside the earth by the slow decay of radioactive particles, a process that happens in all rocks. The earth has a number of different layers:

- The core has a solid iron core and an outer core made of hot melted rock called magma.
- The mantle surrounds the core and is about 1,800 miles thick. It is made up of magma and rock.
- The crust is the outermost layer of the earth. The crust forms the continents and ocean floors. It can be 3 to 5 miles thick under the oceans and 15 to 35 miles thick on the continents.

The earth's crust is broken into pieces called plates. Magma comes close to the earth's surface near the edges of these plates. This is where volcanoes occur. The lava that erupts from volcanoes is partly magma. The rocks and water absorb the heat from this magma deep underground (Figure 15.25). The rocks and water found deeper underground have higher temperatures.

People around the world use geothermal energy to heat their homes and to produce electricity by digging deep wells and pumping the heated underground water or steam to the surface. People can also make use of the stable temperatures near the surface of the earth to heat and cool buildings.

Geothermal reservoirs are naturally occurring areas of hydrothermal resources. They are deep underground and are largely undetectable above ground. Geothermal energy finds its way to the earth's surface in three ways:

1. Volcanoes and fumaroles (holes where volcanic gases are released)
2. Hot springs
3. Geysers

When magma comes near the earth's surface, it heats ground water trapped in porous rock or water running along fractured rock surfaces and faults. Hydrothermal features have two common ingredients, water (hydro) and heat (thermal).

The most active geothermal resources are usually found along major tectonic plate boundaries where earthquakes and volcanoes are located. One of the most active geothermal areas in the world is called the Ring of Fire (Figure 15.26). This area encircles the Pacific Ocean.

Geologists use various methods to find geothermal reservoirs. Drilling a well and testing the temperature deep underground is the most reliable method for locating a geothermal reservoir.

Most of the geothermal power plants in the United States are located in the western states and in Hawaii, where geothermal energy resources are close to the earth's surface (Figure 15.27). California generates the most electricity from geothermal energy. The Geysers dry steam reservoir in northern California is the largest known dry steam field in the world and has been producing electricity since 1960 (Figure 15.28).

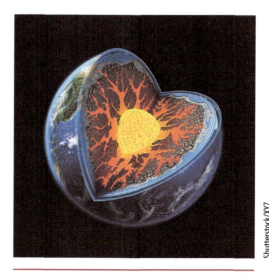

FIGURE 15.25   Magma is closest to the surface at plate boundaries, and the underground heat can be harnessed for heat and to produce electricity.

Shutterstock/XYZ

FIGURE 15.26   The Ring of Fire.

Courtesy of United States Energy Information Administration

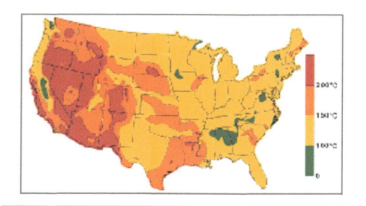

FIGURE 15.27   U.S. geothermal resource map.

Courtesy of United States Energy Information Administration

Some applications of geothermal energy use the Earth's temperatures near the surface, while others require drilling miles into the Earth. There are three main types of geothermal energy systems:

1. Direct use and district heating systems use hot water from springs or reservoirs near the surface.
2. Electricity generation power plants require water or steam at high temperature (300° to 700°F). Geothermal power plants are generally built where geothermal reservoirs are located within a mile or two of the surface.
3. Geothermal heat pumps use stable ground or water temperatures near the Earth's surface to control building temperatures above ground.

Geothermal energy is also used to heat buildings through district heating systems. Hot water near the Earth's surface can be piped directly into buildings and industries for heat. A district heating system provides heat for most of the buildings in Reykjavik, Iceland.

According to the U.S. Environmental Protection Agency (EPA), geothermal heat pumps (Figure 15.29) are the most energy efficient, environmentally clean, and cost effective systems for temperature control. Although most homes still use traditional furnaces and air conditioners, geothermal heat pumps are becoming more popular. In recent years, the U.S. Department of Energy and the EPA have partnered with industry to promote the use of geothermal heat pumps.

Industrial applications of geothermal energy include food dehydration, gold mining, and milk pasteurizing. Dehydration, or the drying of vegetable and fruit products, is the most common industrial use of geothermal energy.

Geothermal Energy in the U.S. and Around the World

The United States leads the world in the amount of electricity generated with geothermal energy. In 2013, U.S. geothermal power plants produced about 17 billion kilowatthours (kWh), 0.4% of total U.S. electricity generation. In 2013, six states had geothermal power plants.

Twenty four countries including the United States had geothermal power plants in 2011, which generated approximately 67.5 billion kWh. The Philippines was the second largest geothermal power producer after the United States at 9.9 billion kWh, which equaled approximately 15% of the country's total power generation. Iceland, the seventh largest producer at 4.7 billion kWh, produced 28% of its total electricity using geothermal energy.

Geothermal power plants use hydrothermal resources that have both water (hydro) and heat (thermal). Geothermal power plants require high temperature (300°F to 700°F) hydrothermal resources that come from

State rankings for geothermal power in 2014

Courtesy of United States Energy Information Administration

FIGURE 15.28 Map of top 5 U.S. states by amount of electricity generated by geothermal power plants in 2014 include; California (#1), Nevada, Utah, Hawaii, Oregon (#5).

Source: U.S. Energy Information Administration, *Electric Power Monthly* (February 2015), Table 19.1B; preliminary data for 2014.

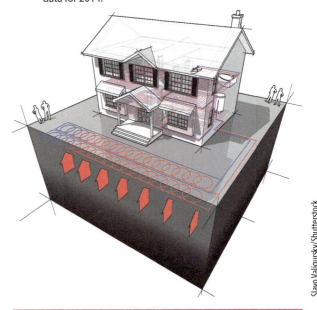

Slavo Valigursky/Shutterstock

FIGURE 15.29 Diagram of a closed loop geothermal heat pump system.

either dry steam wells or hot water wells. We use these resources by drilling wells into the earth and then piping steam or hot water to the surface. The hot water or steam is used to operate a turbine that generates electricity. Some geothermal wells may be as deep as two miles.

Geothermal Energy and the Environment

The environmental impact of geothermal energy depends on how it is used or on how it is converted to useful energy. Direct use applications and geothermal heat pumps have almost no negative impact on the environment. Direct use applications and geothermal heat pumps can actually have a positive effect because they may reduce or avoid the use of other types of energy that may have greater negative impacts on the environment. Geothermal features in national parks like geysers and fumaroles in Yellowstone National Park are protected by law (Figure 15.30).

FIGURE 15.30   **Grand Prismatic Spring, Yellowstone National Park, Wyoming.**

Lorcel/Shutterstock

15.6 Solar

The sun has produced energy for billions of years. Solar energy is the sun's rays (solar radiation) that reach the Earth. This energy can be converted into other forms of energy, such as heat and electricity. In the 1830s, the British astronomer John Herschel famously used a solar thermal collector box (a device that absorbs sunlight to collect heat) to cook food during an expedition to Africa. Today, people use the sun's energy for lots of things.

The amount of solar energy that the earth receives each day is many times greater than the total amount of energy consumed around the world. Indeed, covering 4% of the world's desert areas with photovoltaics could supply the equivalent of all of the world's electricity. The Gobi Desert alone could supply almost all of the world's total electricity demand. However, solar energy is a variable and intermittent energy source. The amount and intensity of sunlight varies by location, and weather and climate conditions affect its availability on a daily and seasonal basis (Figure 15.31). The type and size of a solar energy collection and conversion system determines how much of available solar energy can be converted to useful energy.

When converted to thermal (or heat) energy, solar energy can be used to:

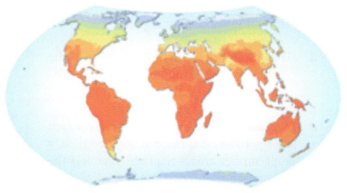

FIGURE 15.31   The red areas of this map indicate regions with the most incident solar energy resources (i.e., near the Equator), while the green and blue shaded areas are regions with low incident solar energy resources (i.e., near the polar regions). The amount and intensity of solar energy varies with location on Earth, season and weather conditions.

Courtesy of United States Energy Information Administration

- Heat water—for use in homes, buildings, or swimming pools
- Heat spaces—inside homes, greenhouses, and other buildings
- Heat fluids—to high temperatures to operate a turbine to generate electricity

Solar energy can be converted to electricity in two ways:

1. Photovoltaic (PV devices) or "solar cells" change sunlight directly into electricity. Individual PV cells are grouped into panels and arrays of panels that can be used in a wide range of applications ranging from single small cells that charge calculator and watch batteries, to systems that power single homes, to large power plants covering many acres.

Solar thermal/electric power plants generate electricity by concentrating solar energy to heat a fluid and produce steam that is used to power a generator. In 2012, solar thermal-power generating units were the main source of electricity at 12 power plants in the United States (11 in California and 1 in Nevada).

In 2010, a 75 Megawatt solar thermal unit was added to Florida Power and Light's Martin plant, a 3,700 MW oil- and gas-fired facility. This innovative system uses a parabolic trough solar array to produce supplemental steam for use with an existing turbine/generator.

Benefits and Drawbacks to Solar

The main benefits of solar energy are:

- Solar energy systems do not produce air pollutants or carbon-dioxide
- When located on buildings, they have minimal impact on the environment

Two limitations of solar energy are:

- The amount of sunlight that arrives at the Earth's surface is not constant. It varies depending on location, time of day, time of year, and weather conditions.
- Because the sun doesn't deliver that much energy to any one place at any one time, a large surface area is required to collect the energy at a useful rate.

Photovoltaic Cells

A photovoltaic cell, commonly called a solar cell or PV, is the technology used to convert solar energy directly into electrical power. A photovoltaic cell is a nonmechanical device usually made from silicon alloys (Figure 15.32).

Sunlight is composed of photons, or particles of solar energy. These photons contain various amounts of energy corresponding to the different wavelengths of the solar spectrum. When photons strike a photovoltaic cell, they may be reflected, pass right through, or be absorbed. Only the absorbed photons provide energy to generate electricity. When enough sunlight (energy) is absorbed by the material (a semiconductor), electrons are

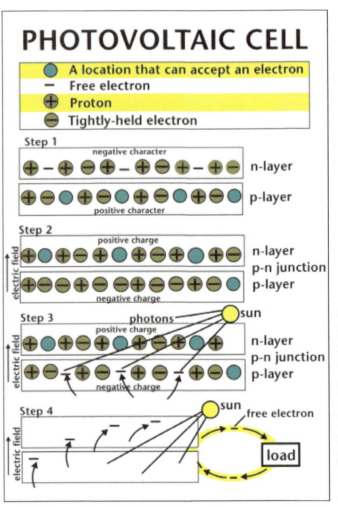

Courtesy of United States Energy Information Administration

FIGURE 15.32   Image of how a photovoltaic cell works.

Source: National Energy Education Development Project (public domain)

dislodged from the material's atoms. Special treatment of the material surface during manufacturing makes the front surface of the cell more receptive to free electrons, so the electrons naturally migrate to the surface.

When the electrons leave their position, holes are formed. When many electrons, each carrying a negative charge, travel toward the front surface of the cell, the resulting imbalance of charge between the cell's front and back surfaces creates a voltage potential like the negative and positive terminals of a battery. When the two surfaces are connected through an external load, such as an appliance, electricity flows.

The photovoltaic cell is the basic building block of a photovoltaic system. Individual cells can vary in size from about 0.5 inches to about 4 inches across. However, one cell only produces 1 or 2 watts, which isn't enough power for most applications. To increase power output, cells are electrically connected into a packaged weather-tight module. Modules can be further connected to form an array. The term array refers to the entire generating plant, whether it is made up of one or several thousand modules. The number of modules connected together in an array depends on the amount of power output needed.

The performance of a photovoltaic array is dependent upon sunlight. Climate conditions (such as clouds or fog) have a significant effect on the amount of solar energy received by a photovoltaic array and, in turn, its performance. The efficiency of most commercially available photovoltaic modules in converting sunlight to electricity ranges from 5% to 15%. Researchers around the world are trying to achieve efficiencies up to 30%.

Some advantages of photovoltaic systems are:

1. Conversion from sunlight to electricity is direct, so bulky mechanical generator systems are unnecessary.
2. PV arrays can be installed quickly and in any size.
3. The environmental impact is minimal, requiring no water for system cooling and generating no by-products.

Photovoltaic cells, like batteries, generate direct current (DC), which is generally used for small loads (electronic equipment). When DC from photovoltaic cells is used for commercial applications or sold to electric utilities using the electric grid, it must be converted to alternating current (AC) using inverters, solid state devices that convert DC power to AC.

Hundreds of thousands of houses and buildings around the world have PV systems on their roofs. Many multi-megawatt (MW) PV power plants have also been built (Figure 15.33) U.S. shipments (includes imports, exports, and domestic shipments) of PV panels (modules) by U.S. industry in 2012 was the equivalent of about 4,655 Megawatts, about 245 times greater than the shipments of about 19 Megawatts in 19941. Since about 2004, most of the PV panels installed in the United States have been in "grid-connected" systems on homes, buildings, and central-station power facilities. There are PV products available that can replace conventional roofing materials.

History of the Photovoltaic Cell

The first practical PV cell was developed in 1954 by Bell Telephone researchers examining the sensitivity of a properly prepared silicon wafer to sunlight. Beginning in the late 1950s, PV cells were used to power U.S. space satellites. PV cells were next widely used for small consumer electronics like calculators and watches and to provide electricity in remote or "off-grid" locations where there were no electric power lines. Technology advances and government financial incentives have helped to greatly expand PV use since the mid-1990s.

The success of PV in outer space first generated commercial applications for this technology. The simplest photovoltaic systems power many of the small calculators

Photovoltaic systems

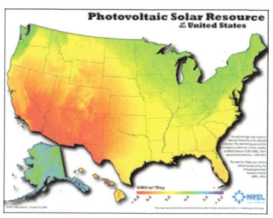

FIGURE 15.33 **Density of PV cells in use across the U.S.** Areas shaded in red and orange have the highest density of this type of solar energy, while areas shaded green and blue on the map indicate a low density.

and wrist watches used every day. More complicated systems provide electricity to pump water, power communications equipment, and even provide electricity to our homes.

Solar Concentrating Collectors

Concentrating solar power technologies use mirrors to reflect and concentrate sunlight onto receivers that collect solar energy and convert it to heat. This thermal energy can then be used to produce high temperature heat or electricity via a steam turbine or heat engine driving a generator.

Solar thermal power plants use the sun's rays to heat a fluid to very high temperatures. The fluid is then circulated through pipes so it can transfer its heat to water to produce steam. The steam, in turn, is converted into mechanical energy in a turbine and into electricity by a conventional generator coupled to the turbine.

So solar thermal power generation works essentially the same as generation from fossil fuels except that instead of using steam produced from the combustion of fossil fuels, the steam is produced by the heat collected from sunlight. Solar thermal technologies use concentrator systems to achieve the high temperatures needed to heat the fluid.

The three main types of solar thermal power systems are:

- Parabolic trough (the most common type of plant)
- Solar dish
- Solar power tower

Courtesy of United States Energy Information Administration

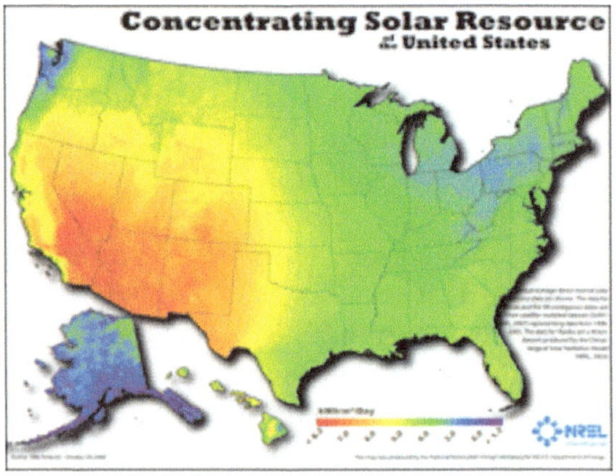

FIGURE 15.34 Concentrating Solar Resource of the United States. Areas shaded in red and orange have the highest density of this type of solar energy, while areas shaded green and blue on the map indicate a low density.

Parabolic troughs are used in the largest solar power facility in the world located in the Mojave Desert at Kramer Junction, California. This facility has operated since the 1980s and accounts for the majority of solar electricity produced by the electric power sector today.

A parabolic trough collector has a long parabolic-shaped reflector that focuses the sun's rays on a receiver pipe located at the focus of the parabola. The collector tilts with the sun as the sun moves from east to west during the day to ensure that the sun is continuously focused on the receiver (Figure 15.35).

Tom Grundy/Shutterstock

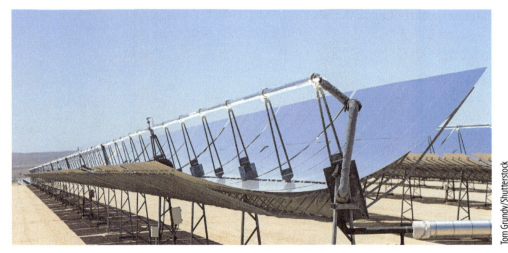

FIGURE 15.35 Parabolic trough power plant.

Because of its parabolic shape, a trough can focus the sun at 30 to 100 times its normal intensity (concentration ratio) on the receiver pipe located along the focal line of the trough, achieving operating temperatures over 750°F.

The "solar field" has many parallel rows of solar parabolic trough collectors aligned on a north-south horizontal axis. A working (heat transfer) fluid is heated as it circulates through the receiver pipes and returns to a series of "heat exchangers" at a central location. Here, the fluid circulates through pipes so it can transfer its heat to water to generate high-pressure, superheated steam. The steam is then fed to a conventional steam turbine and generator to produce electricity. When the hot fluid passes through the heat exchangers, it cools down, and is then recirculated through the solar field to heat up again.

The plant is usually designed to operate at full power using solar energy alone, given sufficient solar energy. However, all parabolic trough power plants can use fossil fuel combustion to supplement the solar output during periods of low solar energy, such as on cloudy days.

Solar Dish

A solar dish/engine system uses concentrating solar collectors that track the sun, so they always point straight at the sun and concentrate the solar energy at the focal point of the dish (Figure 15.36). A solar dish's concentration ratio is much higher than a solar trough's, typically over 2,000, with a working fluid temperature over 1380°F. The power-generating equipment used with a solar dish can be mounted at the focal point of the dish, making it well suited for remote operations or, as with the solar trough, the energy may be collected from a number of installations and converted to electricity at a central point.

The engine in a solar dish/engine system converts heat to mechanical power by compressing the working fluid when it is cold, heating the compressed working fluid, and then expanding the fluid through a turbine or with a piston to produce work. The engine is coupled to an electric generator to convert the mechanical power to electric power.

FIGURE 15.36   Image of a solar dish collector.

topten22photo/Shutterstock

Solar Power Tower

A solar power tower, or central receiver, generates electricity from sunlight by focusing concentrated solar energy on a tower-mounted heat exchanger (receiver). This system uses hundreds to thousands of flat, sun-tracking mirrors called heliostats to reflect and concentrate the sun's energy onto a central receiver tower (Figure 15.37). The energy can be concentrated as much as 1,500 times that of the energy coming in from the sun.

Energy losses from thermal-energy transport are minimized because solar energy is being directly transferred by reflection from the heliostats to a single receiver, rather than being moved through a transfer medium to one central location, as with parabolic troughs.

Power towers must be large to be economical. This is a promising technology for large-scale grid-connected power plants. Power towers are in the early stages of development compared with parabolic trough technology.

The U.S. Department of Energy, along with a number of electric utilities, built and operated a demonstration solar power tower near Barstow, California, during the 1980s and 1990s. Projects from private companies include:

- 5-Megawatt, two-tower project, built in the Mojave Desert in southern California in 2009
- 390-Megawatt, three-tower project being built in the Mojave Desert
- 110-Megawatt project located in Nevada

Nonconcentrating Solar Collectors

Solar Collectors Are Either Nonconcentrating or Concentrating

With nonconcentrating collectors, the collector area (the area that intercepts the solar radiation) is the same as the absorber area (the area absorbing the radiation). Flat-plate collectors are the most common type of nonconcentrating collector and are used when temperatures below about 200°F are sufficient. They are often used for heating buildings.

There are many flat-plate collector designs but generally all consist of:

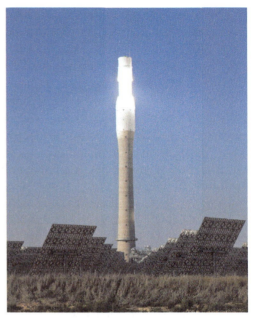

FIGURE 15.37 **Solar power tower.**

- A flat-plate absorber that intercepts and absorbs the solar energy
- A transparent cover(s) that allows solar energy to pass through but reduces heat loss from the absorber
- A heat-transport fluid (air or water) flowing through tubes to remove heat from the absorber, and a heat insulating backing

Solar thermal (heat) energy is often used for heating water used in homes and swimming pools and for heating the insides of buildings ("space heating"). Solar space heating systems can be classified as passive or active.

Passive space heating is what happens to your car on a hot summer day. The sun's rays heat up the inside of your car. In buildings, the air is circulated past a solar heat surface and through the building by convection (meaning that less dense warm air tends to rise while denser cool air moves downward). No mechanical equipment is needed for passive solar heating.

Active heating systems require a collector to absorb and collect solar radiation. Fans or pumps are used to circulate the heated air or heat absorbing fluid. Active systems often include some type of energy storage system.

Solar Energy and the Environment

Using solar energy does not produce air or water pollution and does not produce greenhouse gases, but using solar energy may have some indirect negative impacts on the environment. For example, there are some toxic materials and chemicals that are used in the manufacturing process of photovoltaic (PV) cells, which convert sunlight into electricity. Some solar thermal systems use potentially hazardous fluids to transfer heat. U.S. environmental laws regulate the use and disposal of these types of materials.

As with any type of power plant, large solar power plants can affect the environment where they are located. Clearing land for construction and the placement of the power plant may have long-term impacts on plant and animal life by reducing habitat areas for native plants and animals. Power plants may require water for cleaning solar collectors or concentrators and may require water for cooling turbine-generators. Using ground water or surface water in some arid locations with significant solar potential may affect the ecosystem. In addition, birds and insects can be killed if they fly into a concentrated beam of sunlight created by a solar power tower.

WASTE & POLLUTION

CHAPTER 16

Solid & Hazardous Waste

16.1 Introduction

The terms **trash** and **municipal solid waste** are often used interchangeably. The two terms are also applied interchangeably to mixtures of waste materials containing **hazardous waste**. There are many easily recognizable examples of municipal solid waste materials, such as leftover food, empty bottles, used newspapers, empty pizza boxes, grass clippings, used clothing, damaged furniture, packaging materials, paint cans, and dead batteries. Hazardous waste is different than municipal solid waste. Hazardous waste presents significant challenges for storage and collection, as hazardous waste products are suspected or known to be dangerous to humans and/or the environment. Examples of hazardous waste products include toxic wastes, barrels containing flammable chemicals, large volumes of paint cans, unused lead-based paint containers, and batteries.

Classifying Municipal and Hazardous Waste

It is important to recognize that some discarded materials, like paint and batteries, are considered municipal solid waste and hazardous waste. When the quantity or volume of the paint materials or batteries exceeds what is normally produced by a single- or multi-family home, then these types of waste are deemed inappropriate for treatment as municipal solid waste. These products, due to their volume, must be treated with greater caution and are treated as hazardous waste. Other products, like some very toxic solvents used for degreasing equipment in industrial operations, are only approved for industry use and should never be discarded in any quantity as municipal solid waste. Since no amount of these really toxic chemicals should be in a home environment, no amount is permissible as municipal solid waste, making these chemicals exclusively hazardous waste. For questions about what chemicals or products are acceptable for being discarded as municipal solid waste items, one should always contact the local waste management company or state/local environmental protection agency.

Any waste product deemed hazardous by law has one or more of the following properties:

1. Ignitable (easily flammable)
2. Reactive (unstable, could explode when interacting with chemicals)
3. Corrosive (destroys handles, containers, or even tissues/skin)
4. Toxic (can harm living organisms, including human life)

Waste Management History

Throughout human history, waste management has been part of the human experience. As civilizations became more established in single locations for longer periods of time, there was a greater desire and need to have waste removed from the city streets. Many ancient civilizations, particularly the Greeks, had designated locations for waste dumping outside the residential and business areas of the city. These cities and their residents knew that the buildup of waste in city streets would attract numerous nuisance animal species while also producing odors unpleasant for the residents and city guests.

Although many cities still struggled to keep their streets clean, the issue of waste accumulation peaked in Europe in the fourteenth century with the arrival of the Black Death (bubonic plague). The accumulation

Example Waste
Classification Scheme

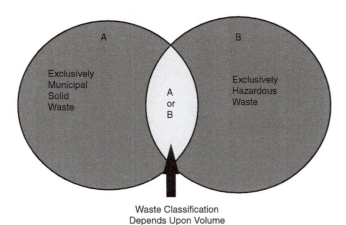

Waste Classification
Depends Upon Volume

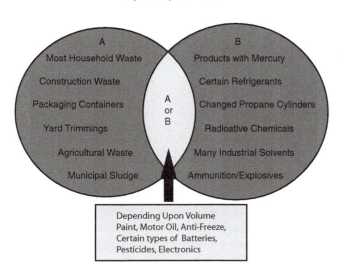

FIGURE 16.1 Classifying Waste Products.

of rubbish in the homes and streets attracted a variety of rodents, including rats. Some of these rats were carrying rat fleas that harbored the bacterium *Yersinia pestis*. When the rat hosts died, the fleas would periodically bite humans, thus infecting some humans with the plague bacteria. In medieval England, 1.5 million people were killed in 12 years by this bacterium, which is believed to have jumped to humans while circulating in rat and flea populations. This population loss in England represented 25% of the total population, which is estimated to have been 4 million. Inadequate sanitation, although possibly not well recognized then, is now widely recognized as a major contributor to this horrible event in human history.

By the 1700s, waste management was still largely underappreciated by many civilizations. However, with the advent of the Industrial Revolution, numerous new products were being produced and consumed in Western Europe and the fledgling United States. Accordingly, there was a tremendous increase in the amount of rubbish being created. Furthermore, the new and growing American democracy was fueled by an Industrial Revolution, and many discoveries contributed to a more enlightened and educated society. During this era, Benjamin Franklin and others advocated for the banning of public dumps in the city of Philadelphia so that it would be free from the frequent disease epidemics, and their "liberty of breathing freely in their own houses" would be restored.[1] Over the next 100 years, many cities wrestled with ideas and policies on how to best manage their growing trash problem.

One remedy for dealing with the looming trash problem was incineration. In 1885, New York City constructed an incinerator on Governors Island for the purpose of destroying trash by burning. Around the United States, hundreds of incinerators were put into operation over the next 80 years in urban areas. In rural areas, trash was burned on-site, and unburned products like glass were discarded at the back of properties or in uninhabited areas. By the late 1950s and early 1960s, U.S. and European scientists were concerned about the impact of air quality on human health. Citing a number of deaths and illnesses related to air quality, scientists believed that pollution controls were needed.[2] By the 1970s and 1980s, many laws, particularly the Clean Air Act, led to the regulation of these facilities for the protection of human health. The cost of compliance with the Clean Air Act was too expensive for many incinerators and public entities. In the interest of public health, and due to the inability of most cities to implement appropriate pollution controls, most incineration facilities were shut down.

Other alternatives were present, such as sanitary and open fills. All of these practices, including hog feeding (Figure 16.2), were gaining tremendous attention by the 1930s and 1940s (Figure 16.2). Hog feeding was even common in many large U.S. cities. The theory was that by feeding the hogs refuse, cities would be able to get rid

FIGURE 16.2 In the 1930s to 1950s, hogs were fed waste as a strategy for getting rid of it.

Shutterstock/Budimir Jevtic

of the garbage, while also being able to produce a valuable and edible product. Unfortunately, by the late 1950s, there was growing evidence that humans consuming pork from garbage-fed hogs were more likely to get ill with trichinosis, which led to the eventual banning of this process from public waste-collection entities.

With incineration and hog feeding deemed unacceptable to public health, the only remaining option for municipal waste was to store or bury it. Consequently, this approach is still used, and the majority of the solid waste generated by the United States is placed in sanitary landfills and is managed at great cost. Due to the expensive nature of collecting and managing the solid waste destined for landfills, there has been an emphasis placed on the three R's, which specifically encourages citizens, businesses, and organizations to *reduce, reuse, and recycle.*

Hazardous Waste History

Lead (Pb) and mercury (Hg) have been with American populations for many years; however, since the 1920s, the United States has observed major increases in its production of synthetic chemicals, many of which can be deemed hazardous waste. For example, the chemical trichloroethylene (TCE) began replacing dry-cleaning solvents in the 1930s. The U.S. military used TCE frequently for degreasing equipment during World War II and the Korean War. Industrial applications also increased substantially as the nation became more industrialized. Like TCE, many other chemicals were developed as America and the world became more industrialized. The health effects associated with human exposure to many of these chemicals were largely unknown or only associated with anecdotal evidence, as opposed to scientific study. Consequently, many of these hazardous waste chemicals were incinerated or placed in landfills just like the waste generated by households.

The Love Canal Disaster

The most historically important hazardous waste event in U.S. history may be the Love Canal disaster. During the 1890s, a plan was proposed by William T. Love to construct a canal to connect the Niagara River to Lake Ontario. The canal was to be constructed originally for hydropower generation, and then for shipping goods by water around the very large Niagara Falls. Plans eventually fell through, but approximately one mile of the canal was dug, 50 feet wide and approximately 25 feet in depth, on average. With little use for this trench, the city of Niagara Falls, New York, decided to dedicate it as a dump site. From the 1920s until the 1940s, only the city dumped waste into the trench; but during World War II, some military wastes associated with the war effort were also dumped in the trench. By the 1950s, the Hooker Chemical Company was granted permission to dump its waste at this location, with much of the waste being in 55-gallon drums, while the city and military stopped using the site. In 1953, the dumping ceased and the canal was buried. Within several years, the school board purchased the property for $1 and constructed two schools in the area. By the 1970s, other development had occurred in the area, including the construction of homes (Figure 16.3). By the mid-1970s media outlets in the area started reporting numerous birth defects, and advocates got the attention of the newly formed U.S.

FIGURE 16.3 **An aerial photograph from 1956 of Love Canal showing a school and houses built over and near a landfill. Niagara Falls, New York.**

Courtesy of EPA

Environmental Protection Administration. EPA investigations acknowledged a high frequency of chromosomal abnormalities in the people and discovered elevated concentrations of hazardous waste chemicals in the environment and even in the basements of homes. By 1978, many Americans viewed Love Canal as the worst public health disaster in American history. President Jimmy Carter declared the area a disaster area. The community in the vicinity of Love Canal was instructed to leave and was compensated by the U.S. government. The immediate site is now abandoned and has been a site of tremendous clean-up activities. The chemical company was recently ordered to pay $129 million in restitution above and beyond the homeowner lawsuits.[4]

The occurrence of events like the Love Canal disaster resulted in numerous health-related problems for the exposed people. This led to tremendous political pressure on elected officials in the Congress and on Presidents Ford and Carter to enact hazardous waste management legislation. Consequently, the Love Canal disaster brought about several major U.S. hazardous waste laws.

Waste Management Regulations

Following the growing acceptance of landfill disposal methods of solid waste, the United States passed the Solid Waste Disposal Act of 1965. This act essentially banned open-air burning of trash and required pollution controls to be utilized when incineration was being performed. It was rightfully presumed that municipalities would consider adopting alternative strategies for waste management when compliance with clean air laws related to incineration made it apparent that incineration would be very costly. In anticipation by the U.S. Environmental Protection Agency of alternative strategies being considered, the Solid Waste Disposal Act required all commercial, industrial, municipal, and household waste to be managed in an environmentally responsible manner for the protection of human health and the environment. Acknowledging some of the limitations of the Solid Waste Act, as brought to light by the Love Canal disaster, there was pressure from the citizenry to reduce waste and ensure that buried waste would be properly managed and safe for public health. Consequently, the new EPA (formed in 1970), working in concert with Congress, the president, and the states, created new legislation passed by Congress in 1976. The name of the new law, the Resource Conservation and Recovery Act, embodied the spirit of reducing waste and protecting the environment.

This new law, commonly called **RCRA** (pronounced rec•ruh), remains the most important law for governing the disposal and management of solid and hazardous waste in the United States. This law empowered and mandated the states to develop and follow plans for managing nonhazardous solid waste and municipal solid waste for the purposes of protecting the environment and human health. RCRA also created a national system for controlling and monitoring hazardous waste products from the time all new hazardous products were created until they were destroyed or ultimately disposed. This provision, pertaining to monitoring hazardous products from their creation until their destruction or ultimate disposal, is commonly called the "cradle to grave" provision (RCRA Subtitle C). RCRA also required underground storage tanks, or USTs, to be regulated. For example, gas stations with underground storage tanks are required by law to have their tanks inspected to ensure they are not leaking and are in good condition (Figure 16.4).

In summary, three programs were created by RCRA:

1. Solid waste program
2. Hazardous waste program
3. Underground storage tank program

One of the limitations of RCRA is that it was solely a preventive tool for preventing any future disasters that could be like Love Canal. RCRA dealt primarily with new waste generation; whereas there was a substantial need and desire to also deal with old hazardous waste sites created by companies,

FIGURE 16.4 **Gas truck filling underground tank.**

Shutterstock/American Spirit

communities, and even our state, local, and federal governments, including the Department of Defense. In 1980 the U.S. Congress, with support from President Carter, passed the Comprehensive Environmental Response, Compensation, and Liability Act, also known as CERCLA (pronounced sir•cluh). Under this law, the EPA has access to and the ability to obtain financial resources for pursuing litigation against negligent parties responsible for degrading the environment and the quality of human life. Using dollars obtained through litigation, known as "Superfund" dollars, EPA can clean-up **orphan sites**. EPA adds dollars to this appropriately named fund (Superfund) through enforcement actions, consent decrees (agreements with the parties brokered while taking action through the courts), and court orders. The Superfund approach authorized through CERCLA makes it possible for to carry out expensive clean-up efforts of dangerous orphan hazardous waste without directly increasing the tax burden on ordinary citizens.

16.2 Nonhazardous Solid Waste

Overview

Congress enacted the Solid Waste Disposal Act of 1965 to address the growing quantity of solid waste generated in the United States and to ensure its proper management. Subsequent amendments to the Solid Waste Disposal Act, such as RCRA, have substantially increased the federal government's involvement in solid waste management.

During the 1980s, solid waste management issues rose to new heights of public concern in many areas of the United States because of increasing solid waste generation, shrinking disposal capacity, rising disposal costs, and public opposition to the siting of new disposal facilities. These solid waste management challenges continue today, as many communities are struggling to develop cost-effective, environmentally protective solutions. The growing amount of waste generated has made it increasingly important for solid waste management officials to develop strategies to manage wastes safely and cost-effectively.

RCRA defines the term **solid waste** as:

- Garbage (e.g., milk cartons and coffee grounds)
- Refuse (e.g., metal scrap, wall board, and empty containers)
- Sludges from waste treatment plants, water supply treatment plants, or pollution control facilities (e.g., scrubber slags)
- Industrial wastes (e.g., manufacturing process wastewaters and nonwastewater sludges and solids)
- Other discarded materials, including solid, semisolid, liquid, or contained gaseous materials resulting from industrial, commercial, mining, agricultural, and community activities (e.g., boiler slags).

The definition of solid waste is not limited to wastes that are physically solid. Many solid wastes are liquid, while others are semisolid or gaseous.

The term solid waste, as defined by the Statute, is very broad, including not only the traditional nonhazardous solid wastes, such as municipal garbage and industrial wastes, but also hazardous wastes. Hazardous waste, a subset of solid waste, is regulated under RCRA Subtitle C. Household hazardous waste will be discussed in the next section.

Municipal Solid Waste

Municipal solid waste is a subset of solid waste and is defined as durable goods (e.g., appliances, tires, batteries), nondurable goods (e.g., newspapers, books, magazines), containers and packaging, food wastes, yard trimmings, and miscellaneous organic wastes from residential, commercial, and industrial nonprocess sources (see Figure 16-5).

Municipal solid waste generation has grown steadily over the past 49 years from 88 million tons per year (2.7 pounds per person per day) in 1960, to 243 million tons per year (4.3 pounds per person per day) in

2009. While generation of waste has grown steadily, recycling has also greatly increased. In 1960, only about 7 percent of municipal solid waste was recycled. By 2009, this figure had increased to 33.8 percent.

To address the increasing quantities of municipal solid waste, EPA recommends that communities adopt "integrated waste management" systems tailored to meet their needs. The term "integrated waste management" refers to the complementary use of a variety of waste management practices to safely and effectively handle the municipal solid waste stream. An integrated waste management system will contain some or all of the following elements: source reduction, recycling (including composting), waste combustion for energy recovery, and/or disposal by landfilling (see Figure 16-6). In designing systems, EPA encourages communities to consider these components in a hierarchical sequence. The hierarchy favors source reduction to reduce both the volume and toxicity of waste and to increase the useful life of manufactured products. The next preferred tier in the hierarchy is recycling, which includes composting of yard and food wastes. Source reduction and recycling are preferred over combustion and/or landfilling, because they divert waste from the third tier and they have positive impacts on both the environment and economy. The goal of EPA's approach is to use a combination of all

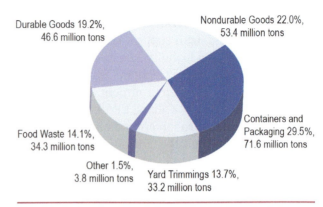

FIGURE 16.5 Products Generated in MSW by Weight, 2009 (total weight – 243 million tons).

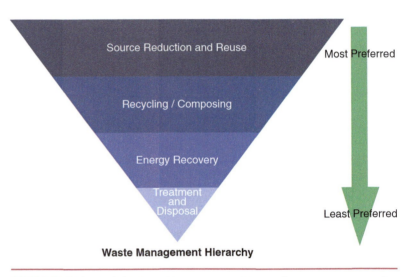

Waste Management Hierarchy

FIGURE 16.6 The Solid Waste Management Hierarchy.

these methods to safely and effectively manage municipal solid waste. EPA recommends that communities tailor their systems from the four components to meet their specific needs, looking first to source reduction, and second to recycling as preferences to combustion and/or landfilling.

Source Reduction

Rather than managing waste after it is generated, **source reduction** changes the way products are made and used in order to decrease waste generation. Source reduction, also called waste prevention, is defined as the design, manufacture, and use of products in a way that reduces the quantity and toxicity of waste produced when the products reach the end of their useful lives. The ultimate goal of source reduction is to decrease the amount and the toxicity of waste generated. Businesses, households, and all levels of government can play an active role in source reduction. Businesses can manufacture products with packaging that is reduced in both volume and toxicity. They also can reduce waste by altering their business practices (e.g., reusing packaging for shipping, making double-sided copies, maintaining equipment to extend its useful life, using reusable envelopes). Community residents can help reduce waste by leaving grass clippings on the lawn or composting them with other yard trimmings in their backyards, instead of bagging such materials for eventual disposal. Figure 16.7 Consumers play a crucial role in an effective source reduction program by purchasing products having reduced packaging or that contain reduced amounts of toxic constituents. This purchasing subsequently increases the demand for products with these attributes.

Recycling

Municipal solid waste **recycling** refers to the separation and collection of wastes, their subsequent transformation or remanufacture into usable or marketable products or materials, and the purchase of products made from recyclable materials. In 2009, 33.8 percent (82.0 million tons) of the municipal solid waste generated in the United States was recycled (see Figure 16-8). Solid waste recycling:

■ Preserves raw materials and natural resources
■ Reduces the amount of waste that requires disposal
■ Reduces energy use and associated pollution
■ Provides business and job opportunities
■ Reduces pollution associated with use of virgin materials.

Solid waste recycling also reduces greenhouse gas (GHG) emissions. For example, using the Waste Reduction Model (WARM), it can be calculated that the GHG savings of recycling 1 short ton of aluminum instead of landfilling it would be 3.71 metric tons of carbon equivalent (MTCE, Figure 16.9).

Communities can offer a wide range of recycling programs to their businesses and residents, such as drop-off centers, curbside collection, and centralized composting of yard and food wastes.

Additional information about recycling of common wastes and materials can be found at www.epa.gov/epawaste/conserve/materials.

Composting processes are designed to optimize the natural decomposition or decay of organic matter, such as leaves and food. Composting may be considered to be source reduction or recycling, as it reduced the amount of solid waste that ends up in the municipal waste system, it is a natural recycling process. Compost is a humus-like material that can be added to soils to increase soil fertility, aeration, and nutrient retention. Composting can serve as a key component of municipal solid waste recycling activities, considering that food and yard wastes accounted for nearly 28 percent of the total amount of municipal solid waste generated in 2009. Some communities are implementing large-scale composting programs in an effort to conserve landfill capacity.

There are three stages to recycling: collecting recyclable materials; manufacturing recycled-content products; and selling those products. Often symbolized by the chasing arrows logo, all three stages of the recycling process must work effectively in order to close the recycling loop. Creating markets for recycled materials—the third arrow—is critical to the success of the recycling process. Without a strong market for recycled materials, there is no incentive to collect recyclables and manufacture recycled-content products. Market development means fostering businesses that manufacture and market recycled-content products and strengthening consumer demand for those products. Market development can include, for example, expanding the processing and remanufacturing capacity of recycling businesses to handle the increasing volume of collected recyclables.

FIGURE 16.7 **Composting yard clippings and food wastes helps to reduce the amount of organic waste that ends up in a landfill.**

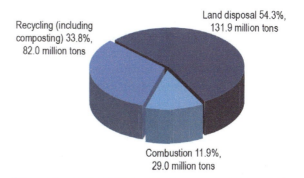

Recycling (including composting) 33.8%, 82.0 million tons

Land disposal 54.3%, 131.9 million tons

Combustion 11.9%, 29.0 million tons

FIGURE 16.8 **Management of MSW in the U.S., 2009 (total weight – 243 million tons).**

FIGURE 16.9 **This symbol is synonymous with recycling.**

Across America, more individuals, organizations, businesses, and government agencies are collecting materials for recycling than ever before, which keeps valuable resources out of landfills. However, resource recovery is only part of the recycling story. Recycling also creates new businesses that haul, process, and broker recovered materials, as well as companies that manufacture and distribute products made with recycled content. These recycling businesses put people to work. The jobs created by recycling businesses draw from the full spectrum of the labor market (ranging from low- and semi-skilled jobs to highly skilled jobs). Materials sorters, dispatchers, truck drivers, brokers, sales representatives, process engineers, and chemists are just some of the jobs needed in the recycling industry. Recycling is actively contributing to America's economic vitality.

Combustion

Confined and controlled burning, known as **combustion**, can not only decrease the volume of solid waste destined for landfills, but can also recover energy from the waste-burning process. Modern waste-to-energy facilities use energy recovered from combustion of solid waste to produce steam and electricity. In 2009, combustion facilities handled 11.9 percent (29.0 million tons) of the municipal solid waste generated (see Figure 16-4). Used in conjunction with source reduction and recycling, combustion can recover energy and materials and greatly reduce the volume of wastes entering landfills.

There are three types of technologies for the combustion of MSW: mass burn facilities, modular systems, and refuse derived fuel systems. Mass burn facilities are by far the most common types of combustion facilities in the United States. The waste used to fuel the mass burn facility may or may not be sorted before it enters the combustion chamber. Many advanced municipalities separate the waste on the front end to pull off as many recyclable products as possible. Modular systems are designed to burn unprocessed, mixed MSW. They differ from mass burn facilities in that they are much smaller and are portable and can be moved from site to site. Refuse derived fuel systems use mechanical methods to shred incoming MSW, separate out non-combustible materials, and produce a combustible mixture suitable as a fuel in a dedicated furnace or as a supplemental fuel in a conventional boiler system.

Additional information about energy recovery from waste can be found at www.epa.gov/waste/nonhaz/municipal/wte.

Landfilling

Landfilling of solid waste still remains the most widely used waste management method. Americans landfilled approximately 54.3 percent (131.9 million tons) of municipal solid waste in 2009 (see Figure 16-8). Many communities are having difficulties siting new landfills, largely as a result of increased citizen concerns about the potential risks and aesthetics associated with having a landfill in their neighborhood. To reduce risks to health and the environment, EPA developed minimum criteria that solid waste landfills must meet.

Climate Change

Solid waste disposal contributes to greenhouse gas emissions in a variety of ways. First, the anaerobic decomposition of waste in landfills produces methane, a greenhouse gas 21 times more potent than carbon dioxide. Second, the incineration of waste produces carbon dioxide as a by-product. In addition, the transportation of waste to disposal sites produces greenhouse gas emissions from the combustion of the fuel used in the equipment. Finally, the disposal of materials indicates that they are being replaced by new products; this production often requires the use of fossil fuels to obtain raw materials and manufacture the items.

Waste prevention and recycling—jointly referred to as waste reduction—help us better manage the solid waste we generate. But preventing waste and recycling also are potent strategies for reducing greenhouse gas emissions. Together, waste prevention and recycling reduce methane emissions from landfills; reduce emissions from incinerators; reduce emissions from energy consumption; and increase storage of carbon in trees.

Looking at each of these benefits in more detail, waste prevention and recycling (including composting) divert organic wastes from landfills, thereby reducing the methane released when these materials decompose. Recycling and waste prevention allow some materials to be diverted from incinerators and thus reduce greenhouse gas emissions from the combustion of waste. Recycling saves energy because manufacturing goods from recycled materials typically requires less energy than producing goods from virgin materials. Waste prevention is even more effective at saving energy. When people reuse things or when products are made with less material, less energy is needed to extract, transport, and process raw materials and to manufacture products. When energy demand decreases, fewer fossil fuels are burned and less carbon dioxide is emitted to the atmosphere. Finally, trees absorb carbon dioxide from the atmosphere and store it in wood, in a process called "carbon sequestration." Waste prevention and recycling of paper products allow more trees to remain standing in the forest, where they can continue to remove carbon dioxide from the atmosphere.

Additional information about the relationship between solid waste and climate change can be found at www.epa.gov/climatechange/wycd/waste.

Industrial Waste

Industrial waste is also a subset of solid waste and is defined as solid waste generated by manufacturing or industrial processes that is not a hazardous waste regulated under Subtitle C of RCRA. Such waste may include, but is not limited to, waste resulting from the following manufacturing processes: electric power generation; fertilizer or agricultural chemicals; food and related products or by-products; inorganic chemicals; iron and steel manufacturing (Figure 16.10); leather and leather products; nonferrous metals manufacturing or foundries; organic chemicals; plastics and resins manufacturing; pulp and paper industry; rubber and miscellaneous plastic products; stone, glass, clay, and concrete products; textile manufacturing; transportation equipment; and water treatment (Figure 16.10). Industrial waste does not include mining waste or oil and gas production waste.

Each year in the United States, approximately 60,000 industrial facilities generate and dispose of approximately 7.6 billion tons of industrial solid waste. Most of these wastes are in the form of wastewaters (97%). EPA has, in partnership with state and tribal representatives and a focus group of industry and public interest stakeholders, developed a set of recommendations and tools to assist facility managers, state and tribal regulators, and the interested public in better addressing the management of land-disposed, nonhazardous industrial wastes.

Similarly to municipal solid waste, EPA recommends considering pollution prevention options when designing an industrial waste management system. Pollution prevention will reduce waste disposal needs and can minimize impacts across all environmental media. Pollution preven-

FIGURE 16.10 **Manufacturing of steel, and many other products, creates solid waste that may not be classified as hazardous, but still must be disposed of properly.**

Shutterstock/zhu difeng

tion can also reduce the volume and toxicity of waste. Lastly, pollution prevention can ease some of the burdens, risks, and liabilities of waste management. As with municipal solid waste, EPA recommends a hierarchical approach to industrial waste management: first, prevent or reduce waste at the point of generation (source reduction); second, recycle or reuse waste materials; third, treat waste; and finally, dispose of remaining waste in an environmentally protective manner. There are many benefits of pollution prevention activities, including protecting human health and the environment, cost savings, simpler design and operating conditions, improved worker safety, lower liability, higher product quality, and improved community relations.

Criteria for Solid Waste Disposal Facilities

One of the initial focuses of the Solid Waste Disposal Act (as amended by RCRA) was to require EPA to study the risks associated with solid waste disposal and to develop management standards and criteria for solid waste disposal units (including landfills) in order to protect human health and the environment. This study resulted in the development of criteria for classifying solid waste disposal facilities and practices.

On September 13, 1979, EPA promulgated criteria to designate solid waste disposal facilities and practices which would not pose adverse effects to human health and the environment (Part 257, Subpart A). Facilities failing to satisfy the criteria are considered **open dumps** requiring attention by state solid waste programs. RCRA prohibits open dumping. As a result, open dumps had to either be closed or upgraded to meet the criteria for sanitary landfills. States were also required to incorporate provisions into their solid waste programs to prohibit the establishment of new open dumps. States have the option of developing standards more stringent than the Part 257, Subpart A criteria.

Solid waste disposal is overseen by the states, and compliance is assured through state-issued permits. EPA does not issue permits for solid waste management. Each state is to obtain EPA approval for their MSWLF permitting program. This approval process assesses whether a state's program is sufficient to ensure each landfill's compliance with the criteria. In states without an approved program, the federal criteria are self-implementing; the owner or operator of a solid waste disposal facility in those states must directly implement the requirements.

The criteria contain provisions designed to ensure that wastes disposed of in solid waste disposal units will not threaten endangered species, surface water, ground water, or flood plains. Further, owners and operators of disposal units are required to implement public health and safety precautions such as disease vector (e.g., rodents, flies, mosquitoes) controls to prevent the spread of disease and restrictions on the open burning of solid waste. In addition, facilities are required to install safety measures to control explosive gases generated by the decomposition of waste, minimize the attraction of birds to the waste disposed in the unit, and restrict public access to the facility. The criteria also restrict the land spreading of wastes with high levels of cadmium and polychlorinated biphenyls (PCBs) in order to adequately protect ground water from these dangerous contaminants.

Criteria for Solid Waste Disposal Facilities

In October 1988, EPA submitted a Report to Congress indicating that the United States was generating an increasing amount of municipal solid waste. The Report revealed that approximately 160 million tons of municipal solid waste were generated each year, 131 million tons of which were landfilled in just over 6,500 MSWLFs. EPA also reported that although these landfills used a wide variety of environmental controls, they may pose significant threats to ground water and surface water resources. For instance, rain water percolating through the landfills can dissolve harmful constituents in the waste and can eventually seep into the ground, potentially contaminating ground water. In addition, improperly maintained landfills can pose other health risks due to airborne contaminants, or the threat of fire or explosion.

To address these environmental and health concerns, and to standardize the technical requirements for these landfills, EPA promulgated revised minimum federal criteria in 1991.

A **municipal solid waste landfill** is defined as a discrete area of land or excavation that receives household waste. In 2009, there were approximately 1,908 MSWLFs in the continental United States.

The revised criteria from the EPA in 1991 addressed seven major aspects of MSWLFs:

- Location
- Operation
- Design
- Ground-water monitoring
- Corrective action
- Closure and post-closure activities
- Financial assurance.

The location criteria restrict where a MSWLF may be located. New landfills must meet minimum standards for placement in or near flood plains, wetlands, fault areas, seismic impact zones, and other unstable areas. Because some bird species are attracted to landfills, the criteria also restrict the placement of landfills near airports to reduce the bird hazards (i.e., collisions between birds and aircraft that may cause damage to the aircraft or injury to the passengers).

The operating criteria establish daily operating standards for running and maintaining a landfill. The standards dictate sound management practices that ensure protection of human health and the environment (Figure 16.11). The provisions require covering the landfill daily, controlling disease vectors, and controlling explosive gases. They also prohibit the open burning of solid waste and require the owner and operator of the landfill to control unauthorized access to the unit.

Leachate is formed when rain water filters through wastes placed in a landfill. When this liquid comes in contact with buried wastes, it leaches, or draws out, chemicals or constituents from those wastes. The design criteria require each new landfill to have a liner consisting of a flexible membrane and a minimum of two feet of compacted soil, as well as a leachate collection system. The liner and collection system prevent the potentially harmful leachate from contaminating the soil and ground water below the landfill. States with EPA-approved MSWLF permit programs can allow the use of an alternative liner design that controls ground-water contamination.

In order to check the performance of system design, MSWLF facility managers must also establish a ground-water monitoring program. Through a series of monitoring wells, the facility owner and operator are alerted if the landfill is leaking and causing contamination. If contamination is detected, the owner and operator of the landfill must perform **corrective action** (i.e., clean up the contamination caused by the landfill).

FIGURE 16.11 Environmental regulations require waste in a landfill to be covered to keep out pests, vented to prevent the buildup of explosive gases and lined to prevent leachate from infiltrating underlying soil and groundwater.

FIGURE 16.12 Pay-As-You-Throw (PAYT) programs are gaining popularity because they encourage people to reduce the amount of waste that they produce by charging them per bag collected.

When landfills reach their capacity and can no longer accept additional waste, the criteria stipulate procedures for properly closing the facility to ensure that the landfill does not endanger human health and the environment in the future. The **closure** activities at the end of a facility's use are often expensive, and the owner and operator must have the ability to pay for them. To this end, the criteria require each owner and operator to prove that they have the financial resources to perform these closure and **post-closure** activities, as well as any known corrective action.

Bioreactor Landfills

EPA is investigating the feasibility of improving how waste is managed in MSWLFs. Projects are being conducted to assess bioreactor landfill technology. A bioreactor landfill operates to more rapidly transform and degrade organic waste. The increase in waste degradation and stabilization is accomplished through the addition of liquid and air to enhance microbial processes. This bioreactor concept differs from the traditional "dry tomb" municipal landfill approach. Thus, decomposition and biological stabilization of the waste in a bioreactor landfill can occur in a shorter time frame than occurs in a traditional landfill. This provides a potential decrease in long-term environmental risks and landfill operating and post-closure costs.

Additional information about bioreactor landfills can be found at www.epa.gov/epawaste/nonhaz/ municipal/ landfill/bioreactors.htm.

16.3 Other Solid Waste Management Initiatives

Along with the Resource Conservation Challenge (which is discussed in Chapter IV), EPA has developed a number of solid waste management initiatives to help facilitate and promote proper waste management, and encourage source reduction by both industry and the public. Several such initiatives are described below.

Materials and Waste Exchanges

Materials and waste exchanges are markets for buying and selling reusable and recyclable commodities. Some are physical warehouses that advertise available commodities through printed catalogs, while others are simply websites that connect buyers and sellers. Some are coordinated by state and local governments. Others are wholly private, for-profit businesses. The exchanges also vary in terms of area of service and the types of commodities exchanged. In general, waste exchanges tend to handle hazardous materials and industrial process waste while materials exchanges handle nonhazardous items.

Typically, the exchanges allow subscribers to post materials available or wanted on a Web page listing. Organizations interested in trading posted commodities then contact each other directly. As more and more individuals recognize the power of this unique tool, the number of Internet-accessible materials exchanges continues to grow, particularly in the area of national commodity-specific exchanges.

A list of international and national exchanges, as well as state-specific exchanges can be found at www.epa. gov/epawaste/conserve/tools/exchange. htm.

Pay-As-You-Throw (PAYT)

Some communities are using economic incentives to encourage the public to reduce solid waste sent to landfills. One of the most successful economic incentive programs used to achieve source reduction and recycling is variable rate refuse pricing, or unit pricing. Unit pricing programs, sometimes referred to as pay-as-you-throw (PAYT) systems, have one primary goal: customers who place more solid waste at the curb for disposal pay more for the collection and disposal service. Thus, customers who recycle more have less solid waste for disposal and pay less. There are a few different types of unit pricing systems. Most require customers to pay a per-can or per-bag fee for refuse collection and require the purchase of a special bag or tag to place on bags or cans. Other systems allow customers to choose between different size containers and charge more for collection of larger containers. The number of PAYT communities in the United States grew to more than 7,133 in 2007, and the program served a population of 75 million today. Based on greenhouse gas calculations, PAYT is attributed with reducing an equivalent of over 10 million metric tons of carbon dioxide annually.

Additional information about unit pricing or pay-as-you-throw programs is available at www.epa.gov/payt.

Leftover household products that contain corrosive, toxic, ignitable, or reactive ingredients are considered to be household hazardous waste (HHW). Products, such as paints, cleaners, oils, batteries, and pesticides, that contain potentially hazardous ingredients require special care when you dispose of them.

Improper disposal of HHW can include pouring them down the drain, on the ground, into storm sewers, or in some cases putting them out with the trash. The dangers of such disposal methods might not be immediately obvious, but improper disposal of these wastes can pollute the environment and pose a threat to human health.

Certain types of HHW have the potential to cause physical injury to sanitation workers, contaminate septic tanks or wastewater treatment systems if poured down drains or toilets, and present hazards to children and pets if left around the house. Federal law allows disposal of HHW in the trash. However, many communities have collection programs for HHW to reduce the potential harm posed by these chemicals. EPA encourages participation in these HHW collection programs rather than discarding the HHW in the trash.

To avoid the potential risks associated with household hazardous wastes, it is important that people always monitor the use, storage, and disposal of products with potentially hazardous substances in their homes. Below are some tips for individuals to follow in their own homes:

- Use and store products containing hazardous substances carefully to prevent any accidents at home. Never store hazardous products in food containers; keep them in their original containers and never remove labels. Corroding containers, however, require special handling. Call your local hazardous materials official or fire department for instructions.
- When leftovers remain, never mix HHW with other products. Incompatible products might react, ignite, or explode, and contaminated HHW might become unrecyclable.
- Remember to follow any instructions for use and disposal provided on product labels.
- Call your local environmental, health, or solid waste agency for instructions on proper use and disposal and to learn about local HHW drop off programs and upcoming collection days.

Used Oil

Used oil is exactly what its name implies, any petroleum-based or synthetic that has been used. During normal use, impurities such as dirt, metal scrapings, water or chemicals, can get mixed in with the oil, so that in time, the oil no longer performs well. Eventually, this used oil must be replaced with virgin or re-refined oil to do the job correctly.

If you are one of the many people who change their own motor oil, you too need to know how to properly dispose of the used oil. Did you know that the used oil from one oil change can contaminate one million gallons of fresh water — a years' supply for 50 people!

- Used motor oil is insoluble, persistent and can contain toxic chemicals and heavy metals.
- It's slow to degrade.
- It sticks to everything from beach sand to bird feathers.
- It's a major source of oil contamination of waterways and can result in pollution of drinking water sources (Figure 16.13).

FIGURE 16.13 **Used motor oil, if not properly disposed, can get washed into rivers, or infiltrate into groundwater, potentially causing contamination of water supplies.**

If all the oil from American do-it-yourself oil changers were recycled, it would be enough motor oil for more than 50 million cars a year. Imagine how much foreign oil that would eliminate. Used motor oil from cars, trucks, boats, motorcycles, farm equipment and lawnmowers can be recycled and re-refined.

- On average, about four million people reuse motor oil as a lubricant for other equipment or take it to a recycling facility.
- Recycled used motor oil can be re-refined into new oil, processed into fuel oils and used as raw materials for the petroleum industry.
- One gallon of used motor oil provides the same 2.5 quarts of lubricating oil as 42 gallons of crude oil.
- Become used motor oil recycler number four million and one!

CFL Bulbs

CFLs, fluorescent bulbs and other bulbs that contain mercury, should be recycled where possible, rather than disposing of them in regular household trash. Recycling prevents the release of mercury into the environment.

CFLs and other fluorescent bulbs often break when thrown into a dumpster, trash can or compactor, or when they end up in a landfill or incinerator (Figure 16.13). Learn more about CFLs and mercury.

Other materials in the bulbs get reused. Recycling CFLs and other fluorescent bulbs allows the reuse of the glass, metals and other materials that make up fluorescent lights. Virtually all components of a fluorescent bulb can be recycled.

Your area may require recycling. Some states and local jurisdictions have more stringent regulations than U.S. EPA does, and may require that you recycle CFLs and other mercury-containing light bulbs. Visit search.Earth911.com to contact your local waste collection agency, which can tell you if such a requirement exists in your state or locality.

Many hardware supply stores and other retailers also offer in-store recycling and these are also listed on search.Earth911.com.

Finally, some bulb manufacturers and other organizations sell pre-labeled recycling kits that allow you to mail used bulbs to recycling centers. The cost of each kit includes shipping charges to the recycling center. You fill up a kit with old bulbs, seal it, and bring it to the post office or leave it for your postal carrier. Websites that provide more information about mail-back services.

FIGURE 16.14 Compact fluorescent lights (CFLs) are a source of mercury contamination if improperly disposed.

Shutterstock/SKChavan

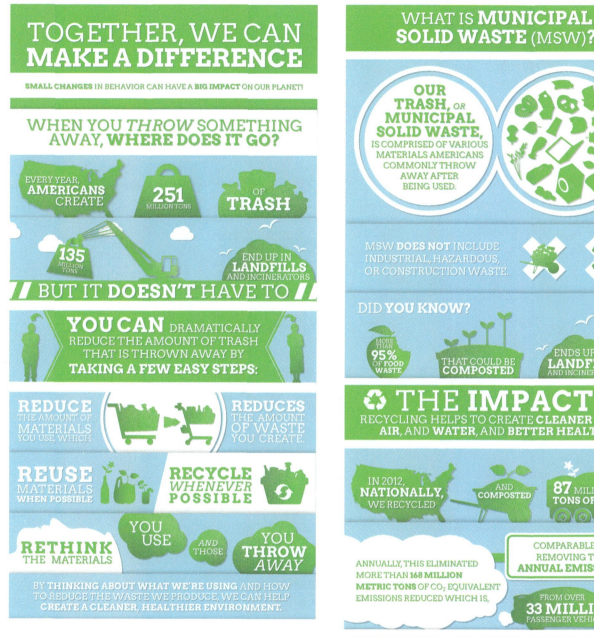

Courtesy of the United States Environmental Protection Agency

Figure 16.15 This infographic is available at http://www.epa.gov/solidwaste/nonhaz/municipal/infographic/index.htm. You can download it and share it with friends, family or colleagues. Also, it provides a nice summary of the topics in this chapter.

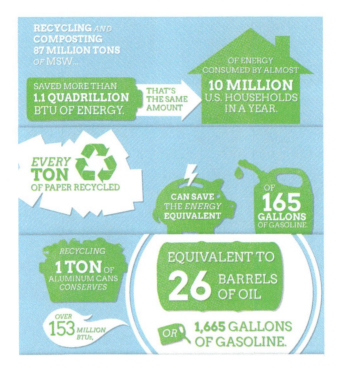

RECYCLING *AND* COMPOSTING 87 MILLION TONS *OF* MSW...

SAVED MORE THAN **1.1 QUADRILLION** BTU OF ENERGY.

THAT'S THE SAME AMOUNT

OF ENERGY CONSUMED BY ALMOST **10 MILLION** U.S. HOUSEHOLDS IN A YEAR.

EVERY **TON** OF PAPER RECYCLED

CAN SAVE THE *ENERGY* EQUIVALENT

OF **165** GALLONS OF GASOLINE.

RECYCLING **1 TON** *OF* ALUMINUM CANS *CONSERVES*

OVER **153** *MILLION* BTUs,

EQUIVALENT TO **26** BARRELS OF OIL

OR **1,665** GALLONS OF GASOLINE.

OUR **PROGRESS**

OVER THE LAST FEW DECADES, **THE RECYCLING, COMPOSTING, AND DISPOSAL** OF MSW HAS CHANGED.

4.38 LBS *IN* 2012

3.66 LBS *IN* 1980

WHILE THE AMOUNT OF **MSW PRODUCED WENT UP** PER PERSON PER DAY, **RECYCLING HAS ALSO INCREASED.**

<10% IN 1980

35% IN 2012

IN 2012, **OVER 19 MILLION TONS** OF YARD TRIMMINGS WERE COMPOSTED...

ALMOST FIVE TIMES AS MUCH AS IN 1990.

1990

THE TOTAL AMOUNT OF **MSW GOING TO LANDFILLS DROPPED**

2012

FROM **145.3 MILLION**

TO **135 MILLION TONS**

HOW **YOU** CAN HELP

IN **STORES:**

- Shop for products **made with recycled materials.**
- Buy items with **less packaging.**
- Buy refillable, **reusable containers.**
- Bring **reusable cloth or canvas bags to the grocery store.**
- Buy only **what you need** or **what you know you will use** (applies to food as well).

AT **HOME:**

- Use **energy-efficient light bulbs** and **rechargeable batteries.**
- **Reuse plastic bags.**
- Ask to **be removed from paper mailing lists.**
- Don't throw anything away that can be **reused or repaired.**
- For unwanted used electronics, **try upgrading the device to continue using it.** Otherwise, **donate or recycle it.**
- **Print on both sides of paper** (and use recycled paper).
- **Compost** your food scraps and yard waste.

MAKE A **DIFFERENCE TODAY!**

If we all take **small steps every day** to reduce the amount of waste we produce, **we can help protect our planet** for generations to come.

For more information, visit **www.epa.gov/recycle**.

United States Environmental Protection Agency

This infographic is based on data from EPA's 2012 MSW Characterization Report. For more information, see **http://go.usa.gov/bPxY**.

CHAPTER 17

Soil Pollution

17.1 Introduction to Soil Contamination

We recognize now that soil, water and atmospheric systems are all connected, however, this was not always the case. After the Clean Air Act (1970) and the Clean Water Act (1972) were established, there remained one frontier for pollution, i.e., one place where it was still legal to dump waste. Indeed as we learned in Chapter 16, it was legal to dump wastes (solid and liquid) onto the surface of soil on your own land until this loophole was closed in 1976 by the Resource Conservation and Recovery Act (RCRA). The topic of soil and water pollution through unsafe disposal practices, and the impacts of this practice on human health, was further publicized when it was featured in the popular movies A Civil Action (1998) and Erin Brockovich (2000).

What kind of contamination is it?

Soil contamination is either solid or liquid hazardous substances mixed with the naturally occurring soil. Usually, contaminants in the soil are physically or chemically attached to soil particles, or, if they are not attached, are trapped in the small spaces between soil particles.

How did it get there?

Soil contamination results when hazardous substances are either spilled or buried directly in the soil or migrate to the soil from a spill that has occurred elsewhere. For example, soil can become contaminated when small particles containing hazardous substances are released from a smokestack and are deposited on the surrounding soil as they fall out of the air. Another source of soil contamination could be water that washes contamination from an area containing hazardous substances and deposits the contamination in the soil as it flows over or through it.

How does it hurt animals, plants and humans?

Contaminants in the soil can hurt plants when they attempt to grow in contaminated soil and take up the contamination through their roots. Contaminants in the soil can adversely impact the health of animals and humans when they ingest, inhale, or touch contaminated soil, or when they eat plants or animals that have themselves been affected by soil contamination. Animals ingest and come into contact with contaminants when they burrow in contaminated soil. Humans ingest and come into contact with contaminants when they play in contaminated soil or dig in the soil as part of their work. Certain contaminants, when they contact our skin, are absorbed into our bodies. When contaminants are attached to small surface soil particles they can become airborne as dust and can be inhaled.

17.2 What Is Superfund?

The RCRA was instrumental in preventing further pollution of soils (and therefore the water systems connected to the soils, and the surrounding atmosphere), and it provided a mechanism to fine liable parties to clean up contaminated sites. However, it did not provide provisions for cleaning up the previously contaminated, and often abandoned, sites. The Comprehensive Environmental Response, Compensation and Liability Act of 1980 has been instrumental in cleaning up many contaminated sites that were abandoned by companies that went out of business or sold the land to unsuspecting developers who then gave up the land when they found out it was contaminated and could not be safely developed, including sites like Love Canal and Times Beach. This act established a cleanup fund called Superfund. It is critical to have a funding mechanism to remediate these contaminated sites, because cleanup costs are often millions of dollars and the process can take years to complete. In some cases, the sites are semi-permanently closed and monitored, when cleanup is deemed to be too expensive and the contamination can be contained on-site.

How Superfund Works

EPA can take three types of actions (known as response actions) to deal with abandoned hazardous waste sites: emergency responses, early actions, and long-term actions.

- An **emergency response** is used at a site that requires immediate action to eliminate serious risks to human health and the environment (for example, cleaning up chemicals spilled from an overturned truck on the highway).
- An **early action** is used at a site posing a threat in the near future by preventing human contact with contaminants such as providing clean drinking water to a neighborhood, removing hazardous materials from the site, or preventing contaminants from spreading. Early actions may last a few days or up to five years.
- A **long-term action** is used at a site where cleanup may take many years or decades (groundwater cleanups are frequently in this category). Often both early and long-term actions are performed at the same time. For example, leaking storage drums may be removed in an early action while contaminated soil is cleaned up under a long-term action.

EPA and state agencies find out about contaminated sites in many ways - a phone call from a citizen, a reported accident, or a planned search to discover sites. EPA first reviews a site to decide what needs to be done. EPA collects information, inspects the area, and talks to people in the community to find out how the site affects them and the environment. Some sites don't require any action; others may be cleaned up by state agencies or other programs. The remaining sites - those that meet certain requirements - call for action by the Federal government.

At sites that require Federal action, EPA conducts tests to find out what hazardous substances are present and how serious the risks may be to people and the environment. To figure out how dangerous a hazardous waste site is, EPA uses a "scorecard" called the **Hazard Ranking System (HRS)**. EPA uses the information it collected to score a site according to the risk it poses to people's health and the environment. Risk is a way of saying how likely it is that someone will be exposed to a hazardous substance, and the chance he or she will be harmed by that exposure. Environmental risk estimates

FIGURE 17.1   Oil containers left out can spill onto soil and contaminate underlying groundwater.

Shutterstock/Seksan Panpinyo

how likely it is that a hazardous substance will harm the environment (water, plants, animals, air, and so forth).

To give an HRS score to a site, EPA looks at **migration pathways** - how contamination moves in the environment. EPA examines four migration pathways:

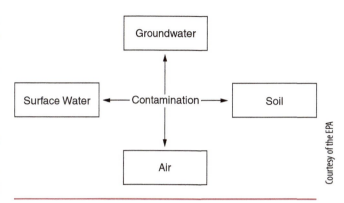

- **Groundwater** that may be used for drinking water
- **Surface water** (like rivers and lakes) used for drinking water, as well as for plant and animal habitats
- **Soil** that people may come in contact with or that can be absorbed lower in the food chain
- **Air** that carries contaminants.

FIGURE 17.2 **Migration pathways for contaminants to get into natural systems and then spread to others.**

Sites that get a high score on the HRS can be put on the National Priorities List (NPL). The NPL is a list of the nation's worst hazardous waste sites that qualify for extensive, long-term cleanup action under Superfund. Once a site is placed on the NPL, a more detailed study further pinpoints the cause and extent of contamination, as well as the risks posed to people and the environment nearby. This information helps identify different ways to clean up the site. EPA lists these cleanup options in a proposed plan for long-term cleanup. The proposed plan describes different ways to clean up the site and the choice EPA prefers. The public has at least 30 days to comment on the plan.

After EPA answers the public's concerns, it publishes a Record of Decision (ROD) that describes how it will clean up the site. The cleanup method is designed to address the unique conditions at the site. The design and actual cleanup is conducted either by EPA, a state, or the people responsible for contaminating the site.

Who pays for the cleanup?

The law says EPA can make the people responsible for contamination pay for site studies and cleanup work. EPA negotiates with these **Potentially Responsible Parties (PRPs)** to reach an agreement. Sometimes EPA pays for the cleanup out of a pool of money called the Superfund and then tries to make PRPs pay back the costs. Superfund money comes mainly from taxes on chemical and petroleum industries.

Who's involved in the cleanup?

Like any team, EPA works with many other groups to clean up a Superfund site:

- **Communities** provide important information about the site and surrounding area. They ensure that citizens' concerns are addressed during the cleanup process. They also help determine what cleanup method should be used and how the site will be used in the future.
- **States** work with EPA on making cleanup decisions, pay for 10 percent of cleanup costs in their state, and make sure sites are maintained after cleanup. They may also lead the cleanup activities. In addition, states address other sites on their own.
- **PRPs** are responsible for and are encouraged to participate in all aspects of the cleanup. If PRPs refuse or are unable to pay for a cleanup, EPA may either legally require them to perform certain cleanup tasks or conduct the cleanup itself and try to make the PRPs pay EPA back.
- **Federal agencies** can be involved in site cleanup either as site owners, as PRPs, or as EPA's partners in conducting the cleanup (the Department of Justice, for example, provides legal help).
- **Contractors** can be hired by the PRP or EPA, and usually perform much of the actual cleanup work at a Superfund site.

Over the past 20+ years, U.S. agencies (acting under the authority of CERCLA regulations) have located and analyzed tens of thousands of hazardous waste sites, protected people and the environment from contamination at the worst sites, and involved others in cleanup.

EPA's Office of Solid Waste and Emergency Response (OSWER) in Washington, D.C. oversees the Superfund program.

17.3 Cleaning Up the Soil

There are three general approaches to cleaning up contaminated soil: 1) soil can be excavated from the ground and be either treated or disposed; 2) soil can be left in the ground and treated in place; or 3) soil can be left in the ground and contained to prevent the contamination from becoming more widespread and reaching plants, animals, or humans. Containment of soil in place is usually done by placing a large plastic cover over the contaminated soil to prevent direct contact and keep rain water from seeping into the soil and spreading the contamination. Treatment approaches can include: air stripping, capping the site, precipitation of metal wastes, excavation, incineration of contaminated soil, pump-and-treat, phytoremediation, bioremediation. These will all be described in further detail.

Air Stripping

What Is it?

Air stripping removes volatile organic compounds from contaminated groundwater or surface water. Volatile organic compounds, or VOCs, are chemicals that quickly vaporize when heated or disturbed. For example, the gasoline fumes you smell at the gas station are VOCs volatilized in the air. In air stripping, these vapors are transferred from the water in which they were dissolved into a passing air stream. This air stream can be further processed to collect and reuse or destroy the VOCs.

How does it work?

The process starts when contaminated surface water or groundwater is pumped from large storage tanks into the top of a "packed tower" attached to an air blower. This packed tower is simply a large metal cylinder packed with material. While the stream of contaminated water is released into the tower, an air stream is pumped up from the bottom. The material in the tower forces the water stream to trickle down through various channels and air spaces. As the air stream flows upward, the contact of the two streams, called the "counter-current" flow, vaporizes the VOCs out of the water stream and collects them in the air stream, which exits the top of the tower.

How does the tower's packing material work? Inside the packed tower, the water stream forms a thin film on the material, which allows much more of the air stream to come into contact with the water stream. Using smaller packing material increases the surface area available for air stripping and improves the transfer process.

Why is it used?

Air stripping is useful for removing VOCs like trichloroethylene (TCE), dichloroethylene, chlorobenzene, and vinyl chloride. These are all hazardous substances. Equipment used in air stripping is relatively simple, allowing for quick start-up and shutdown and easy maintenance. This makes air stripping well-suited for hazardous waste site operations.

An important factor to consider in using air stripping is its impact on air pollution. Moving VOCs from water to air can mean just transferring pollution. Gases generated during air stripping may need to be collected and treated before they can be released into the air to avoid damaging the atmosphere.

How well does it work?

Air stripping can remove up to 98 percent of VOCs and up to 80 percent of certain semivolatile compounds. It does not work well for removing metals or inorganic contaminants.

Capping

What Is it?

Capping, often used in combination with other cleanup methods, covers buried wastes to prevent contaminants from spreading. Spreading, or migration, can be caused by rainwater or surface water moving through the site or by wind blowing dust off a site. Caps are usually made of a combination of materials like synthetic fibers, heavy clays, and sometimes concrete. Caps should minimize water movement through the wastes using efficient drainage; resist damage caused by settling; prevent standing water by funneling away as much water as the underlying filter or soils can handle; and allow easy maintenance.

How does it work?

The primary purpose of a cap is to minimize contact between rain or surface water and the buried waste. Two types of caps, multilayered and single-layer, serve this purpose.

- Multilayered caps have three layers: vegetation, drainage, and water-resistant. The vegetation layer prevents erosion of the cap's soils; the drainage layer channels rainwater away from the cap and keeps water from collecting on the water-resistant layer, which covers the waste.
- Single-layer caps are made of any material that resists water penetration. The most effective single-layer caps are made of concrete or asphalt, but single-layer caps are usually not acceptable unless there are valid reasons for not using a multilayer cap.

Why is it used?

Capping is required when contaminated materials are left in place at a site. It is used when the underground contamination is so extensive that excavating and removing it isn't practical, or when removing wastes would be more dangerous to human health and the environment than leaving them in place. Wells are often used to monitor groundwater where a cap has been installed to detect any movement of the wastes.

How well does it work?

Capping works well for sealing off contamination from the above-ground environment and reducing underground waste migration. Caps can be put over virtually any site, and can be completed relatively quickly. Capping materials and equipment are readily available. A multilayered cap will usually last for at least 20 years. Proper maintenance will make it last even longer.

Precipitation (NOT rain/snowfall) of Metal Wastes

What Is it?

Precipitation separates heavy metals from the water they contaminate. How does it work? Precipitation changes dissolved heavy metal contaminants into a solid form that can be separated from the water. Water contaminated with heavy metals is treated with chemicals, which cause the metal molecules to stick together and separate from the water. The solids are removed from the water. The clean water is then pumped back into the ground

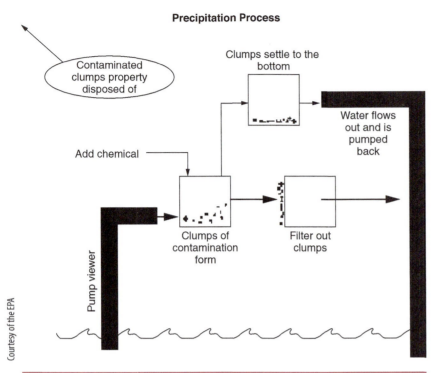

Precipitation Process

Contaminated clumps property disposed of

Clumps settle to the bottom

Water flows out and is pumped back

Add chemical

Clumps of contamination form

Filter out clumps

Pump viewer

Courtesy of the EPA

FIGURE 17.3   Precipitation Process.

and the collected metals are properly disposed of (Figure 17.3).

Why is it used?

Precipitation is easy to perform and can be used in many areas. It efficiently treats contaminated groundwater for reuse, and is one of the main methods of treating industrial wastewaters.

How well does it work?

Precipitation can be costly and is difficult to use if the water is contaminated with many types of metals, since different metals may interfere with one another and the cleanup process. Precipitation is very successful in treating wastewaters and is becoming the most widely selected cleanup method for removing heavy metals from groundwater.

Excavation

What Is it?

Excavation removes contaminated material from a hazardous waste site using heavy construction equipment, such as backhoes, bulldozers, and front loaders. At certain sites, specially designed equipment may be used to prevent the spread of contaminants.

How does it work?

The first step in excavation involves sampling and mapping the contaminated area to identify the contaminated area to be excavated. Samples are taken at several different depths in the same location so that a vertical, as well as horizontal, map of the contamination may be developed. Historical records, such as photographs or eye-witness accounts from past employees, and the contamination's effects on vegetation can also be used to pinpoint the area to be excavated.

Once the contamination is fully mapped, it can be removed. When hazardous waste has been buried in the ground a layer of soil may need to be removed before the waste is excavated. This layer, called overburden, is set aside and is later replaced in its original location. Contaminated materials are then dug up and loaded onto trucks for hauling. After it is cleaned up, excavated soil may be returned to its original location for use as backfill. Soil in the walls and bottom of the excavated area is tested to ensure that all contamination has been removed.

Excavation proceeds until cleanup goals are met. Excavation of hazardous waste or contaminated materials must be carefully planned to make sure contamination doesn't spread to clean areas of the site. For example, once excavation equipment is in a contaminated area, it must stay there until the work is completed, then thoroughly cleaned and decontaminated prior to leaving the site. In the event that contaminants have seeped into the groundwater, additional treatment may be necessary.

Why is it used?

Hazardous wastes can generally be excavated without exposing people near the site to contamination. Wastes can be removed for further treatment or disposal at an approved landfill. Excavation uses common construction equipment and is a widely used and accepted method of dealing with hazardous waste. Excavation is also relatively inexpensive compared to other, more complicated treatment technologies.

How well does it work?

Excavation is very effective in removing contamination and is commonly used at remediation sites. There are no strict limits on the types of wastes that can be excavated and removed. Concern for worker Human Health & Ecological Risk, however, may prevent excavation of explosive, reactive, or highly toxic waste material.

Incineration

What Is it?

Incineration involves burning hazardous wastes to destroy such organic compounds as dioxins and PCBs. Incinerators can handle many forms of waste, including contaminated soils, sludges, solids, and liquids. Incineration is not effective in treating inorganic substances such as hydrochloric acid, salts, and metals.

EPA establishes and specifies the conditions under which each incinerator can operate by issuing permits. A permit defines how the incinerator must operate, such as:

- Maximum carbon monoxide level in stack gases (gases from the combustion process which exit the stack after treatment by air pollution control devices)
- Maximum feed rates (how fast hazardous wastes are fed into the incinerator)
- Minimum burning temperature. The permit conditions are designed to deliver a "complete burn" of the hazardous waste. For example, a permit requires the waste feed to be cut off if burning conditions are not optimal.

How does it work?

Incineration uses high temperatures (between 1600°F and 2500°F) to degrade contaminants into nontoxic substances, such as water, carbon dioxide, and nitrogen oxides (nitrogen and oxygen). Properly done, high-temperature incineration can be an effective, odorless, and smokeless process. The process is illustrated in Figure 17.4

EPA incinerator regulations assume that all leftover ash and material removed from the incinerator are hazardous. Accordingly, they must be disposed of at a facility that has a permit to handle hazardous waste. In addition, water used in the incineration process must meet strict standards before it can be discharged to surface waters.

Why is it used?

Incineration can be a permanent waste disposal solution because it destroys wastes that would otherwise take up space in a landfill. Incineration effectively destroy over 99 percent of all organic compounds.

Incineration Process

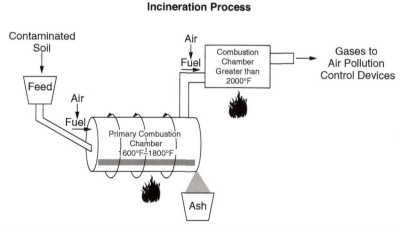

FIGURE 17.4 Incineration Process. To view a larger version of this diagram, visit http://www.epa.gov/superfund/students/clas_act/haz-ed/ff_08.htm

A common misconception is that the more toxic the chemical, the more difficult it is to burn. EPA's research shows that how toxic a chemical is does not relate to how easily it breaks down under heat during incineration.

How well does it work?

No incinerator can destroy 100 percent of the hazardous waste fed into it. Small amounts are released into the atmosphere through the incinerator stack or are mixed with the ash. EPA requires that each incinerator destroy and remove 99.99 percent of all hazardous waste it processes. For PCBs and dioxin wastes, the standard is 99.999 percent. When operated properly, incinerators can meet or exceed these standards. Operating at this level of efficiency, however, is a complex, highly technical task.

Pump-and-Treat

When soils are contaminated, a primary issue is that these pollutants will then contaminate the water in the soil (i.e., groundwater). As you will recall from our earlier chapter about water, **groundwater** is fresh water (from rain or melting ice and snow) that soaks into the soil and is stored in the tiny spaces (pores) between rocks and particles of soil. Groundwater accounts for nearly 95 percent of the nation's fresh water resources. It can stay underground for hundreds of thousands of years, or it can come to the surface and help fill rivers, streams, lakes, ponds, and wetlands. Groundwater can also come to the surface as a spring or be pumped from a well (Figure 17.5). Both of these are common ways we get groundwater to drink. About 50 percent of our municipal, domestic, and agricultural water supply is groundwater.

How does the ground store water?

Groundwater is stored in the tiny open spaces between rock and sand, soil, and gravel. How well loosely arranged rock (such as sand and gravel) holds water depends on the size of the rock particles. Layers of loosely arranged particles of uniform size (such as sand) tend to hold more water than layers of rock with materials of different sizes. This is because smaller rock materials settle in the spaces between larger rock materials, decreasing the amount of open space that can hold water. Porosity (how well rock material holds water) is also affected by the shape of rock particles. Round particles will pack more tightly than particles with sharp edges. Material with angular-shaped edges has more open space and can hold more water.

Groundwater is found in two zones. The **unsaturated zone**, immediately below the land surface, contains water and air in the open spaces, or pores. The **saturated zone**, a zone in which all the pores and rock fractures are filled with water, underlies the unsaturated zone. The top of the saturated zone is called the water table (Figure 17.6). The water table may be just below or hundreds of feet below the land surface.

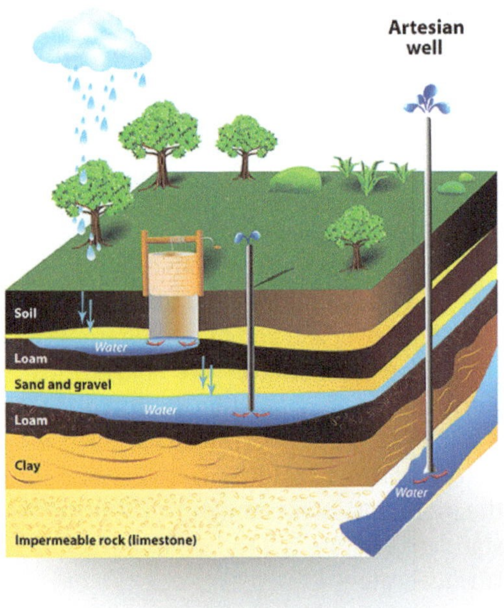

What is an aquifer?

Where groundwater can move rapidly, such as through gravel and sandy deposits, an **aquifer** can form. In an aquifer, there is enough groundwater that it can be pumped to the surface and used for drinking water, irrigation, industry, or other uses.

FIGURE 17.5 **Groundwater is stored in porous soils and rocks underground, called aquifers, which can act as steady sources of water.**

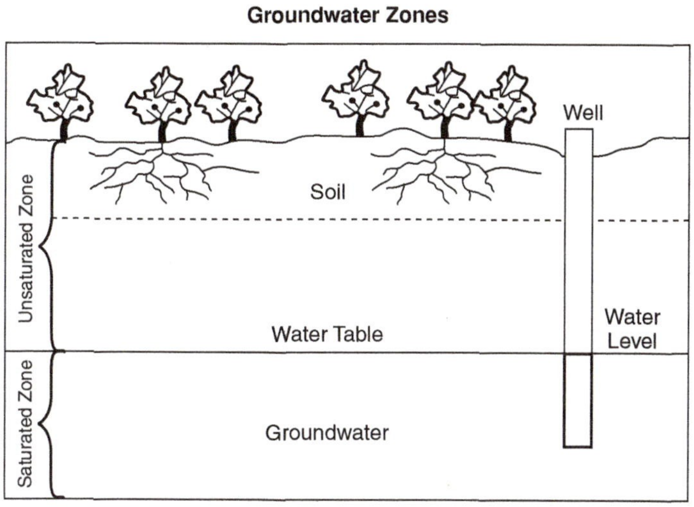

Groundwater Zones

Courtesy of the EPA

FIGURE 17.6   Groundwater Zones.

For water to move through underground rock, pores or fractures in the rock must be connected. If rocks have good connections between pores or fractures and water can move freely through them, we say that the rock is **permeable**. **Permeability** refers to how well a material transmits water. If the pores or fractures are not connected, the rock material cannot produce water and is therefore not considered an aquifer. The amount of water an aquifer can hold depends on the volume of the underground rock materials and the size and number of pores and fractures that can fill with water.

An aquifer may be a few feet to several thousand feet thick, and less than a square mile or hundreds of thousands of square miles in area. For example, the High Plains Aquifer underlies about 280,000 square miles in 8 states- Colorado, Kansas, Nebraska, New Mexico, Oklahoma, South Dakota, Texas, and Wyoming.

How does water fill an aquifer?

Aquifers get water from precipitation (rain and snow) that filters through the unsaturated zone. Aquifers can also receive water from surface waters like lakes and rivers. When the aquifer is full, and the water table meets the surface of the ground, water stored in the aquifer can appear at the land surface as a spring or seep. **Recharge** areas are where aquifers take in water; **discharge** areas are where groundwater flows to the land surface. Water moves from higher-elevation areas of recharge to lower-elevation areas of discharge through the saturated zone.

How is groundwater contaminated?

Groundwater can become contaminated in many ways. If surface water that recharges an aquifer is polluted, the groundwater will also become contaminated. Contaminated groundwater can then affect the quality of surface water at discharge areas. Groundwater can also become contaminated when liquid hazardous substances soak down through the soil into groundwater.

Contaminants that can dissolve in groundwater will move along with the water, potentially to wells used for drinking water. If there is a continuous source of contamination entering moving groundwater, an area of contaminated groundwater, called a **plume**, can form (Figure 17.7). A combination of moving groundwater and a continuous source of contamination can, therefore, pollute very large volumes and areas of groundwater. Some plumes at Superfund sites are several miles long. More than 88 percent of current Superfund sites have some groundwater contamination.

How do liquids contaminate groundwater?

Some hazardous substances dissolve very slowly in water. When these substances seep into groundwater faster than they can dissolve, some of the contaminants will stay in liquid form. If the liquid is less dense than water, it will float on top of the water table, like oil on water. Pollutants in this form are called **light non-aqueous phase liquids (LNAPLs)**. If the liquid is more dense than water, the pollutants are called **dense non-aqueous phase liquids (DNAPLs)**. DNAPLs sink to form pools at the bottom of an aquifer. These pools continue to contaminate the aquifer as they slowly dissolve and are carried away by moving groundwater. As DNAPLs flow

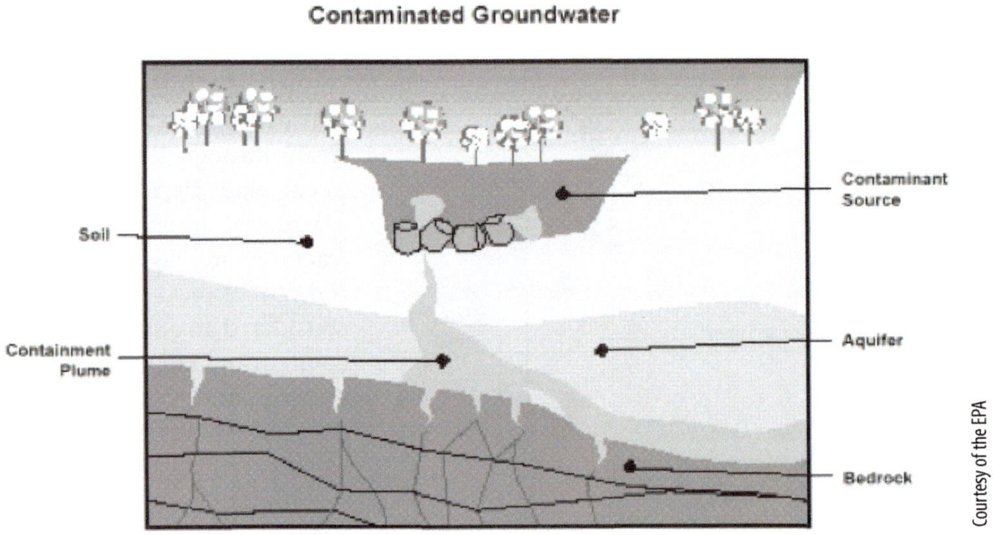

Contaminated Groundwater

FIGURE 17.7   Contaminants near the soil surface can infiltrate soils and spread to aquifers deep underground.

downward through an aquifer, tiny globs of liquid become trapped in the spaces between soil particles. This form of groundwater contamination is called **residual contamination**.

What affects groundwater contamination?

Many processes can affect how contamination spreads and what happens to it in the groundwater, potentially making the contaminant more or less harmful, or toxic. Some of the most important processes affecting hazardous substances in groundwater are advection, sorption, and biological degradation.

- **Advection** occurs when contaminants move with the groundwater. This is the main form of contaminant migration in groundwater.
- **Sorption** occurs when contaminants attach themselves to soil particles. Sorption slows the movement of contaminants in groundwater, but also makes it harder to clean up contamination.
- **Biological degradation** happens when microorganisms, such as bacteria and fungi, use hazardous substances as a food and energy source. In the process, contaminants break down and hazardous substances often become less harmful.

Why is cleaning up groundwater so hard?

Cleaning up contaminated groundwater often takes longer than expected because groundwater systems are complicated and the contaminants are invisible to the naked eye. This makes it more difficult to find contaminants and to design a treatment system that either destroys the contaminants in the ground or takes them to the surface for cleanup. Groundwater contamination is the reason for most of Superfund's long-term cleanup actions. Figure 17.8 illustrates the groundwater treatment strategy of "pump-and-treat" in action. Contaminated water is pumped out of a contaminant plume (often from the most concentrated area of the plume, so that the pumping does not cause the contamination to spread further).

Phytoremediation

Phytoremediation is the direct use of green plants and their associated microorganisms to stabilize or reduce contamination in soils, sludges, sediments, surface water, or ground water. First tested actively at waste sites in the early 1990s, phytoremediation has been tested at more than 200 sites nationwide. Because it is a natural

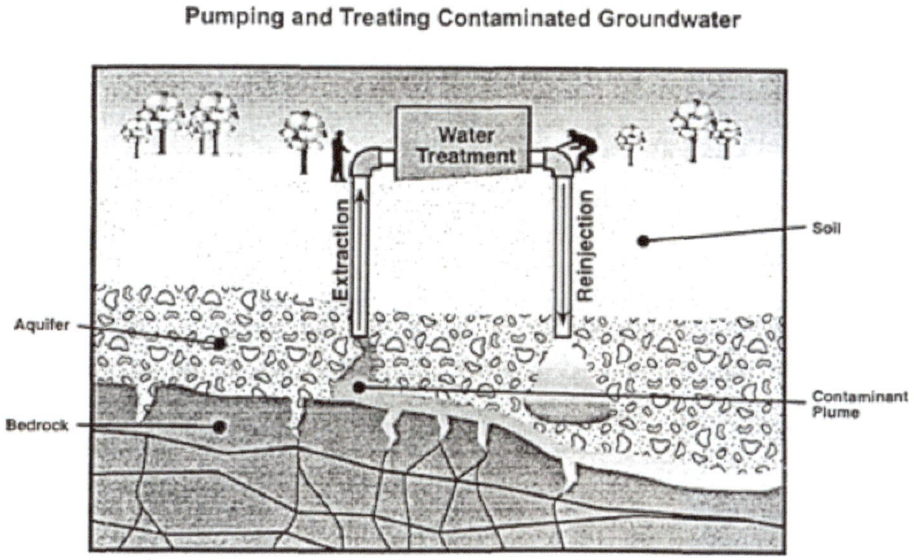

Pumping and Treating Contaminated Groundwater

Water Treatment

Extraction

Reinjection

Soil

Aquifer

Contaminant Plume

Bedrock

Courtesy of the EPA

FIGURE 17.8   Pumping and Treating Contaminated Groundwater.

FIGURE 17.9   Trees, grasses and other plants can be used to extract contaminants from soil. These plants can then be "harvested" and incinerated to collect the contaminants. Some contaminants are also converted into less toxic compounds during this process.

process, phytoremediation can be an effective remediation method at a variety of sites and on numerous contaminants. However, sites with low concentrations of contaminants over large cleanup areas and at shallow depths present especially favorable conditions for phytoremediation. Plant species are selected for use based on factors such as ability to extract or degrade the contaminants of concern, adaptation to local climates, high biomass, depth root structure, compatibility with soils, growth rate, ease of planting and maintenance, and ability to take up large quantities of water through the roots.

Oregon Poplar Site

The Oregon Poplar site, located in Clackamas, Oregon, comprises three to four acres within a vacant parcel located parallel to the small Mt. Scott Creek stream. The site had been an abandoned grassy field in a primarily commercial and light industrial area. Contaminants of concern at the site were primarily volatile organic compounds (VOCs), resulting most likely from illegal dumping activities. The ground water beneath the site is shallow (two to ten feet below the ground surface), locally confined, and in hydraulic contact with the Mt. Scott Creek stream. These characteristics along with low concentration of contaminants and little to no risk to human health make the site a good candidate for phytoremediation.

Hybrid poplar trees were planted on site in 1998 to remediate the ground water contaminated with VOCs. By July 30, 2002, the trees had not only survived, but shown considerable growth. Four of the larger trees were selected as the focus of sampling because their roots most likely be in contact with contaminated ground water. Although the water and soil samples proved inconclusive, tissue samples taken from the four trees indicated that the trees were actively removing VOCs from the ground water and soil. Although tissue samples from all sections of the trees revealed contaminant uptake, higher contaminant concentrations seemed to be found in the trunk rather than the leaf tissue. The success of the trees at the Oregon Poplar site supports the notion that phytoremediation may be an innovative technology worthy of nationwide consideration.

Bioremediation

Bioremediation uses microorganisms to degrade organic contaminants in soil, groundwater, sludge, and solids. The microorganisms break down contaminants by using them as an energy source or cometabolizing them with an energy source. To stimulate and enhance microbial activity, microorganisms (**bioaugmentation**) or amendments (**biostimulation**), such as air, organic substrates or other electron donors/acceptors, nutrients, and other compounds that affect and can limit treatment in their absence can be added. Biostimulation can be used where the bacteria necessary to degrade the contaminants are present but conditions do not favor their growth (e.g., anaerobic bacteria in an aerobic aquifer, aerobic bacteria in an anaerobic aquifer, lack of appropriate nutrients or electron donors/acceptors). Bioaugmentation can be used when the bacteria necessary to degrade the contaminants do not occur naturally at a site or occur at too low of a population to be effective. Biostimulation and bioaugmentation can be used to treat soil and other solids, groundwater, or surface water (EPA 2006).

Bioremediation may be conducted *in situ* or *ex situ*. In situ processes treat soil and groundwater in place, without removal or transportation offsite. This approach may be advantageous since the costs of materials handling and some environmental impacts may be reduced. However, *in situ* processes may be limited by the ability to control or manipulate the physical and chemical environment during bioremediation. *Ex situ processes*, on the other hand, involve the removal of the contaminated media to a treatment area (EPA 2006).

The Superfund cleanup process begins with site discovery or notification to EPA of possible releases of hazardous substances. Sites are discovered by various parties, including citizens, State agencies, and EPA Regional offices. Once discovered, sites are entered into the Comprehensive Environmental Response, Compensation, and Liability Information System (CERCLIS), EPA's computerized inventory of potential hazardous substance release sites (search CERCLIS for hazardous waste sites). Some sites may be cleaned up under other authorities. EPA then evaluates the potential for a release of hazardous substances from the site through these steps in the Superfund cleanup process. Community involvement, enforcement, and emergency response can occur at any time in the process. A wide variety of characterization, monitoring, and remediation technologies are used through the cleanup process. This process is summarized in the table below.

TABLE 17.1   The Superfund cleanup process

PA/SI	Preliminary Assessment/Site Inspection Investigations of site conditions. If the release of hazardous substances requires immediate or short-term response actions, these are addressed under the Emergency Response program of Superfund.
NPL Listing	National Priorities List (NPL) Site Listing Process A list of the most serious sites identified for possible long-term cleanup.
RI/FS	Remedial Investigation/Feasibility Study Determines the nature and extent of contamination. Assesses the treatability of site contamination and evaluates the potential performance and cost of treatment technologies.
ROD	Records of Decision Explains which cleanup alternatives will be used at NPL sites. When remedies exceed 25 million, they are reviewed by the National Remedy Review Board.
RD/RA	Remedial Design/Remedial Action Preparation and implementation of plans and specifications for applying site remedies. The bulk of the cleanup usually occurs during this phase. All new fund-financed remedies are reviewed by the National Priorities Panel.
Construction Completion	Construction Completion Identifies completion of physical cleanup construction, although this does not necessarily indicate whether final cleanup levels have been achieved.
Post Construction Completion	Post Construction Completion Ensures that Superfund response actions provide for the long-term protection of human health and the environment. Included here are Long-Term Response Actions (LTRA), Operation and Maintenance, Institutional Controls, Five-Year Reviews, Remedy Optimization.
NPL Delete	National Priorities List Deletion Removes a site from the NPL once all response actions are complete and all cleanup goals have been achieved.
Reuse	Site Reuse/Redevelopment Information on how the Superfund program is working with communities and other partners to return hazardous waste sites to safe and productive use without adversely affecting the remedy.

CHAPTER 18

Water Pollution & Wastewater

18.1 Measuring Water Quality

In the U.S., the U.S. Geological Survey and other state and municipal agencies have been measuring water for decades. Millions of measurements and analyses have been made (Figure 18.1). Some measurements are taken almost every time water is sampled and investigated, no matter where in the U.S. the water is being studied. Even these simple measurements can sometimes reveal something important about the water and the environment around it.

The results of a single measurement of a water's properties are actually less important than looking at how the properties vary over time. For example, if you take the pH of the creek behind your school and find that it is 5.5, you might say "Wow, this water is acidic!" But, a pH of 5.5 might be "normal" for that creek. It is similar to how my normal body temperature (when I'm not sick) is about 97.5 degrees, but my third-grader's normal temperature is "really normal"—right on the 98.6 mark. As with our temperatures, if the pH of your creek begins to change, then you might suspect that something is going on somewhere that is affecting the water, and possibly, the water quality. So, often, the *changes* in water measurements are more important than the actual measured values.

pH is only one measurement of a water body's health; there are others, too.

Water Temperature

Water temperature is not only important to swimmers and fisherman, but also to industries and even fish and algae. A lot of water is used for cooling purposes in power plants that generate electricity (Figure 18.2). They need cool water to start with, and they generally release warmer water back to the environment. The temperature of the released water can affect downstream habitats. Temperature also can affect the ability of water to hold oxygen as well as the ability of organisms to resist certain pollutants.

pH

pH is a measure of how acidic/basic water is. The range goes from 0–14, with 7 being neutral. pH values of less than 7 indicate acidity, whereas a pH of greater than 7 indicates a base. pH is really a measure of the relative amount of free hydrogen and hydroxyl ions in the water. Water that has more free hydrogen ions is acidic, whereas water that has more free hydroxyl ions is basic. Since pH can be affected by chemicals in the water, pH is an important indicator of water that is changing chemically. pH is reported in "logarithmic units," like the Richter scale, which measures earthquakes. Each number represents a 10-fold change in the acidity/basicness of the water. Water with a pH of five is ten times more acidic than water having a pH of six.

Pollution can change a water's pH, which in turn can harm animals and plants living in the water. For instance, water coming out of an abandoned coal mine can have a pH of 2, which is very acidic and would definitely affect any fish who were present in that water body! By using the logarithm scale, this mine-drainage water would be 100,000 times more acidic than neutral water—so stay out of abandoned mines.

Sections 18.1 and 18.6 courtesy of the United States Geologic Survey
Sections 18.4 and 18.5 Copyright © 2013 Bridgepoint Education. Reprinted by permission.
Sections 18.2 and 18.3 courtesy of the United States Environmental Protection Agency

Courtesy of United States Geologic Survey

FIGURE 18.1   A USGS employee is measuring water quality properties from a stream sample.

Courtesy of United States Geologic Survey

FIGURE 18.2   Cooling towers from power plants can release water to a stream or lake that is warmer than the resident water, elevating the temperature in the body of water.

Specific Conductance

Specific conductance is a measure of the ability of water to conduct an electrical current. It is highly dependent on the amount of dissolved solids (such as salt) in the water. Pure water, such as distilled water, will have a very low specific conductance, and sea water will have a high specific conductance. Rainwater often dissolves airborne gasses and airborne dust while it is in the air, and thus often has a higher specific conductance than distilled water. Specific conductance is an important water-quality measurement because it gives a good idea of the amount of dissolved material in the water (Figure 18.3).

High specific conductance indicates high dissolved-solids concentration; dissolved solids can affect the suitability of water for domestic, industrial, and agricultural uses. At higher levels, drinking water may have an unpleasant taste or odor or may even cause gastrointestinal distress. Additionally, high dissolved-solids concentration can cause deterioration of plumbing fixtures and appliances. Relatively expensive water-treatment processes, such as reverse osmosis, are needed to remove excessive dissolved solids from water.

Agriculture also can be adversely affected by high-specific-conductance water, as crops cannot survive if the water they use is too saline, for instance. Agriculture can also be the cause of increases in the specific conductance of local waters. When water is used for irrigation, part of the water evaporates or is consumed by plants, concentrating the original amount of dissolved solids in less water; thus, the dissolved-solids concentration and the specific conductance in the remaining water is increased. The remaining higher specific-conductance water reenters the river as irrigation-return flow. In a USGS study in Colorado, USA, specific conductance was found to vary during the year as a result of the temporal variability of streamflow. As this chart shows, specific conductance generally was lowest in the Arkansas RIver near Avondale, Colorado, in May to August, when streamflow generally was largest, and increased with decreasing streamflow in the fall, winter, and spring.

Often in school, students do an experiment where they connect a battery to a light bulb and run two wires from the battery into a beaker of water. When the wires are put into a beaker of distilled water, the light will not light. But, the bulb does light up when the beaker contains salt water (saline). In the saline water, the salt has dissolved, releasing free electrons, and the water will conduct an electrical current.

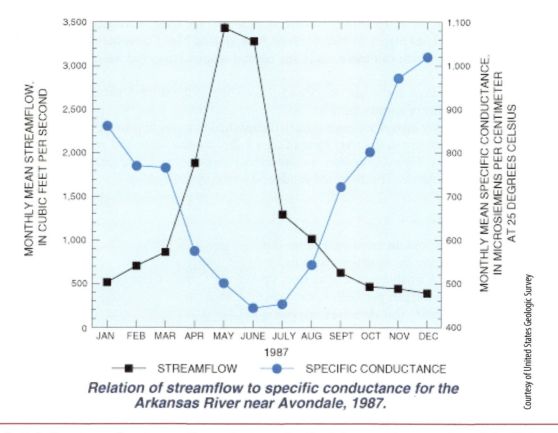

Courtesy of United States Geologic Survey

Relation of streamflow to specific conductance for the
Arkansas River near Avondale, 1987.

FIGURE 18.3   **Relating streamflow to specific conductance for the Arkansas River near Avondale, CO in 1987.**

Turbidity

Turbidity is the amount of particulate matter that is suspended in water. Turbidity measures the scattering effect that suspended solids have on light: the higher the intensity of scattered light, the higher the turbidity. Material that causes water to be turbid include:

- clay
- silt
- finely divided organic and inorganic matter
- soluble colored organic compounds
- plankton
- microscopic organisms

Courtesy of United States Geologic Survey

FIGURE 18.4   **The murky river in this picture is very turbid, meaning that light cannot reach very far through the water profile.** Aquatic plants depend on light in order to grow.

Turbidity makes the water cloudy or opaque. Figure 18.4 shows highly turbid water from a tributary (where construction was probably taking place) flowing into the less turbid water of the Chattahoochee River in Georgia. Turbidity is measured by shining a light through the water and is reported in nephelometric turbidity units (NTU). During periods of low flow (base flow), many rivers are a clear green color, and turbidities are low, usually less than 10 NTU. During a rainstorm, particles from the surrounding land are washed into the river making the water a muddy brown color, indicating water that has higher turbidity values. Also, during high flows, water velocities are faster and water volumes are higher, which can more easily stir up and suspend material from the stream bed, causing higher turbidities.

Turbidity can be measured in the laboratory and also on-site in the river. A handheld turbidity meter can be used to measure turbidity of a water sample.

State-of-the-art turbidity meters are beginning to be installed in rivers to provide an instantaneous turbidity reading. They read turbidity in the river by shining a light into the water and reading how much light is reflected back to the sensor. They also contain a conductivity sensor to measure electrical conductance of the water, which is strongly influenced by dissolved solids and a temperature gauge.

Dissolved Oxygen

Although water molecules contain an oxygen atom, this oxygen is not what is needed by aquatic organisms living in our natural waters. A small amount of oxygen, up to about ten molecules of oxygen per million of water, is actually dissolved in water. This dissolved oxygen is breathed by fish and zooplankton and is needed by them to survive. You can't tell by looking at water that there are oxygen molecules dissolved in it. Likewise, if you look at a closed bottle of a soft drink, you don't see the carbon dioxide dissolved in that until you shake it up and open the top.

Rapidly moving water, such as in a mountain stream or large river, tends to contain a lot of dissolved oxygen, while stagnant water contains little. Bacteria in water can consume oxygen as organic matter decays. Thus, excess

FIGURE 18.5 **Eutrophic conditions, Hartbees River, South Africa.**

National Eutrophication Monitoring Programme

organic material in our lakes and rivers can cause an oxygen-deficient situation to occur. Aquatic life can have a hard time in stagnant water that has a lot of rotting, organic material in it, especially in summer, when dissolved-oxygen levels are at a seasonal low.

As the amount of dissolved oxygen drops below normal levels in water bodies, the water quality is harmed and creatures begin to die off. Indeed, a water body can "die", a process called **eutrophication**.

Hardness

The amount of dissolved calcium and magnesium in water determines its "hardness." Water hardness varies throughout the United States. If you live in an area where the water is "soft," then you may never have even heard of water hardness. But, if you live in Florida, New Mexico, Arizona, Utah, Wyoming, Nebraska, South Dakota, Iowa, Wisconsin, or Indiana, where the water is relatively hard, you may notice that it is difficult to get a lather up when washing your hands or clothes. And, industries in your area might have to spend money to soften their water, as hard water can damage equipment. Hard water can even shorten the life of fabrics and clothes! Does this mean that students who live in areas with hard water keep up with the latest fashions since their clothes wear out faster?

Suspended Sediment

Suspended sediment is the amount of soil moving along in a stream. It is highly dependent on the speed of the water flow, as fast-flowing water can pick up and suspend more soil than calm water. During storms, soil is washed from the stream banks into the stream. The amount that washes into a stream depends on the type of land in the river's watershed and the vegetation surrounding the river.

If land is disturbed along a stream and protection measures are not taken, then excess sediment can harm the water quality of a stream. You›ve probably seen those short, plastic fences that builders put up on the edges of the property they are developing. These silt fences are supposed to trap sediment during a rainstorm and keep it from washing into a stream, as excess sediment can harm the creeks, rivers, lakes, and reservoirs.

Sediment coming into a reservoir is always a concern; once it enters it cannot get out—most of it will settle to the bottom. Reservoirs can "silt in" if too much sediment enters them. The volume of the reservoir is reduced, resulting in less area for boating, fishing, and recreation, as well as reducing the power-generation capability of the power plant in the dam.

18.2 Types of Water Pollution

General water quality is impacted when water quality properties change from the "normal" properties to which life is adapted in that water body. These changes can occur through natural processes, such as a 100-year flood that washes high levels of sediments into a river. It can also be human caused, for example, when pesticides are washed off fields into nearby rivers, or when warm water is discharged into a river from an electrical power plant. Water is polluted when harmful compounds or microorganisms are added to it though natural or human activities. The three main types of water pollution are the addition of (1) pathogenic microorganisms or parasites, (2) toxic compounds, and (3) excessive nutrients. Other forms include thermal and sediment pollution. Microorganisms and parasites are a concern for human health because of the diseases they cause. These diseases are called **waterborne** because they involve contact with or consumption of infected water. Waterborne diseases are estimated to cause 1.8 million human deaths per year and about 88% of the disease burden in the world.

Pollution can enter a body of water from a single **point source** or from multiple **nonpoint sources** (Figure 18.6). A point source is a pollution source that can be traced to its origin, such as a discharge pipe from a factory sending pollutants into the water. Nonpoint sources are much harder to trace and locate, and thus reducing the pollution from them can be very difficult. Take, for example, the pollution from a roadway that runs along a river. As cars drive along this road they deposit rubber, oil, gasoline, and other wastes on the road surface. At times it rains, and the rainwater flows from the road to the river carrying the pollution with it into

FIGURE 18.6 Point source pollution is when we can identify a single location where the pollution originates, such as the pipes dumping waste into the water body in the photo on the left. Nonpoint source pollution is when we cannot identify exactly where pollutants came from. For example, in the photo on the right, if excess nutrients were identified in the river, those polluting nutrients may have originated from fertilized yards, parks, or golf courses in the city on the left side of the river, or they may have originated in the farm fields on the right side of the river. Left photo courtesy of Environmental Protection Agency, right photo courtesy of USDA Natural Resources Conservation Service.

the river. This problem can be minimized if a barrier of plants, known as a riparian buffer, is grown between the road and the river. The buffer slows water flow, allowing the plants and soil to absorb and hopefully mitigate the pollution. The pollution may just collect in another area, but what happens depends on the pollutants present as well as the pollution load. Nonpoint source pollution can also be a result of rainfall or snowmelt moving over and through soil. As the water moves, it picks up and carries pollutants, finally depositing them into lakes, rivers, wetlands, coastal waters, and groundwaters.

All water carries some level of impurities. Trying to purify water completely is almost impossible due to the extremely strong attraction of water to other polar compounds. Even the best water purification systems will still produce water with some impurities in the range of 1–5 parts per billion (ppb). These systems are very expensive to operate, only produce limited quantities of water, and are used in laboratories and for the production of items requiring high purity like drugs. This means that water quality is a relative term. Definitions of water quality usually refer to changes in the water content due to human activities and whether the materials added change the way the water can be used. If you look at a swamp, the water is full of organic material and living organisms varying from bacteria to alligators. The addition of heavy metals from a mining site or oil from a leaky well could seriously harm life in the swamp. Wetlands are important in cleaning and recycling water, and pollution in either of these forms would severely damage the swamp environment. Therefore, when we talk about pollution of water bodies, we must consider whether those bodies of water will still perform the same functions they did prior to being polluted. Is this lake still producing fish, do any of the fish contain high levels of a pollutant like mercury, can we still safely consume the fish? Can we drink the water from the river or is it now toxic? The water does not have to have obvious problems, such as being cloudy or smelly, to be a health hazard to humans, and the health problem does not need to affect everyone who drinks the water. Birth control drugs (made of the female hormones estrogen and progesterone) have been discovered to pass through women's bodies and out in their urine. These hormones, even after passing through a functional wastewater treatment facility, are still functional and act on the organisms living in water where these pollutants are introduced. They could possibly affect people who use the water for their drinking source. These hormones and other polluting compounds that act like hormones have been shown to cause a feminization of male reptiles, fish, and amphibians living in the water. This has reduced some of these populations by causing the males to become nonreproductive.

Almost anything can be a water pollutant if it can dissolve in water, stay suspended in the water column, or cover the surface of the water and prevent air from interacting with the water. The first types of pollutants are those that dissolve in water. They have some polar properties and can fit into the hydrogen bonding of the water molecules. Nonpolar compounds like oils and fats do not dissolve in water. These form droplets or a layer on top of the water. A layer of oil from an oil spill can form a barrier that stops oxygen from diffusing into the water. Pollutants can also be solids that cannot dissolve like sand, silt, or clay particles. These can cloud the water or clog fish gills and reduce the penetration of light, thereby reducing photosynthesis. Once a pollutant enters the water, it can be distributed throughout the hydrologic system. It can enter soil or rock formations and be left there. Generally, the evaporation process will clean the evaporated water molecules of a pollutant, but most of the pollutant remains in and contaminates the area where the evaporation occurred.

Nutrient Pollution

Nutrient pollution is one of America's most widespread, costly and challenging environmental problems, and is caused by excess nitrogen and phosphorus in the air and water.

Nitrogen and phosphorus are nutrients that are natural parts of aquatic ecosystems. Nitrogen is also the most abundant element in the air we breathe. Nitrogen and phosphorus support the growth of algae and aquatic plants, which provide food and habitat for fish, shellfish and smaller organisms that live in water.

But when too much nitrogen and phosphorus enter the environment—usually from a wide range of human activities—the air and water can become polluted. Nutrient pollution has impacted many streams, rivers, lakes, bays and coastal waters for the past several decades, resulting in serious environmental and human health issues, and impacting the economy.

Too much nitrogen and phosphorus in the water causes algae to grow faster than ecosystems can handle. Significant increases in algae harm water quality, food resources and habitats, and decrease the oxygen that fish and other aquatic life need to survive (Figure 18.7). Large growths of algae are called algal blooms and they can severely reduce or eliminate oxygen in the water, leading to illnesses in fish and the death of large numbers of fish (a process called eutrophication, as discussed earlier). Some algal blooms are harmful to humans because they produce elevated toxins and bacterial growth that can make people sick if they come into contact with polluted water, consume tainted fish or shellfish, or drink contaminated water.

Nutrient pollution in ground water—which millions of people in the United States use as their drinking water source—can be harmful, even at low levels. Infants are vulnerable to a nitrogen-based compound called nitrates in drinking water. Excess nitrogen in the atmosphere can produce pollutants such as ammonia and ozone, which can impair our ability to breathe, limit visibility and alter plant growth. When excess nitrogen comes back to earth from the atmosphere, it can harm the health of forests, soils and waterways.

FIGURE 18.7 Too much nitrogen and phosphorus in the water can have diverse and far-reaching impacts on public health, the environment and the economy. Photo credit: Bill Yates.

Sources and Solutions: Nutrient Pollution

Excessive nitrogen and phosphorus that washes into water bodies and is released into the air are often the direct result of human activities. The primary sources of nutrient pollution are:

- **Agriculture:** Animal manure (Figure 18.8), excess fertilizer applied to crops and fields, and soil erosion make agriculture one of the largest sources of nitrogen and phosphorus pollution in the country.
- **Stormwater:** When precipitation falls on our cities and towns, it runs across hard surfaces—like rooftops, sidewalks and roads—and carries pollutants, including nitrogen and phosphorus, into local waterways.
- **Wastewater:** Our sewer and septic systems are responsible for treating large quantities of waste, and these systems do not always operate properly or remove enough nitrogen and phosphorus before discharging into waterways.
- **Fossil Fuels:** Electric power generation, industry, transportation and agriculture have increased the amount of nitrogen in the air through use of fossil fuels.
- **In and Around the Home:** Fertilizers, yard and pet waste, and certain soaps and detergents contain nitrogen and phosphorus, and can contribute to nutrient pollution if not properly used or disposed of. The amount of hard surfaces and type of landscaping can also increase the runoff of nitrogen and phosphorus during wet weather (Figure 18.9).

FIGURE 18.8 Animal waste contributes excess nutrients to our waterways when manure is improperly managed.

Agriculture Is the Primary Source of Nutrient Pollution

Farming operations can contribute to nutrient pollution when not properly managed. Fertilizers and animal manure, which are both rich in nitrogen and phosphorus, are the primary sources of nutrient pollution from agricultural sources. Excess nutrients can impact water quality when it rains or when water and soil containing nitrogen and phosphorus wash into nearby waters or leach into ground waters.

Fertilized soils and livestock can be significant sources of gaseous, nitrogen-based compounds like ammonia and nitrogen oxides. Ammonia can be harmful to aquatic life if large amounts are deposited to surface waters. Nitrous oxide is a potent greenhouse gas.

There are many ways that agricultural operations can reduce nutrient pollution, including:

- **Watershed efforts:** The collaboration of a wide range of people and organizations often across an entire watershed is vital to reducing nutrient pollution. State governments, farm organizations, conservation groups, educational institutions, non-profit organizations, and community groups all play a part in successful efforts to improve water quality.
- **Nutrient management:** Applying fertilizers in the proper amount, at the right time of year and with the right method can significantly reduce the potential for pollution.
- **Cover crops:** Planting certain grasses, grains or clovers can help keep nutrients out of the water by recycling excess nitrogen and reducing soil erosion.
- **Buffers:** Planting trees, shrubs and grass around fields, especially those that border water bodies, can help by absorbing or filtering out nutrients before they reach a water body.
- **Conservation tillage:** Reducing how often fields are tilled reduces erosion and soil compaction, builds soil organic matter, and reduces runoff.
- **Managing livestock waste:** Keeping animals and their waste out of streams, rivers and lakes keeps nitrogen and phosphorus out of the water and restores stream banks.
- **Drainage water management:** Reducing nutrient loadings that drain from agricultural fields helps prevent degradation of the water in local streams and lakes.

FIGURE 18.9 **Our homes, yards and streets contribute to nitrogen pollution in a variety of ways, but solutions exist to address this pollution at its source.**

FIGURE 18.10 **Applying fertilizers in the proper amount, at the right time of year and with the right method can significantly reduce how much fertilizer reaches water bodies.**

Urban Water Pollution

The influence of humans on the earth's waters is huge. Where people are concentrated, there is a double problem of providing drinking water and disposing of wastes and wastewater. The wastewater from an urban area can include sewage from homes and commercial building restrooms, factory or industrial wastes, and stormwater runoff.

Homes produce wastewater not only from toilets but also from kitchens and laundry. Shower and handwashing water is not very polluted and in green homes is returned to be used in the toilets. One of the major problems with domestic wastewater is the pathogenic bacteria, viruses, and parasites that are viable in this water. Some of these pathogens are easily killed or do not survive for long periods outside of the human body. Others are quite persistent and are not killed or inactivated by wastewater treatment facilities. Even after going through municipal drinking water processing some will survive. An example of this is the parasite *Cryptosporidium*, which passed though the water treatment system in Milwaukee, Wisconsin, in 1993 and sickened over 400,000 people, killing at least 100.

catolla/Shutterstock

FIGURE 18.11 Keeping animals and their waste out of streams keeps nitrogen and phosphorus out of the water and protects stream banks.

The stormwater runoff from city streets and buildings may have automobile wastes, acid rain, lawn fertilizer, and fecal wastes from pets and wild animals in it. Many cities flow their stormwater into their wastewater treatment systems, which can overload their systems during a large rainstorm or spring snow melt. This causes untreated wastewater, including some raw sewage, to exit the facility and enter whatever natural waterway receives the water. Other cities send their stormwater straight to the nearest lake or stream, introducing polluted water directly into the surface water system (Figure 18.12). Cities that have addressed their stormwater runoff problems have separate treatment facilities or store the water and slowly mix it into their wastewater system. Stormwater runoff can be used for certain types of irrigation because it is typically low in pollutants, so some cities store and use the water.

FIGURE 18.12 Signs such as this one can be placed at the inlets to storm sewers to educate people about where the wastes that flow through the storm sewer system end up. These types of signs have become common in many communities around the United States. Picture courtesy of NOAA.

Industrial Pollution and Thermal Pollution

Most industries are located in urban areas, but many industries produce large volumes of polluted water or types of pollutants that typical municipal facilities cannot handle, so they need their own treatment facilities. If the industries do dispose of their wastes into the municipal treatment system, there can be any type of waste product or pollutant in this water. Extra care must be taken to ensure that the industrial waste does not contain organic waste that can cause depletion of oxygen as it is decomposed by resident microbes (i.e., a high Biological Oxygen Demand, or BOD), does not contain heavy metals that can kill the beneficial bacteria used to clean the wastewater, and does not contain pollutants that cannot be handled by the treatment process. Pollutants that are not successfully treated will pass out into the environment where they can damage ecosystems.

Mining and metal smelting are unique industries that have numerous serious water pollution problems. Water flowing through mines or waste dump materials, both abandoned and active, can become highly acidic. Water is also commonly used in processing mining materials. Mined materials often contain pyrite and other sulfur compounds. When these compounds weather and react with oxygen and water they form sulfuric acid and create low pH water (Figure 18.13). This acidified water (often referred to as acid mine drainage) has the added problem of dissolving metals from mined materials, most importantly the heavy metals, which are extremely toxic, and carrying these metals with the flowing water. There are abandoned mines in many places around the world that have water flowing from them that is too acidic to touch, and if drunk, the lead, copper, zinc, or other metal would kill a person. These water sources usually mix with larger water flows and get diluted, but they can also pollute the entire water drainage system all the way to and including parts of the ocean.

Water is not only adversely affected by the addition of chemicals, it is also affected by temperature changes. There are few if any examples of bodies of water being cooled by human activities, but humans do cause increases in water temperature in many locations around the world. With the exception of global warming, which is warming the oceans, this thermal pollution is mostly in freshwater. There are two main ways of causing thermal pollution. The first has

Daniel Cain, USGS.

FIGURE 18.13 **A USGS scientist collects a water sample for analysis of mineral particles know as colloids.** Toxic metals (such as copper in excess), which can come from old or existing mine sites, bind to the particles and are then ingested by aquatic animals. The metals in this creek have stained the rocks red along its edges.

already been mentioned in this chapter, and that is the storage of water in reservoirs, which allows more heating from the sun and evaporation of the water. Domestic and industrial uses of water also heat the water, and when the heated water returns to any natural body of water it raises its temperature.

Many industries use water to cool their operations. This ranges from cooling metal during a smelter operation to using steam during the generation of electricity. The steam is usually cooled by a local body of water and then the water is returned to its source. The addition of heat to the water causes the amount of available oxygen to decline (in other words, the solubility of oxygen decreases with increasing water temperatures). The increase may be only a few degrees, but it can alter the aquatic environment. Many cold-water fish species, such as trout, cannot survive in the warmer waters and reduced oxygen levels. In other fish, a few degrees can change their reproductive cycle and either trigger or stop spawning. These changes in temperature are usually not constant but vary over time with the needs of the industry, making the thermal pollution even more disruptive. This heating can prevent lakes from freezing over and increase the growth of algae and aquatic plants, similar to the addition of excessive nutrients.

Thermal pollution can be remediated and controlled just like the other types of water pollution. Cooling waters do not have to be returned to local water sources. The industry can maintain its own separate water supply and use shallow ponds to cool the water before it is reused. Another method, the cooling tower, is used in many industrial applications. In the cooling tower the water is cooled by evaporation as it is sprayed from the top of the tower, transferring heat from the water to the atmosphere. Both of these methods lose a lot of water to evaporation. To prevent evaporation loss, water can also be cooled using a heat exchanger where the water flows though pipes with air moving over them. The heat is transferred to the atmosphere much as a radiator works in a car. Water can also be pumped though pipes underground where the heat is lost to the soil. All of these methods can cause thermal pollution in other areas (e.g., the atmosphere, soil) or overload the local atmosphere with humidity which can cause local fog or freezing rain in colder periods.

Agricultural Water Pollution

A single farm or ranch can pollute local waters in the same way a city or industry can. The average-size hog farm in the United States produces the same fecal waste output as a moderate-size city. Farms have several unique pollution problems that affect the air, water, and soil. Farms apply huge amounts of pesticides and fertilizers to the soil and crops to maximize their food production. This production cannot be reduced or serious food shortages would occur, and therefore we need to monitor and control the use of agricultural chemicals carefully. Many of the chemicals used can wash from the soil or move into the water table below the fields. These chemicals are almost entirely water soluble, and most are applied mixed with water. When used in excess, they move easily into local water sources. This movement is increased when irrigation is used and can be extreme when nonsprinkler forms of irrigation are used. The loss of agricultural chemicals is an economic hardship on the farmer and an environmental problem for everyone. Farmers can control both the application and method of application of agricultural chemicals and can influence the flow of water from their fields. They can leave strips of natural vegetation between their fields and waterways (Figure 18.14) and leave residues from the harvested crops on their fields (Figure 18.15).

Eutrophication is a common result of farm chemical runoff into bodies of water. This problem can even extend to the oceans when rivers that have been overenriched with nutrients from agriculture reach the ocean. This is seen at the mouth of the Mississippi River, where it discharges into the Gulf of Mexico and causes a large area of "dead water" (Figure 18.16). The overloading of nutrients is not the only problem. Many farm chemicals are toxic and persistent in the environment, which means they do not break down and are not detoxified by organisms in the soil or water. Many of these can accumulate in living tissue, especially fat tissues, and increase in concentration as they move up the food chain. An example is the chlorinated hydrocarbons that have been or are still used as insecticides; these reach toxic levels in the top predators after being at almost undetectable levels in the water or soil. DDT (1,1,1-trichloro-2,2-bis(p-chlorophenyl) ethane) has liposoluble properties, it was readily absorbed and built up to toxic levels in fish, mammals,

FIGURE 18.14 **This Iowa stream has rows of trees, shrubs, and grasses, known as a riparian buffer, planted between it and the surround agricultural fields.** The riparian buffer serves to remove particulate material as well as dissolved nutrients and other chemicals from water coming off the fields, meaning that the water entering the stream is cleaner than if the buffer was not in place. Photo courtesy of USDA-NRCS.

FIGURE 18.15 **Corn growing through soybean residues in Iowa.** The residues slow down water flow across the field and protect the soil structure, allowing more infiltration and reducing erosion. Photo courtesy of USDA-NRCS.

birds, and humans as it passed up the food chain. DDT's toxic action was first seen as a disruption of hormone systems. DDT was banned in the United States in 1972 but is highly persistent and is still globally distributed; it is also still used in many parts of the world. It accumulates in animals tissues because it is extremely stable, and it persists in soils and in plants. DDT has reached most groundwater and surface water sources. DDT has a reported **half-life** of > 60 years outside of organisms. A half-life is the length of time it takes for half of the compound to break down or change, but many of the breakdown products of DDT are still toxic.

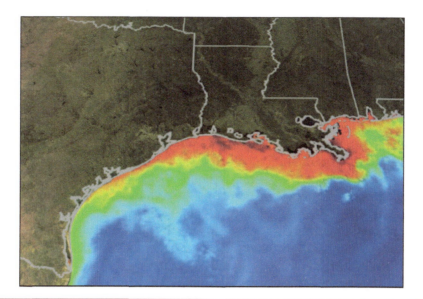

FIGURE 18.16 **Gulf of Mexico Dead Zone.** Less oxygen dissolved in the water is often referred to as a "dead zone" (in red above) because most marine life either dies or, if they are mobile, such as fish, leave the area. Habitats that would normally be teeming with life become, essentially, biological deserts. The Gulf of Mexico dead zone typically ranges from a low of approximately 1,197 square miles to as much as 6,213 square miles. Figure courtesy of NOAA.

Sediment Pollution and Contaminated Sediment

Sediments are fragmented materials that originate from weathering and erosion of rocks or unconsolidated deposits and are transported by, suspended in, or deposited by water. Suspended and bedded sediments (SABS) are defined by EPA as particulate organic and inorganic matter that suspend in or are carried by the water, and/or accumulate in a loose, unconsolidated form on the bottom of natural water bodies. This includes the frequently used terms of clean sediment, suspended sediment, total suspended solids, bedload, turbidity, or in common terms, dirt, soils or eroded materials.

Contaminated Sediments

Many of the sediments in our rivers, lakes, and oceans have been contaminated by pollutants. Some of these pollutants, such as the pesticide DDT and the industrial chemicals known as polychlorinated biphenyls (PCBs), were released into the environment long ago. The use of DDT and PCBs in the United States was banned in the 1970s, but these chemicals persist for many years.

Other contaminants enter our waters every day. Some flow directly from industrial and municipal waste dischargers, while others come from polluted runoff in urban and agricultural areas. Still other contaminants are carried through the air, landing in lakes and streams far from the factories and other facilities that produced them. In cases like this, the sediment may serve as a contaminant reservoir or source of contamination. Experts believe that contaminated sediments are a widespread and serious problem. Areas of concern are found on the Atlantic and Pacific coasts, in the Gulf of Mexico and the Great Lakes, and along inland waterways.

Contaminated sediments affect small creatures such as worms, crustaceans, and insect larvae that inhabit the bottom of a water body in what is known as the benthic environment. Some kinds of toxic sediments kill benthic organisms, reducing the food available to larger animals such as fish.

Some contaminants in the sediment are taken up by benthic organisms in a process called bioaccumulation. When larger animals feed on these contaminated organisms, the toxins are taken into their bodies, moving up the food chain in increasing concentrations in a process known as biomagnification. As a result, fish and shellfish, waterfowl, and freshwater and marine mammals, as well as benthic organisms, are affected by contaminated sediments.

Species that cannot tolerate the toxic contaminants found in some sediments simply die, reducing the variety of organisms, also known as biodiversity, in the affected environment. Animals that survive exposure to contaminated sediments may develop serious health problems, including fin rot, tumors, and reproductive effects.

When contaminants bioaccumulate in trout, salmon, ducks, and other food sources, they pose a threat to human health. In 1998, fish consumption advisories were issued for more than 2,506 bodies of water in the United States. *Possible long-term effects of eating contaminated fish include cancer and neurological defects.*

Contaminated sediments do not always remain at the bottom of a water body. Anything that stirs up the water, such as a storm or a boat's propeller, can resuspend some sediments. Resuspension may mean that all of the animals in the water, and not just the bottom-dwelling organisms, will be directly exposed to toxic contaminants.

Every year, approximately 300 million cubic yards of sediment are dredged to deepen harbors and clear shipping lanes in the United States. Roughly 3 – 12 million cubic yards of these sediments are so contaminated they require special, and sometimes costly, handling. If dredging to improve navigation cannot be conducted because sediments are contaminated, the volume of shipping on these waterways will decline.

No single government agency is completely responsible for addressing the problem of contaminated sediments. A variety of laws give federal, state, and tribal agencies authority to address sediment quality issues. Private industry and the public also have roles to play in contaminated sediment prevention. Increasing public awareness of the problem is crucial to developing an effective solution.

18.3 Clean Water Act & TMDLs

Background

The goal of the Clean Water Act (CWA) is "to restore and maintain the chemical, physical, and biological integrity of the Nation's waters" states, territories, and authorized tribes are required to develop lists of impaired waters. These are waters for which technology-based regulations and other required controls are not stringent enough to meet the water quality standards set by states. The law requires that states establish priority rankings for waters on the lists and develop Total Maximum Daily Loads (TMDLs), for these waters. A **TMDL** is a calculation of the maximum amount of a pollutant that a water body can receive and still safely meet water quality standards.

The CWA includes two basic approaches for protecting and restoring the nation's waters. One is a technology-based, end-of-pipe approach, whereby EPA promulgates effluent guidelines that rely on technologies available to remove pollutants from waste streams. These guidelines are used to derive individual, technology-based National Pollutant Discharge Elimination System (NPDES) permit limits. The other approach is water-quality based and is designed to achieve the desired uses of a water. This approach may ultimately result in more stringent NPDES permit limits.

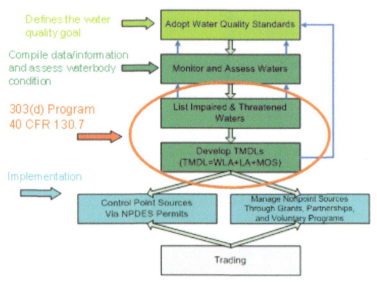

FIGURE 18.17 The process by which EPA and state permitting agencies implement the Clean Water Act.

Water Quality-Based Approach of the Clean Water Act

Water quality standards are the foundation of the water-quality based control program mandated by the Clean Water Act. Water quality standards define the goals for a waterbody by designating its uses, setting criteria to protect those uses, and establishing provisions to protect water quality from pollutants. A water quality standard consists of four basic elements:

1. designated uses of the waterbody (*e.g.* recreation, water supply, aquatic life, agriculture),
2. water quality criteria to protect designated uses (numeric pollutant concentrations and narrative requirements),
3. an antidegradation policy to maintain and protect existing uses and high quality waters, and
4. general policies addressing implementation issues (*e.g.*, low flows, variances, mixing zones).

By adopting water quality standards, states are able to determine which healthy waters need protection, which waters must be restored and how much pollutant reductions are needed. Consequently, these water quality standards set a goal for restoring and protecting a watershed over the long term.

Water quality monitoring provides the data to characterize waters and identify changes or trends in water quality over time. The collection of monitoring data enables states to identify existing or emerging water quality problems and determine whether current pollution control mechanisms are effective in complying with the regulations. The CWA requires that each state monitor and assess the health of all their waters and report their findings every two years to EPA.

Monitoring data, as well as other information, must be used by the states to develop a list of water-quality limited segments, i.e., waters that will not meet water quality standards for a particular pollutant even after a technology-based permit is in place. States must develop TMDLs, or Total Maximum Daily Loads, for every water body/pollutant combination that is of concern. The TMDL allocates that load to point sources, (Wasteload Allocation or WLA), and nonpoint sources (Load Allocation or LA) which include both anthropogenic and natural background sources of the pollutant.

In many cases, the TMDL analysis is the trigger for determining the source(s) of pollutants. A TMDL may contain WLAs only, LAs only, or a combination of both. The EPA cannot enforce implementation of a TMDL once the analysis is complete. If the TMDL identifies nonpoint sources of pollutants as a major cause of impairment, states can apply for EPA funded grants to fund state programs for nonpoint source assessment and control as well as individual projects.

TMDL in Detail

Point sources include all sources subject to regulation under the National Pollutant Discharge Elimination System (NPDES)program, e.g. wastewater treatment facilities, some stormwater discharges and concentrated animal feeding operations (CAFOs). Nonpoint sources include all remaining sources of the pollutant as well as anthropogenic and natural background sources. TMDLs must also account for seasonal variations in water quality, and include a margin of safety(MOS) to account for uncertainty in predicting how well pollutant reductions will result in meeting water quality standards.

TMDLs must also include a margin of safety (MOS) to account for the uncertainty in predicting how well pollutant reduction will result in meeting water quality standards, and account for seasonal variations.

The TMDL calculation is:

$$\text{TMDL} = \Sigma\text{WLA} + \Sigma\text{LA} + \text{MOS}$$

where **WLA** is the sum of wasteload allocations (point sources), **LA** is the sum of load allocations (nonpoint sources and background), and **MOS** is the margin of safety.

By regulation, each pollutant causing a waterbody to be impaired or threatened is referred to as a water-body/pollutant combination, and a TMDL is developed for each waterbody/pollutant combination. For example, if one waterbody is impaired or threatened by three pollutants three TMDLs will be developed for the waterbody. However, in many cases, the word TMDL is used to describe a document that addresses several waterbody/pollutants combinations (i.e., several TMDLs exist in one TMDL document). More and more states are bundling TMDLs on a watershed scale.

18.4 National Water Quality Assessment Program

In 1991, the U.S. Geological Survey (USGS, part of the Department of the Interior) began a systematic, long-term program to monitor watersheds. The National Water-Quality Assessment Program (NAWQA), established to help manage surface and groundwater supplies, has involved the collection and analysis of water quality data in over 50 major river basins and aquifer systems in nearly all 50 states.

The program has encompassed three principal categories of investigation: (1) the current conditions of surface water and groundwater; (2) changes in those conditions over time; and (3) major factors—such as climate, geography, and land use—that affect water quality. For each of these categories, the water and sediment have been tested for such pollutants as pesticides, plant nutrients, volatile organic compounds, and heavy metals.

The NAWQA findings were disturbing. Water quality is most affected in **watersheds** with highest population density and urban development. In agricultural areas, 95 percent of tested streams and 60 percent of shallow wells contained herbicides, insecticides, or both. In urban areas, 99 percent of tested streams and 50 percent of shallow wells had herbicides, especially those used on lawns and golf courses. Insecticides were found more frequently in urban streams than in agricultural ones.

The study also found large amounts of plant nutrients in water supplies. For instance, 80 percent of agricultural streams and 70 percent of urban streams were found to contain phosphorus at concentrations that exceeded EPA guidelines.

Moreover, in agricultural areas, one out of five well-water samples had nitrate concentrations higher than EPA standards for drinking water. Nitrate contamination can result from nitrogen fertilizers or material from defective septic systems leaching into the groundwater, or it may reflect defects in the wells.

18.5 Prevention and Remediation

As the old saying goes, an ounce of prevention is worth a pound of cure. This is especially true when it comes to controlling water pollution. Several important steps taken since the passage of the Clean Water Act have made surface waters today cleaner in many ways than they were 30 years ago.

For example, industrial wastes are mandated to be neutralized or broken down before being discharged to streams, lakes, and harbors. Moreover, the U.S. government has banned the production and use of certain dangerous pollutants such as DDT and PCBs.

In addition, two major changes have been introduced in the handling of sewage. First, smaller, less efficient sewage treatment plants are being replaced with modern, regional plants that include biological treatment, in which microorganisms are used to break down organic matter in the sewage. The newer plants are releasing much cleaner discharges into the receiving bodies of water (rivers, lakes, and ocean).

watersheds: land area that delivers water, sediment, and dissolved substances via small streams to a major stream (river)

Second, many jurisdictions throughout the United States are building separate sewer lines for storm water and sanitary wastes. These upgrades are needed because excess water in the older, "combined" sewer systems would simply bypass the treatment process, and untreated sewage would be discharged directly into receiving bodies of water.

To minimize pollutants from nonpoint sources, the EPA is requiring all municipalities to address the problem of runoff from roads and parking lots. At the same time, the use of fertilizers and pesticides needs to be reduced. Toward this end, county extension agents are educating farmers and homeowners about their proper application and the availability of nutrient testing.

To curtail the use of expensive and potentially harmful pesticides, the approach known as integrated pest management can be implemented. It involves the identification of specific pest problems and the use of nontoxic chemicals and chemical-free alternatives whenever possible. For instance, aphids can be held in check by ladybug beetles and caterpillars can be controlled by applying neem oil to the leaves on which they feed.

Moreover, new urban development projects in many areas are required to implement storm-water management practices. They include such features as: oil and grease traps in storm drains; swales to slow down runoff, allowing it to infiltrate back into groundwater; "wet" detention basins (essentially artificial ponds) that allow solids to settle out of runoff; and artificial wetlands that help break down contaminants in runoff. While such additions may be costly, they significantly improve water quality.

18.6 Wastewater and Its Treatment

We consider wastewater treatment as a water use because it is so interconnected with the other uses of water. Much of the water used by homes, industries, and businesses must be treated before it is released back to the environment.

If the term "wastewater treatment" is confusing to you, you might think of it as "sewage treatment." Nature has an amazing ability to cope with small amounts of water wastes and pollution, but it would be overwhelmed if we didn't treat the billions of gallons of wastewater and sewage produced every day before releasing it back to the environment. Treatment plants reduce pollutants in wastewater to a level nature can handle.

Wastewater is used water. It includes substances such as human waste, food scraps, oils, soaps and chemicals. In homes, this includes water from sinks, showers, bathtubs, toilets, washing machines and dishwashers. Businesses and industries also contribute their share of used water that must be cleaned.

Wastewater also includes storm runoff. Although some people assume that the rain that runs down the street during a storm is fairly clean, but it isn't. Harmful substances that wash off roads, parking lots, and rooftops can harm our rivers and lakes.

Effects of Wastewater Pollutants

If wastewater is not properly treated, then the environment and human health can be negatively impacted. These impacts can include harm to fish and wildlife populations, oxygen depletion, beach closures and other restrictions on recreational water use, restrictions on fish and shellfish harvesting and contamination of drinking water. Some examples of pollutants that can be found in wastewater and the potentially harmful effects these substances can have on ecosystems and human health include:

- decaying organic matter and debris can use up the dissolved oxygen in a lake so fish and other aquatic biota cannot survive;
- excessive nutrients, such as phosphorus and nitrogen (including ammonia), can cause eutrophication, or over-fertilization of receiving waters, which can be toxic to aquatic organisms, promote excessive plant growth, reduce available oxygen, harm spawning grounds, alter habitat and lead to a decline in certain species;
- chlorine compounds and inorganic chloramines can be toxic to aquatic invertebrates, algae and fish;
- bacteria, viruses and disease-causing pathogens can pollute beaches and contaminate shellfish populations, leading to restrictions on human recreation, drinking water consumption and shellfish consumption;

- metals, such as mercury, lead, cadmium, chromium and arsenic can have acute and chronic toxic effects on species.
- other substances such as some pharmaceutical and personal care products, primarily entering the environment in wastewater effluents, may also pose threats to human health, aquatic life and wildlife.

Wastewater Treatment

The major aim of wastewater treatment is to remove as much of the suspended solids as possible before the remaining water, called effluent, is discharged back to the environment. As solid material decays, it uses up oxygen, which is needed by the plants and animals living in the water.

The 3 main steps in the wastewater treatment process are primary, secondary and tertiary treatment (pretty straightforward, right?). During primary treatment, a majority of solids are separated from the liquid waste. Oil and grease are skimmed off the top, as they are allowed to float to the surface as they move slowly through the first separation tank. The dense solids settle to the bottom, allowing the rest of the water to move to secondary treatment. During secondary treatment, microbes in the treatment tanks consume the dissolved and small pieces of organic matter. The tanks are aerated to provide oxygen for the microbes. The water flows through a series of these long aerated tanks and mostly clean water emerges. If this water is to be discharged to a natural water body, or used by people, it must be disinfected (i.e., insure there are no remaining pathogenic organisms). This process can include sand filtering, the addition of chlorine, ozonation of the water, UV light exposure and other options. During advanced, or tertiary treatment, the water may be treated to remove excess nitrogen or phosphorous, for example.

Here's a step-by-step guide describing what happens at each stage of the treatment process and how pollutants are removed to help keep our waterways clean. This information is courtesy of the Greater Vancouver Regional District.

The Primary Treatment Process

1. Screening:

Wastewater entering the treatment plant includes items like wood, rocks, and even dead animals. Unless they are removed, they could cause problems later in the treatment process. Most of these materials are sent to a landfill.

FIGURE 18.18   During secondary treatment, the wastewater travels through a series of long concrete tanks that are bubbled with oxygen to allow microbes in the tanks to consume the organic waste.

2. Pumping:

The wastewater system relies on the force of gravity to move sewage from your home to the treatment plant. So wastewater-treatment plants are located on low ground, often near a river into which treated water can be released. If the plant is built above the ground level, the wastewater has to be pumped up to the aeration tanks (item 3). From here on, gravity takes over to move the wastewater through the treatment process.

3. Removing Sludge:

Wastewater then enters the second section or sedimentation tanks. Here, the sludge (the organic portion of the sewage) settles out of the wastewater and is pumped out of the tanks. Some of the water is removed in a step called thickening and then the sludge is processed in large tanks called digesters.

4. Removing Scum:

As sludge is settling to the bottom of the sedimentation tanks, lighter materials are floating to the surface. This 'scum' includes grease, oils, plastics, and soap. Slow-moving rakes skim the scum off the surface of the waste-water. Scum is thickened and pumped to the digesters along with the sludge.

5. Aerating:

One of the first steps that a water treatment facility can do is to just shake up the sewage and expose it to air. This causes some of the dissolved gases (such as hydrogen sulfide, which smells like rotten eggs) that taste and smell bad to be released from the water. Wastewater enters a series of long, parallel concrete tanks. Each tank is divided into two sections. In the first section, air is pumped through the water.

As organic matter decays, it uses up oxygen. Aeration replenishes the oxygen. Bubbling oxygen through the water also keeps the organic material suspended while it forces 'grit' (coffeegrounds, sand and other small, dense particles) to settle out. Grit is pumped out of the tanks and taken to landfills.

6. Filtration:

Many cities also use filtration in sewage treatment. After the solids are removed, the liquid sewage is filtered through a substance, usually sand, by the action of gravity. This method gets rid of almost all bacteria, reduces turbidity and color, removes odors, reduces the amount of iron, and removes most other solid particles that remained in the water. Water is sometimes filtered through carbon particles, which removes organic particles. This method is used in some homes, too.

7. Disinfection:

There are a number of ways to further disinfect the water after it is filtered. One method is to use chlorination. The wastewater flows into a 'chlorine contact' tank, where the chemical chlorine is added to kill bacteria, which could pose a health risk, just as is done in swimming pools. The chlorine is mostly eliminated as the bacteria are destroyed, but sometimes it must be neutralized by adding other chemicals. This protects fish and other marine organisms, which can be harmed by the smallest amounts of chlorine.

Other strategies include pumping water past a strong UV light, or pumping ozone into the water. The treated water (called effluent) can then discharged to a local river or the ocean

8. Wastewater Residuals:

Another part of treating wastewater is dealing with the solid-waste material. These solids are kept for 20 to 30 days in large, heated and enclosed tanks called 'digesters.' Here, bacteria break down (digest) the material, reducing its volume, odors, and getting rid of organisms that can cause disease. The finished product is mainly sent to landfills, but sometimes can be used as fertilizer.

In-home Septic Systems to Treat Wastewater

Common in rural areas without centralized sewer systems, septic systems are underground wastewater treatment structures that use a combination of nature and time-tested technology to treat wastewater from household plumbing produced by bathrooms, kitchen drains, and laundry.

A typical septic system consists of a septic tank and a drainfield, or soil absorption field. Below is a brief overview of how septic systems work. For an animated, interactive model of a household septic system, visit the Guadalupe-Blanco River Authority website (http://www.gbra.org/septic.swf) on how a septic system works!

FIGURE 18.19 **An infographic showing the basic workings of a septic system.**

1. All water runs out of your house from one main drainage pipe into a septic tank.
2. The septic tank is a buried, water-tight container usually made of concrete, fiberglass, or polyethylene. Its job is to hold the wastewater long enough to allow solids to settle down to the bottom (forming sludge), while the oil and grease floats to the top (as scum). Compartments and a T-shaped outlet prevent the sludge and scum from leaving the tank and traveling into the drainfield area.
3. The liquid wastewater (effluent) then exits the tank into the drainfield. If the drainfield is overloaded with too much liquid, it will flood, causing sewage to flow to the ground surface or create backups in toilets and sinks.
4. Finally, the wastewater percolates into the soil, naturally removing harmful coliform bacteria, viruses, and nutrients.

You may already know you have a septic system. If you don't know, here are tell-tale signs that you probably do:

- You use well water.
- The waterline coming into your home doesn't have a meter.
- You show a "$0.00 Sewer Amount Charged" on your water bill.
- Your neighbors have a septic system.

Once you've determined that you have a septic system, you can find it by:

- Looking on your home's "as built" drawing.
- Checking your yard for lids and manhole covers.
- Contacting a septic inspector/pumper to help you locate it.

Septic systems can get clogged or stop working correctly. A foul odor isn't always the first sign of a malfunctioning septic system. Call a septic professional if you notice any of the following:

- Wastewater backing up into household drains.
- Bright green, spongy grass on the drainfield, even during dry weather.
- Pooling water or muddy soil around your septic system or in your basement.
- A strong odor around the septic tank and drainfield.
- Mind the signs of a failing system.

One call to a septic professional could save you thousands of dollars! Homeowners can contact their local or state health department for more information about onsite wastewater practices in their community at Contacts.

CHAPTER 19

Air Pollution

19.1 Air Pollution

Air pollution is the addition of a gas, vapor, or particle to the atmosphere that causes harm or discomfort to a living organism. There are many sources of pollution that occur naturally, such as volcanoes and forest fires, but since these have always occurred and we have little control over them, this chapter will discuss only anthropogenic impacts. There are five major air-polluting substances released by humans. These are carbon monoxide, volatile organic compounds (hydrocarbons), particulate matter, sulfur dioxide, and oxides of nitrogen. These are called the **primary air pollutants**. In the presence of each other and sunlight, these substances can react to form **secondary air pollutants**, and these can further react with substances that occur naturally and other air pollutants to form even more compounds.

The Environmental Protection Agency (EPA) has established air quality standards for six principal air pollutants known as the **criteria air pollutants**. The criteria pollutants are carbon monoxide (CO), particulate matter (PM), sulfur dioxide (SO_2), nitrogen dioxide (NO_2), lead (Pb), and ozone (O_3). All of these are emitted directly except ozone, which is formed when nitrogen dioxide, other oxides of nitrogen, sulfur oxides (SO_x), and volatile organics react in the presence of sunlight. Increased ozone is seen more in the afternoons in areas with high automobile traffic. Particulate matter is either emitted directly or formed when nitrogen oxides, sulfur oxides, ammonia, and organic compounds react in the atmosphere. There are some compounds that are highly toxic and are emitted in small amounts in limited areas; an example is mercury (Hg) from coal-fired electric power plants. The EPA has established an **Air Quality Index (AQI)** for reporting daily air quality. The EPA calculates the AQI for five major air pollutants regulated by the Clean Air Act: ground-level ozone, particle pollution, carbon monoxide, sulfur dioxide, and nitrogen dioxide (Figure 19.1).

- "Good" AQI is 0–50. Air quality is considered satisfactory, and air pollution poses little or no risk. This region is colored green.

- "Moderate" AQI is 51–100. Air quality is acceptable; however, for some pollutants there may be a moderate health concern for a very small number of people. For example, people who are unusually sensitive to ozone may experience respiratory symptoms. This region is colored yellow.

- "Unhealthy for Sensitive Groups" AQI is 101–150. Although general public is not likely to be affected at this AQI range, people with lung disease, older adults and children are at a greater risk from exposure to ozone, whereas persons with heart and lung disease, older adults, and children are at greater risk from the presence of particles in the air. This region is colored orange.

- "Unhealthy" AQI is 151–200. Everyone may begin to experience some adverse health effects, and members of the sensitive groups may experience more serious effects. This region is colored red.

- "Very Unhealthy" AQI is 201–300. This would trigger a health alert signifying that everyone may experience more serious health effects. This region is colored purple.

- "Hazardous" AQI greater than 300. This would trigger a health warning of emergency conditions. The entire population is more likely to be affected. This region is colored maroon.

FIGURE 19.1 **The Air Quality Index (AQI) is an important way for people to determine if air conditions are favorable.** People with lung diseases or disabilities especially need to be aware of air quality conditions before exercising outside.

19.2 Six Principle Pollutants

The source, and amounts, of each pollutant that goes into the atmosphere in the U.S. every year, are shown in Figure 19.2.

Each of these pollutants has a unique set of sources, as shown in Figure 19.2. These sources include agriculture (fertilizer application, livestock waste and dust), mobile (aircrafts, boats and barges, cars and trucks), industrial uses, fuel combustion (for electricity generation, institutional building heat), miscellaneous (gas stations, solid waste off-gassing, commercial cooking), dust (construction and roads), fires (such as wildfires and prescribed burns), and biogenics (produced by living organisms, such as people and livestock).

Carbon Monoxide

Carbon monoxide (CO) is a colorless, odorless gas emitted from combustion processes. Nationally and, particularly in urban areas, the majority of CO emissions to ambient air come from mobile sources. CO can cause harmful health effects by reducing oxygen delivery to the body's organs (like the heart and brain) and tissues. At extremely high levels, CO can cause death.

EPA first set air quality standards for CO in 1971. For protection of both public health and welfare, EPA set a 8-hour primary standard at 9 parts per million (ppm) and a 1-hour primary standard at 35 ppm.

Particulate Matter

Particle pollution (also called particulate matter or PM) is the term for a mixture of solid particles and liquid droplets found in the air. Some particles, such as dust, dirt, soot, or smoke, are large or dark enough to be seen with the naked eye. Others are so small they can only be detected using an electron microscope.

Particle pollution includes "inhalable coarse particles," with diameters larger than 2.5 micrometers and smaller than 10 micrometers and "fine particles," with diameters that are 2.5 micrometers and smaller. How small is 2.5 micrometers? Think about a single hair from your head. The average human hair is about 70 micrometers in diameter—making it 30 times larger than the largest fine particle (Figure 19.3).

These particles come in many sizes and shapes and can be made up of hundreds of different chemicals. Some particles,

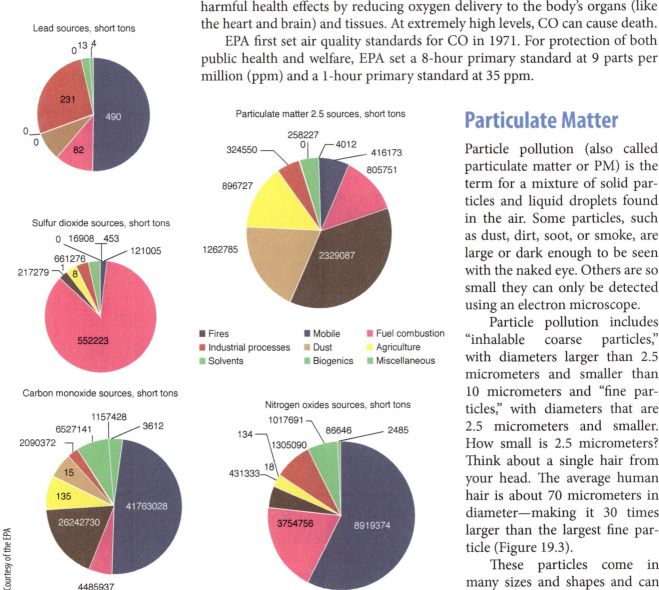

Courtesy of the EPA

FIGURE 19.2   Sources and amounts of 6 principal air pollutants.

known as *primary particles* are emitted directly from a source, such as construction sites, unpaved roads, fields, smokestacks or fires. Others form in complicated reactions in the atmosphere of chemicals such as sulfur dioxides and nitrogen oxides that are emitted from power plants, industries and automobiles. These particles, known as *secondary particles*, make up most of the fine particle pollution in the country.

EPA regulates inhalable particles (fine and coarse). Particles larger than 10 micrometers (sand and large dust) are not regulated by EPA.

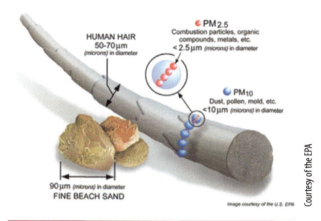

FIGURE 19.3

Health

The size of particles is directly linked to their potential for causing health problems. Small particles less than 10 micrometers in diameter pose the greatest problems, because they can get deep into your lungs, and some may even get into your bloodstream.

Exposure to such particles can affect both your lungs and your heart. Small particles of concern include "inhalable coarse particles" (such as those found near roadways and dusty industries), which are larger than 2.5 micrometers and smaller than 10 micrometers in diameter; and "fine particles" (such as those found in smoke and haze), which are 2.5 micrometers in diameter and smaller.

The Clean Air Act requires EPA to set air quality standards to protect both public health and the public welfare (e.g. visibility, crops and vegetation). Particle pollution affects both.

Numerous scientific studies have linked particle pollution exposure to a variety of problems, including:

- premature death in people with heart or lung disease,
- nonfatal heart attacks,
- irregular heartbeat,
- aggravated asthma,
- decreased lung function, and
- increased respiratory symptoms, such as irritation of the airways, coughing or difficulty breathing.

People with heart or lung diseases, children and older adults are the most likely to be affected by particle pollution exposure. However, even if you are healthy, you may experience temporary symptoms from exposure to elevated levels of particle pollution. For more information about asthma, visit www.epa.gov/asthma.

Environmental Effects

Visibility impairment Fine particles (PM2.5) are the main cause of reduced visibility (haze) in parts of the United States, including many of our treasured national parks and wilderness areas. For more information about visibility, visit www.epa.gov/visibility.

Environmental damage Particles can be carried over long distances by wind and then settle on ground or water. The effects of this settling include: making lakes and streams acidic; changing the nutrient balance in coastal waters and large river basins; depleting the nutrients in soil; damaging sensitive forests and farm crops; and affecting the diversity of ecosystems. For more information about the effects of particle pollution and acid rain, visit http://www.epa.gov/acidrain/.

Aesthetic damage Particle pollution can stain and damage stone and other materials, including culturally important objects such as statues and monuments.

Lead

Lead (Pb) is a metal found naturally in the environment as well as in manufactured products. The major sources of lead emissions have historically been from fuels in on-road motor vehicles (such as cars and trucks) and industrial sources. As a result of EPA's regulatory efforts to remove lead from on-road motor vehicle gasoline, emissions of lead from the transportation sector dramatically declined by 95 percent between 1980 and 1999, and levels of lead in the air decreased by 94 percent between 1980 and 1999. Today, the highest levels of lead in air are usually found near lead smelters. The major sources of lead emissions to the air today are ore and metals processing and piston-engine aircraft operating on leaded aviation gasoline.

Health

In addition to exposure to lead in air, other major exposure pathways include ingestion of lead in drinking water and lead-contaminated food as well as incidental ingestion of lead-contaminated soil and dust. Lead-based paint remains a major exposure pathway in older homes. Learn more about lead in paint, dust and soil by visiting http://www2.epa.gov/lead.

Once taken into the body, lead distributes throughout the body in the blood and is accumulated in the bones. Depending on the level of exposure, lead can adversely affect the nervous system, kidney function, immune system, reproductive and developmental systems and the cardiovascular system. Lead exposure also affects the oxygen carrying capacity of the blood. The lead effects most commonly encountered in current populations are neurological effects in children and cardiovascular effects (e.g., high blood pressure and heart disease) in adults. Infants and young children are especially sensitive to even low levels of lead, which may contribute to behavioral problems, learning deficits and lowered IQ.

Lead is persistent in the environment and accumulates in soils and sediments through deposition from air sources, direct discharge of waste streams to water bodies, mining, and erosion. Ecosystems near point sources of lead demonstrate a wide range of adverse effects including losses in biodiversity, changes in community composition, decreased growth and reproductive rates in plants and animals, and neurological effects in vertebrates.

Sources of Lead at Home

Robert Crum/Shutterstock

FIGURE 19.4 **Before 1978, it was not uncommon to use lead-based paints in homes.**

If your home was built before 1978, there is a good chance it has lead-based paint (Figure 19.4 and 19.5). In 1978, the federal government banned consumer uses of lead-containing paint, but some states banned it even earlier. Lead from paint, including lead-contaminated dust, is one of the most common causes of lead poisoning.

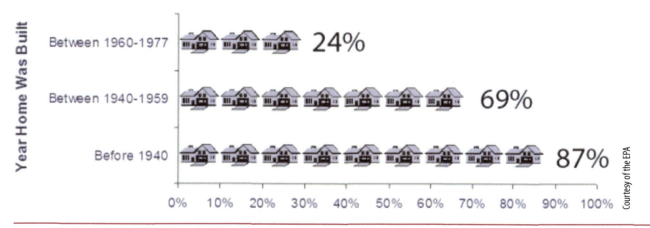

FIGURE 19.5   **Older homes are more likely to contain lead-based paint.**

■ Lead paint is still present in millions of homes, sometimes under layers of newer paint. If the paint is in good shape, the lead paint is usually not a problem. Deteriorating lead-based paint (peeling, chipping, chalking, cracking, damaged, or damp) is a hazard and needs immediate attention.
■ It may also be a hazard when found on surfaces that children can chew or that get a lot of wear-and-tear, such as:
   ❑ Windows and window sills
   ❑ Doors and door frames
   ❑ Stairs, railings, banisters, and porches
■ Be sure to keep all paint in excellent shape and clean up dust frequently.
■ Lead in household dust results from indoor sources such as deteriorating lead-based paint.
■ Lead dust can also be tracked into the home from soil outside that is contaminated by deteriorated exterior lead-based paint and other lead sources, such as industrial pollution and past use of leaded gasoline.
■ Renovation, repair or painting activities can create toxic lead dust when painted surfaces are disturbed or demolished.
■ Pipes and solder—Lead is used in some water service lines and household plumbing materials. Lead can leach, or enter the water, as water flows through the plumbing. Lead pipes and lead solder were commonly used until 1986.

Sulfur Dioxide

Sulfur dioxide (SO_2) is one of a group of highly reactive gasses known as "oxides of sulfur." The largest sources of SO_2 emissions are from fossil fuel combustion at power plants (73%) and other industrial facilities (20%). Smaller sources of SO_2 emissions include industrial processes such as extracting metal from ore, and the burning of high sulfur containing fuels by locomotives, large ships, and non-road equipment. SO_2 is linked with a number of adverse effects on the respiratory system.

Health

Current scientific evidence links short-term exposures to SO_2, ranging from 5 minutes to 24 hours, with an array of adverse respiratory effects including bronchoconstriction and increased asthma symptoms. These effects are particularly important for asthmatics at elevated ventilation rates (e.g., while exercising or playing.)

Studies also show a connection between short-term exposure and increased visits to emergency departments and hospital admissions for respiratory illnesses, particularly in at-risk populations including children, the elderly, and asthmatics.

EPA's National Ambient Air Quality Standard for SO_2 is designed to protect against exposure to the entire group of sulfur oxides (SO_x). SO_2 is the component of greatest concern and is used as the indicator for the larger group of gaseous sulfur oxides (SO_x). Other gaseous sulfur oxides (e.g. SO_3) are found in the atmosphere at concentrations much lower than SO_2.

Emissions that lead to high concentrations of SO_2 generally also lead to the formation of other SO_x. Control measures that reduce SO_2 can generally be expected to reduce people's exposures to all gaseous SO_x. This may have the important co-benefit of reducing the formation of fine sulfate particles, which pose significant public health threats.

SO_x can react with other compounds in the atmosphere to form small particles. These particles penetrate deeply into sensitive parts of the lungs and can cause or worsen respiratory disease, such as emphysema and bronchitis, and can aggravate existing heart disease, leading to increased hospital admissions and premature death.

Ground Level Ozone

Ozone is found in two regions of the Earth's atmosphere—at ground level and in the upper regions of the atmosphere. Both types of ozone have the same chemical composition (O_3). While upper atmospheric ozone protects the earth from the sun's harmful rays, ground level ozone is the main component of smog.

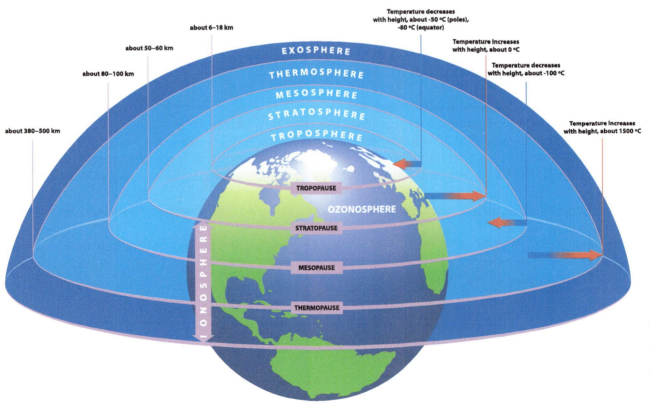

FIGURE 19.6   Ozone near ground level (troposphere) is a health hazard. Ozone in the stratosphere helps to block out UV radiation, and so it is desirable there!.

Troposheric, or ground level ozone, is not emitted directly into the air, but is created by chemical reactions between oxides of nitrogen (NO_x) and volatile organic compounds (VOC). Ozone is likely to reach unhealthy levels on hot sunny days in urban environments. Ozone can also be transported long distances by wind. For this reason, even rural areas can experience high ozone levels.

High ozone concentrations have also been observed in cold months, where a few high elevation areas in the Western U.S. with high levels of local VOC and NO_x emissions have formed ozone when snow is on the ground and temperatures are near or below freezing. Ozone contributes to what we typically experience as "smog" or haze, which still occurs most frequently in the summertime, but can occur throughout the year in some southern and mountain regions.

Under the Clean Air Act, EPA has established health and environmentally protective standards for ozone in the air we breathe. EPA and others have instituted a variety of multi-faceted programs to meet these standards. Learn more about EPA's ozone standards and regulatory actions. Throughout the country, additional programs are being put into place to cut NO_x and VOC emissions from vehicles, industrial facilities, and electric utilities. Programs are also aimed at reducing pollution by reformulating fuels and consumer/commercial products, such as paints and chemical solvents that contain VOC. Voluntary and innovative programs encourage communities to adopt practices, such as carpooling, to reduce harmful emissions.

Emissions from industrial facilities and electric utilities, motor vehicle exhaust, gasoline vapors, and chemical solvents are some of the major sources of NO_x and VOC.

Health Effects

Ozone in the air we breathe can harm our health—typically on hot, sunny days when ozone can reach unhealthy levels. Even relatively low levels of ozone can cause health effects. People with lung disease, children, older adults, and people who are active outdoors may be particularly sensitive to ozone.

Children are at greatest risk from exposure to ozone because their lungs are still developing and they are more likely to be active outdoors when ozone levels are high, which increases their exposure. Children are also more likely than adults to have asthma.

Breathing ozone can trigger a variety of health problems including chest pain, coughing, throat irritation, and congestion. It can worsen bronchitis, emphysema, and asthma. Ground level ozone also can reduce lung function and inflame the linings of the lungs. Repeated exposure may permanently scar lung tissue.

Ozone can:

- Make it more difficult to breathe deeply and vigorously.
- Cause shortness of breath and pain when taking a deep breath.
- Cause coughing and sore or scratchy throat.
- Inflame and damage the airways.
- Aggravate lung diseases such as asthma, emphysema, and chronic bronchitis.
- Increase the frequency of asthma attacks.
- Make the lungs more susceptible to infection.
- Continue to damage the lungs even when the symptoms have disappeared.

These effects may lead to increased school absences, medication use, visits to doctors and emergency rooms, and hospital admissions. Research also indicates that ozone exposure may increase the risk of premature death from heart or lung disease.

The AIRNow Web site provides daily air quality reports for many areas. These reports use the Air Quality Index (or AQI) to tell you how clean or polluted the air is. EnviroFlash, a free service, can alert you via email when your local air quality is a concern. Sign up at www.enviroflash.info.

Airway Inflammation. With airway inflammation, there is an influx of white blood cells, increased mucous production, and fluid accumulation and retention. This causes the death and shedding of cells that line the airways and has been compared to the skin inflammation caused by sunburn.

Human Lung Anatomy and Function

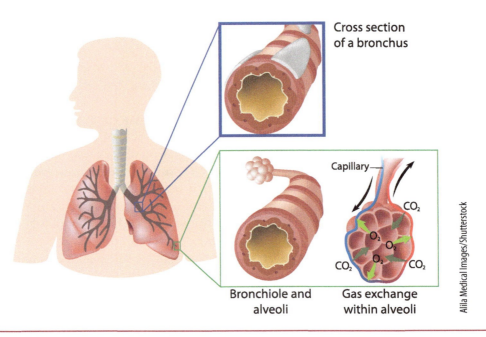

Cross section of a bronchus

Capillary

CO_2

O_2 O_2

CO_2 O_2 CO_2

Bronchiole and alveoli

Gas exchange within alveoli

Alila Medical Images/Shutterstock

FIGURE 19.7   Effects on the Airways. Ozone is a powerful oxidant that can irritate the air ways causing coughing, a burning sensation, wheezing and shortness of breath and it can aggravate asthma and other lung diseases.

Normal alveoli

Asthma

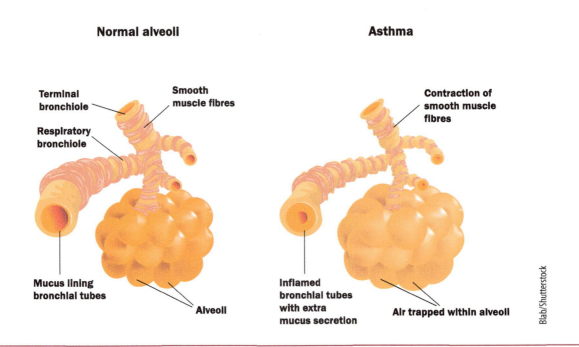

Terminal bronchiole

Smooth muscle fibres

Respiratory bronchiole

Mucus lining bronchial tubes

Alveoli

Contraction of smooth muscle fibres

Inflamed bronchial tubes with extra mucus secretion

Air trapped within alveoli

Blab/Shutterstock

FIGURE 19.8   Alveoli filled with trapped air. Ozone can cause the muscles in the airways to constrict, trapping air in the alveoli. This leads to wheezing and shortness of breath. In people with asthma it can result in asthma attacks.

Ecosystem Effects

Ozone Effects on Plants

Recent systematic surveys by the USDA US Forest Service continue to document widespread injury on sensitive bioindicator plants.

Many studies have documented diagnostic leaf injury due to ozone in the field, including remote wilderness areas. However, damage from ozone is not always visible on leaves and growth decline can occur without these symptoms.

Ground level ozone is absorbed by the leaves of plants, where it can reduce photosynthesis, damage leaves and slow growth. It can also make sensitive plants more susceptible to certain diseases, insects, harsh weather and other pollutants.

Ozone also affects sensitive vegetation and ecosystems, including forests, parks, wildlife refuges and wilderness areas. In particular, ozone harms sensitive vegetation, including trees and plants during the growing season.

Plant species that are sensitive to ozone and potentially at an increased risk from exposure include trees such as black cherry, quaking aspen, ponderosa pine and cottonwood. These trees are found in many areas of the country.

Ground level ozone can have harmful effects on sensitive vegetation and ecosystems. When sufficient ozone enters the leaves of a plant, it can:

■ Interfere with the ability of sensitive plants to produce and store food.
■ Visibly damage the leaves of trees and other plants, harming the appearance of vegetation in urban areas, national parks, and recreation areas.

In addition to reduced tree growth and visible injury to leaves, continued ozone exposure over time can lead to increased susceptibility of sensitive plant species to disease, damage from insects, effects of other pollutants, competition, and harm from severe weather. These effects can also have adverse impacts on ecosystems, including loss of species diversity and changes to habitat quality and water and nutrient cycles.

Nitrogen Oxides

Nitrogen dioxide (NO_2) is one of a group of highly reactive gasses known as "oxides of nitrogen," or "nitrogen oxides (NOx)." Other nitrogen oxides include nitrous acid and nitric acid. EPA's National Ambient Air Quality Standard uses NO_2 as the indicator for the larger group of nitrogen oxides. NO_2 forms quickly from emissions from cars, trucks and buses, power plants, and off-road equipment. In addition to contributing to the formation of ground-level ozone, and fine particle pollution, NO_2 is linked with a number of adverse effects on the respiratory system.

EPA first set standards for NO_2 in 1971, setting both a primary standard (to protect health) and a secondary standard (to protect the public welfare) at 0.053 parts per million (53 ppb), averaged annually. The Agency has reviewed the standards twice since that time, but chose not to revise the annual standards at the conclusion of each review. In January 2010, EPA established an additional primary standard at 100 ppb, averaged over one hour. Together the primary standards protect public health, including the health of sensitive populations - people with asthma, children, and the elderly. No area of the country has been found to be out of compliance with the current NO_2 standards.

Health

Current scientific evidence links short-term NO_2 exposures, ranging from 30 minutes to 24 hours, with adverse respiratory effects including airway inflammation in healthy people and increased respiratory symptoms in people with asthma.

Also, studies show a connection between breathing elevated short-term NO_2 concentrations, and increased visits to emergency departments and hospital admissions for respiratory issues, especially asthma.

NO_2 concentrations in vehicles and near roadways are appreciably higher than those measured at monitors in the current network. In fact, in-vehicle concentrations can be 2–3 times higher than measured at nearby area-wide monitors. Near-roadway (within about 50 meters) concentrations of NO_2 have been measured to be approximately 30 to 100% higher than concentrations away from roadways.

Individuals who spend time on or near major roadways can experience short-term NO_2 exposures considerably higher than measured by the current network. Approximately 16% of U.S housing units are located within 300 ft of a major highway, railroad, or airport (approximately 48 million people). This population likely includes a higher proportion of non-white and economically-disadvantaged people.

NO_2 exposure concentrations near roadways are of particular concern for susceptible individuals, including people with asthma asthmatics, children, and the elderly

The sum of nitric oxide (NO) and NO_2 is commonly called nitrogen oxides or NOx. Other oxides of nitrogen including nitrous acid and nitric acid are part of the nitrogen oxide family. While EPA's National Ambient Air Quality Standard (NAAQS) covers this entire family, NO_2 is the component of greatest interest and the indicator for the larger group of nitrogen oxides.

NOx react with ammonia, moisture, and other compounds to form small particles. These small particles penetrate deeply into sensitive parts of the lungs and can cause or worsen respiratory disease, such as emphysema and bronchitis, and can aggravate existing heart disease, leading to increased hospital admissions and premature death.

Ozone is formed when NOx and volatile organic compounds react in the presence of heat and sunlight. Children, the elderly, people with lung diseases such as asthma, and people who work or exercise outside are at risk for adverse effects from ozone. These include reduction in lung function and increased respiratory symptoms as well as respiratory-related emergency department visits, hospital admissions, and possibly premature deaths.

Emissions that lead to the formation of NO_2 generally also lead to the formation of other NOx. Emissions control measures leading to reductions in NO_2 can generally be expected to reduce population exposures to all gaseous NOx. This may have the important co-benefit of reducing the formation of ozone and fine particles both of which pose significant public health threats.

19.3 Secondary Pollutant: Acid Rain

"Acid rain" is a broad term referring to a mixture of wet and dry deposition (deposited material) from the atmosphere containing higher than normal amounts of nitric and sulfuric acids. The precursors, or chemical forerunners, of acid rain formation result from both natural sources, such as volcanoes and decaying vegetation, and man-made sources, primarily emissions of **sulfur dioxide** (**SO₂**) and **nitrogen oxides** (**NOₓ**) resulting from fossil fuel combustion (Figure 19.9). In the United States, roughly 2/3 of all SO₂ and 1/4 of all NOₓ come from electric power generation that relies on burning fossil fuels, like coal. Acid rain occurs when these gases react in the atmosphere with water, oxygen, and other chemicals to form various acidic compounds. The result is a mild solution of sulfuric acid and nitric acid. When sulfur dioxide and nitrogen oxides are released from power plants and other sources, prevailing winds blow these compounds across state and national borders, sometimes over hundreds of miles.

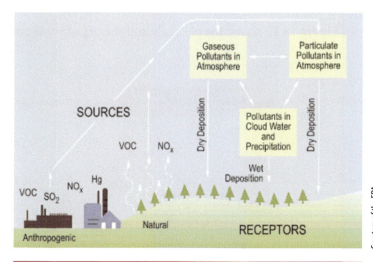

FIGURE 19.9 **The primary mechanisms for man-made sources of acid rain.**

Wet deposition refers to acidic rain, fog, and snow. If the acid chemicals in the air are blown into areas where the weather is wet, the acids can fall to the ground in the form of rain, snow, fog, or mist. As this acidic water flows over and through the ground, it affects a variety of plants and animals. The strength of the effects depends on several factors, including how acidic the water is; the chemistry and buffering capacity of the soils involved; and the types of fish, trees, and other living things that rely on the water.

In areas where the weather is dry, the acid chemicals may become incorporated into dust or smoke and fall to the ground through **dry deposition**, sticking to the ground, buildings, homes, cars, and trees. Dry deposited gases and particles can be washed from these surfaces by rainstorms, leading to increased runoff. This runoff water makes the resulting mixture more acidic. About half of the acidity in the atmosphere falls back to earth through dry deposition.

Measuring Acid Rain

Acid rain is measured using a scale called "pH." The lower a substance's pH, the more acidic it is.

Pure water has a pH of 7.0. However, normal rain is slightly acidic because carbon dioxide (CO₂) dissolves into it forming weak carbonic acid, giving the resulting mixture a pH of approximately 5.6 at typical atmospheric concentrations of CO₂. As of 2000, the most acidic rain falling in the U.S. has a pH of about 4.3.

Acidic and basic are two extremes that describe chemicals, just like hot and cold are two extremes that describe temperature. Mixing acids and bases can cancel out their extreme effects; much like mixing hot and cold water can even out the water temperature. A substance that is neither acidic nor basic is neutral.

The pH scale measures how acidic or basic a substance is. It ranges from 0 to 14. A pH of 7 is neutral. A pH less than 7 is acidic, and a pH greater than 7 is basic. Each whole pH value below 7 is ten times more acidic than the next higher value. For example, a pH of 4 is ten times more acidic than a pH of 5 and 100 times (10 times 10) more acidic than a pH of 6. The same holds true for pH values above 7, each of which is ten times more alkaline—another way to say basic—than the next lower whole value. For example, a pH of 10 is ten times more alkaline than a pH of 9.

Pure water is neutral, with a pH of 7.0. When chemicals are mixed with water, the mixture can become either acidic or basic. Vinegar and lemon juice are acidic substances, while laundry detergents and ammonia are basic.

Chemicals that are very basic or very acidic are called "reactive." These chemicals can cause severe burns. Automobile battery acid is an acidic chemical that is reactive. Automobile batteries contain a stronger form of some of the same acid that is in acid rain. Household drain cleaners often contain lye, a very alkaline chemical that is reactive.

Figure 19.10 shows the pH scale and the pH of some common items:

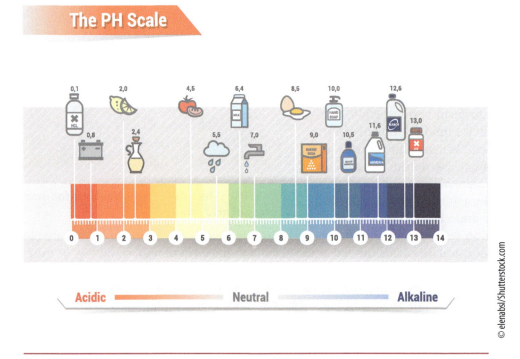

FIGURE 19.10   **Items with pH of less than 7 are acidic and more than a pH of 7 are basic.**

Effects of Acid Rain

Acid rain causes acidification of lakes and streams and contributes to the damage of trees at high elevations (for example, red spruce trees above 2,000 feet) and many sensitive forest soils. In addition, acid rain accelerates the decay of building materials and paints, including irreplaceable buildings, statues, and sculptures that are part of our nation's cultural heritage. Prior to falling to the earth, sulfur dioxide (SO_2) and nitrogen oxide (NO_x) gases and their particulate matter derivatives—sulfates and nitrates—contribute to visibility degradation and harm public health.

The ecological effects of acid rain are most clearly seen in the aquatic, or water, environments, such as streams, lakes, and marshes. Acid rain flows into streams, lakes, and marshes after falling on forests, fields, buildings, and roads. Acid rain also falls directly on aquatic habitats. Most lakes and streams have a pH between 6 and 8, although some lakes are naturally acidic even without the effects of acid rain. Acid rain primarily affects sensitive bodies of water, which are located in watersheds whose soils have a limited ability to neutralize acidic compounds (called "buffering capacity"). Lakes and streams become acidic (i.e., the pH value goes down) when the water itself and its surrounding soil cannot buffer the acid rain enough to neutralize it. In areas where buffering capacity is low, acid rain releases aluminum from soils into lakes and streams; aluminum is highly toxic to many species of aquatic organisms.

Many lakes and streams examined in a National Surface Water Survey (NSWS) suffer from chronic acidity, a condition in which water has a constant low pH level. The survey investigated the effects of acidic deposition

FIGURE 19.11   **A beautiful stream.**

in over 1,000 lakes larger than 10 acres and in thousands of miles of streams believed to be sensitive to acidification. Of the lakes and streams surveyed, acid rain caused acidity in 75 percent of the acidic lakes and about 50 percent of the acidic streams. Several regions in the U.S. were identified as containing many of the surface waters sensitive to acidification. They include the Adirondacks and Catskill Mountains in New York state, the mid-Appalachian highlands along the east coast, the upper Midwest, and mountainous areas of the Western United States. In areas like the Northeastern United States, where soil-buffering capacity is poor, some lakes now have a pH value of less than 5. One of the most acidic lakes reported is Little Echo Pond in Franklin, New York. Little Echo Pond has a pH of 4.2.

Acidification is also a problem in lakes that were not surveyed in federal research projects. For example, although lakes smaller than 10 acres were not included in the NSWS, there are from one to four times as many of these small lakes as there are larger lakes. In the Adirondacks, the percentage of acidic lakes is significantly higher when it includes smaller lakes.

Streams flowing over soil with low buffering capacity are as susceptible to damage from acid rain as lakes. Approximately 580 of the streams in the Mid-Atlantic Coastal Plain are acidic primarily due to acidic deposition. In the New Jersey Pine Barrens, for example, over 90 percent of the streams are acidic, which is the highest rate of acidic streams in the nation. Over 1,350 of the streams in the Mid-Atlantic Highlands (mid-Appalachia) are acidic, primarily due to acidic deposition.

The acidification problem in both the U.S. and Canada grows in magnitude if "episodic acidification" is taken into account. Episodic acidification refers to brief periods during which pH levels decrease due to run-off from melting snow or heavy downpours. Lakes and streams in many areas throughout the U.S. are sensitive to episodic acidification. In the Mid-Appalachians, the Mid-Atlantic Coastal Plain, and the Adirondack Mountains, many additional lakes and streams become temporarily acidic during storms and spring snowmelt. For example, approximately 70 percent of sensitive lakes in the Adirondacks are at risk of episodic acidification. This amount is over three times the amount of chronically acidic lakes. In the mid-Appalachians, approximately 30 percent of sensitive streams are likely to become acidic during an episode. This level is seven times the number of chronically acidic streams in that area. Episodic acidification can cause "fish kills."

Emissions from U.S. sources also contribute to acidic deposition in eastern Canada, where the soil is very similar to the soil of the Adirondack Mountains, and the lakes are consequently extremely vulnerable to chronic acidification problems. The Canadian government has estimated that 14,000 lakes in eastern Canada are acidic.

How Does Acid Rain Affect Fish and Other Aquatic Organisms?

Acid rain causes a cascade of effects that harm or kill individual fish, reduce fish population numbers, completely eliminate fish species from a waterbody, and decrease biodiversity. As acid rain flows through soils in a watershed, aluminum is released from soils into the lakes and streams located in that watershed. So, as pH in a lake or stream decreases, aluminum levels increase. Both low pH and increased aluminum levels are directly toxic to fish. In addition, low pH and increased aluminum levels cause chronic stress that may not kill individual fish, but leads to lower body weight and smaller size and makes fish less able to compete for food and habitat.

	pH 6.5	pH 6.0	pH 5.5	pH 5.0	pH 4.5	pH 4.0
TROUT						
BASS						
PERCH						
FROGS						
SALAMANDERS						
CLAMS						
CRAYFISH						
SNAILS						
MAYFLY						

Courtesy of the EPA

FIGURE 19.12 **Tolerance limits to pH change for a number of aquatic species.** Younger (juvenile) members of a species will be more sensitive to pH changes than adults of the same species.

BMJ/Shutterstock

FIGURE 19.13 **Forests are sensitive to acid rain, especially when they are stressed by other environmentally factors.**

Some types of plants and animals are able to tolerate acidic waters. Others, however, are acid-sensitive and will be lost as the pH declines. Generally, the young of most species are more sensitive to environmental conditions than adults. At pH 5, most fish eggs cannot hatch. At lower pH levels, some adult fish die. Some acid lakes have no fish. The chart below shows that not all fish, shellfish, or the insects that they eat can tolerate the same amount of acid; for example, frogs can tolerate water that is more acidic (i.e., has a lower pH) than trout (Figure 19.12).

Together, biological organisms and the environment in which they live are called an ecosystem. The plants and animals living within an ecosystem are highly interdependent. For example, frogs may tolerate relatively high levels of acidity, but if they eat insects like the mayfly, they may be affected because part of their food supply may disappear. Because of the connections between the many fish, plants, and other organisms living in an aquatic ecosystem, changes in pH or aluminum levels affect biodiversity as well. Thus, as lakes and streams become more acidic, the numbers and types of fish and other aquatic plants and animals that live in these waters decrease.

Effects of Acid Rain on Forests and Plants

Over the years, scientists, foresters, and others have noted a slowed growth of some forests. Leaves and needles turn brown and fall off when they should be green and healthy. In extreme cases, individual trees or entire areas of the forest simply die off without an obvious reason.

After much analysis, researchers now know that acid rain causes slower growth, injury, or death of forests. Acid rain has been implicated in forest and soil degradation in many areas of the eastern U.S., particularly high elevation forests of the Appalachian Mountains from Maine to Georgia that include areas such as the Shenandoah and Great Smoky Mountain National Parks. Of course, acid rain is not the only cause of such conditions. Other factors contribute to the overall stress of these areas, including air pollutants, insects, disease, drought, or very cold weather. In most cases, in fact, the impacts of acid rain on trees are due to the combined effects of acid rain and these other environmental stressors. After many years of collecting information on the chemistry and biology of forests, researchers are beginning to understand how acid rain works on the forest soil, trees, and other plants.

A spring shower in the forest washes leaves and falls through the trees to the forest floor below. Some trickles over the ground and runs into streams, rivers, or lakes, and some of the water soaks into the soil. That soil may neutralize some or all of the acidity of the acid rainwater. This ability is called buffering capacity, and without it, soils become more acidic. Differences in soil buffering capacity are an important reason why some areas that receive acid rain show a lot of damage, while other areas that receive about the same amount of acid rain do not appear to be harmed at all. The ability of forest soils to resist, or buffer, acidity depends on the thickness and composition of the soil, as well as the type of bedrock beneath the forest floor. Midwestern states like Nebraska and Indiana have soils that are well buffered. Places in the mountainous northeast, like New York's Adirondack and Catskill Mountains, have thin soils with low buffering capacity.

Acid rain does not usually kill trees directly. Instead, it is more likely to weaken trees by damaging their leaves, limiting the nutrients available to them, or exposing them to toxic substances slowly released from the soil. Quite often, injury or death of trees is a result of these effects of acid rain in combination with one or more additional threats.

Scientists know that acidic water dissolves the nutrients and helpful minerals in the soil and then washes them away before trees and other plants can use them to grow. At the same time, acid rain causes the release of substances that are toxic to trees and plants, such as aluminum, into the soil. Scientists believe that this combination of loss of soil nutrients and increase of toxic aluminum may be one way that acid rain harms trees. Such substances also wash away in the runoff and are carried into streams, rivers, and lakes. More of these substances are released from the soil when the rainfall is more acidic.

However, trees can be damaged by acid rain even if the soil is well buffered. Forests in high mountain regions often are exposed to greater amounts of acid than other forests because they tend to be surrounded by acidic clouds and fog that are more acidic than rainfall. Scientists believe that when leaves are frequently bathed in this acid fog, essential nutrients in their leaves and needles are stripped away. This loss of nutrients in their foliage makes trees more susceptible to damage by other environmental factors, particularly cold winter weather.

Acid rain can harm other plants in the same way it harms trees. Although damaged by other air pollutants such as ground level ozone, food crops are not usually seriously affected because farmers frequently add fertilizers to the soil to replace nutrients that have washed away. They may also add crushed limestone to the soil. Limestone is an alkaline material and increases the ability of the soil to act as a buffer against acidity.

What Is the Role of Nitrogen in Acid Rain and Other Environmental Problems?

The impact of nitrogen on surface waters is also critical. Nitrogen plays a significant role in episodic acidification and new research recognizes the importance of nitrogen in long-term chronic acidification as well. Furthermore, the adverse impact of atmospheric nitrogen deposition on estuaries and near-coastal water bodies is significant. Scientists estimate that 10 to 45 percent of the nitrogen produced by various human activities that reaches estuaries and coastal ecosystems is transported and deposited via the atmosphere. For example, about 30 percent of the nitrogen in the Chesapeake Bay comes from atmospheric deposition. Nitrogen is an important factor in causing eutrophication (oxygen depletion) of water bodies. The symptoms of eutrophication include blooms of algae (both toxic and non-toxic), declines in the health of fish and shellfish, loss of seagrass beds and coral reefs, and ecological changes in food webs. According to the National Oceanic and Atmospheric Administration (NOAA), these conditions are common in many of our nation's coastal ecosystems. These ecological changes impact human populations by changing the availability of seafood and creating a risk of consuming contaminated fish or shellfish, reducing our ability to use and enjoy our coastal ecosystems, and causing economic impact on people who rely on healthy coastal ecosystems, such as fishermen and those who cater to tourists.

How Is EPA's Acid Rain Program Addressing These Issues?

Acid rain control will produce significant benefits in terms of lowered surface water acidity. If acidic deposition levels were to remain constant over the next 50 years (the time frame used for projection models), the acidification rate of lakes in the Adirondack Mountains that are larger than 10 acres would rise by 50 percent or more. Scientists predict, however, that the decrease in SO_2 emissions required by the Acid Rain Program will significantly reduce acidification due to atmospheric sulfur. Without the reductions in SO_2 emissions, the proportions of acidic aquatic ecosystems would remain high or dramatically worsen.

19.4 Indoor Air Quality

Most of us spend much of our time indoors. The air that we breathe in our homes, in schools, and in offices can put us at risk for health problems. Some pollutants can be chemicals, gases, and living organisms like mold and pests.

© Opka/Shutterstock.com

FIGURE 19.14 **There are many sources of air pollutants in a typical house, as shown by this infographic.**

- "Living Room: a well-used room in the home, may harbor pet hair and dander, secondhand smoke, carbon monoxide from fireplace with a leaking chimney."
- "Bathroom: This is the dampest room in the home, and needs adequate ventilation to prevent the growth of molds that can trigger asthma and other respiratory ailments."
- "Bedroom: This room has lots of surfaces that can collect dust, so clean fabrics in here and vacuum regularly."
- "Kitchen: Often thought of as the 'heart of the home,' it also is a source for many indoor air pollutants. Appliances may leak natural gas, and many people store cleaning chemicals or apply pesticides in this room. Carbon monoxide may also build up if stoves are not ventilated properly."
- "Basement: Radon is a major concern in basements. Also, mold can build up in damp basements, and often fumes are present from chemicals stored in the basement. Carbon monoxide is also an issue with appliances like furnaces and water heaters."

Several sources of air pollution are in homes, schools, and offices. Some pollutants cause health problems such as sore eyes, burning in the nose and throat, headaches, or fatigue. Other pollutants cause or worsen allergies, respiratory illnesses (such as asthma), heart disease, cancer, and other serious long-term conditions. Sometimes individual pollutants at high concentrations, such as carbon monoxide, cause death.

Learn about Pollutants

Understanding and controlling some of the common pollutants found in homes, schools, and offices may help improve your indoor air and reduce your family's risk of health concerns related to indoor air quality (IAQ).

Radon is a radioactive gas that is formed in the soil. It can enter indoors through cracks and openings in floors and walls that are in contact with the ground. When building in a region with risks of radon in the ground, builders should include radon-reducing features in the house (such as vents and fans that blow air from basement). There are also do-it-yourself radon testing kits for people who want to test their houses. Radon mitigation specialists can put in features to reduce radon levels, if it is warranted.

■   *Radon is the leading cause of lung cancer among nonsmokers, and the second leading cause of lung cancer overall.*

Secondhand smoke comes from burning tobacco products. It can cause cancer and serious respiratory illnesses.

■   Children are especially vulnerable to secondhand smoke. It can cause or worsen asthma symptoms and is linked to increased risks of ear infections and Sudden Infant Death Syndrome (SIDS).

Combustion Pollutants are gases or particles that come from burning materials. In homes, the major source of combustion pollutants are improperly vented or unvented fuel-burning appliances such as space heaters, woodstoves, gas stoves, water heaters, dryers, and fireplaces. The types and amounts of pollutants produced depends on the type of appliance, how well the appliance is installed, maintained, and vented, and the kind of fuel it uses. Common combustion pollutants include:

■   Carbon monoxide (CO) which is a colorless, odorless gas that interferes with the delivery of oxygen throughout the body. Carbon monoxide causes headaches, dizziness, weakness, nausea, and even death.
■   Nitrogen dioxide (NO_2) which is a colorless, odorless gas that causes eye, nose and throat irritation, shortness of breath, and an increased risk of respiratory infection.

Volatile organic compounds (VOCs) are chemicals found in paints and lacquers, paint strippers, cleaning supplies, varnishes and waxes, pesticides, building materials and furnishings, office equipment, moth repellents, air fresheners, and dry-cleaned clothing. VOCs evaporate into the air when these products are used or sometimes even when they are stored. Volatile organic compounds irritate the eyes, nose and throat, and cause headaches, nausea, and damage to the liver, kidneys, and central nervous system. Some of them can cause cancer.

■   Read and follow all directions and warnings on common household products.
■   Make sure there is plenty of fresh air and ventilation (e.g., opening windows and using extra fans) when painting, remodeling, or using other products that may release VOCs.
■   Never mix products, such as household cleaners, unless directed to do so on the label.
■   Store household products that contain chemicals according to manufacturers' instructions.
■   Keep all products away from children!

Asthma triggers are commonly found in homes, schools, and offices and include mold, dust mites, secondhand smoke, and pet dander. A home may have *mold* growing on a shower curtain, *dust mites* in pillows, blankets or stuffed animals, *secondhand smoke* in the air, and *cat and dog hairs* on the carpet or floors. Other common asthma triggers include some foods and pollutants in the air.

■   Asthma triggers cause symptoms including coughing, chest tightness, wheezing, and breathing problems. An asthma attack occurs when symptoms keep getting worse or are suddenly very severe.

Asthma attacks can be life threatening. However, asthma is controllable with the right medicines and by reducing asthma triggers.

Molds are living things that produce spores. Molds produce spores that float in the air, land on damp surfaces, and grow. Molds can grow anywhere there is moisture in a house or building.

- Inhaling or touching molds can cause hay fever-type symptoms such as sneezing, runny nose, red eyes, and skin rashes. Molds can also trigger asthma attacks.
- The key to mold control is moisture control.
- If mold is a problem in your home, you should clean up the mold promptly and fix the water problem.
- It is important to dry water-damaged areas and items within 24–48 hours to prevent mold growth.

Take steps to help improve your air quality and reduce your IAQ-related health risks at little or no cost by:

Controlling the sources of pollution: Usually the most effective way to improve indoor air is to eliminate individual sources or reduce their emissions.

Ventilating: Increasing the amount of fresh air brought indoors helps reduce pollutants inside. When weather permits, open windows and doors, or run an air conditioner with the vent control open. Bathroom and kitchen fans that exhaust to the outdoors also increase ventilation and help remove pollutants. Always ventilate and follow manufacturers' instructions when you use products or appliances that may release pollutants into the indoor air.

Changing filters regularly: Central heaters and air conditioners have filters to trap dust and other pollutants in the air. Make sure to change or clean the filters regularly, following the instructions on the package.

Adjusting humidity: The humidity inside can affect the concentrations of some indoor air pollutants. For example, high humidity keeps the air moist and increases the likelihood of mold. Keep indoor humidity between 30 and 50 percent. Use a moisture or humidity gauge, available at most hardware stores, to see if the humidity in your home is at a good level. To increase humidity, use a vaporizer or humidifier. To decrease humidity, open the windows if it is not humid outdoors. If it is warm, turn on the air conditioner or adjust the humidity setting on the humidifier.

Asthma is a serious, sometimes life-threatening respiratory disease that affects the quality of life for millions of Americans. Environmental asthma triggers are found around the home and can be eliminated with simple steps.

- Don't allow smoking in your home or car.
- Dust and clean your home regularly.
- Clean up mold and fix water leaks.
- Wash sheets and blankets weekly in hot water.
- Use allergen-proof mattress and pillow covers.
- Keep pets out of the bedroom and off soft furniture.
- Control pests—close up cracks and crevices and seal leaks; don't leave food out.

Children are especially sensitive to secondhand smoke, which can trigger asthma and other respiratory illnesses. Secondhand smoke is smoke comes from burning tobacco products such as cigarettes, pipes, and cigars.

- To help protect children from secondhand smoke, do not smoke or allow others to smoke inside your home or car.

Schools

With nearly 56 million people, or 20 percent of the U.S. population, spending their days inside elementary and secondary schools, IAQ problems can be a significant concern. All types of schools—whether new or old, big or small, elementary or high school—can experience IAQ problems. School districts are increasingly experiencing budget shortfalls and many are in poor condition, leading to a host of IAQ problems.

- EPA's voluntary *Indoor Air Quality Tools for Schools* Program provides district-based guidance to schools about best practices, industry guidelines, and practical management actions to help school personnel identify, solve, and prevent IAQ problems.
- Children may be more sensitive to pollution, and children with asthma are especially sensitive. Asthma is responsible for millions of missed school days each year. Parents' and caregivers' involvement helps daycare facilities become aware of asthma triggers and the need to reduce them.

Office Buildings

Many office buildings have poor IAQ because of pollution sources and poorly designed, maintained, or operated ventilation systems.

- Office workers help to improve the indoor air in their buildings by paying attention to environmental conditions including ventilation, temperature, and the presence of odors. Report any problems to facility managers immediately.
- To improve IAQ, be careful not to block air vents or grilles, keep your space clean and dry, and do not bring in products that may pollute the indoor air.

Indoor Air Quality in Developing Nations

The World Health Organization estimates that 4.3 million people die annually from unhealthy indoor air conditions. In developing countries, and even in some "off the beaten path" places in developed countries, approximately 3 billion people rely on burning wood, charcoal or coal, dried animal dung, or crop residues in open pits, stoves or fireplaces to heat their homes and cook foods. The gasses produced in the combustion of these fuels produce particulate matter, carbon monoxide, and other carcinogenic compounds. For example, the concentration of fine particulate matter indoors when these fires are burning can reach 100 times the acceptable levels.

FIGURE 19.15 **Unvented cooking fires can severely impair lung health.**

Indoor air pollution from indoor fires leads to many different health issues. In children under 5, it is the leading cause of pneumonia. Indeed, it causes more than half of all pneumonia deaths in this young population. In adults, indoor air pollution causes chronic obstructive pulmonary disease (COPD), a progressive lung disease that includes diseases like asthma, emphysema and lung cancer. It also is a leading cause of a number of different cancers, heart disease and stroke, and even leads to blindness by cataract formation.

These statistics about indoor air pollutino in developing countries are from the World Health Organization websites http://www.who.int/indoorair/en/ and http://www.who.int/indoorair/health_impacts/disease/en/

CHAPTER 20

Urban Issues

20.1 Smart Growth in Developed Countries

Urban Sprawl

The issue of urban sprawl is one that most Americans are familiar with. **Sprawl** can be defined as the spreading of urban or suburban areas into undeveloped lands. In this summary report by the Natural Resources Defense Council (NRDC) the impacts of sprawl on farmland, forests, air pollution, climate change, and water quality are examined. An alternative to sprawl, known as "**Smart Growth,**" is also introduced. Although some of the data quoted in this report from 1999 is dated, many of the same trends continue today. Fortunately, many cities are embracing "Smart Growth" policies and initiatives to revitalize downtown areas and reduce sprawl.

> **sprawl:** the spreading of urban or suburban areas into undeveloped lands

> **smart growth:** form of urban planning that recognizes urban growth will occur but uses zoning laws and other tools to prevent sprawl, direct growth to certain areas, protect ecologically sensitive and important lands and waterways, and develop urban areas that are more environmentally sustainable

Because the land area of the United States is so large, sprawl may only be affecting a small percentage of the total. However, many of the areas that are most impacted by sprawl development are losing land that is highly valuable for farming, forests, or other purposes. In addition, sprawl is happening in and around areas where most Americans already live, and so the visual impacts on the landscape and the negative side effects of increased traffic as well as air and water pollution are getting a lot of attention. Whereas it seems natural to expect some sprawl in areas with growing populations, the fact that it's happening to such an extent in urban areas with stagnant or even declining populations (such as Cleveland, Ohio) indicates that this is a problem that stems from poor planning and a failure to consider the broader environmental and social impacts of this form of development.

In addition to the obvious loss of open space and farmland, this article also draws attention to how sprawl is causing other environmental problems. Because sprawl developments are so spread out and designed around the use of private automobiles, far more energy is used in transportation, and this increases regional air pollution problems and the emission of greenhouse gases linked to climate change. Likewise, converting farms, forests, and open spaces to homes, strip malls, and parking lots involves paving over porous ground. This causes at least two problems. First, as rain runs off parking lots and other nonporous surfaces, it picks up pollutants such as leaked motor oil and washes them into nearby streams and rivers. Sprawl is thus a major cause of what is known as nonpoint water pollution. Second, this increased runoff aggravates flooding problems and lowers groundwater supplies since, instead of soaking into the ground, rain that falls on rooftops, driveways, parking lots, and roads often runs straight into nearby streams and rivers.

Addressing the problems caused by sprawl is a challenging issue since many Americans appear to favor a life in the suburbs and sprawl development can often provide that option. One approach, known as smart growth, attempts to minimize some of the negative impacts of sprawl through better planning and design of new developments. Smart

FIGURE 20.1 **Urban sprawl, especially the spread of subdivisions, has consumed vast areas of land over the past century.**

© Cameron Whitman/iStock/Thinkstock

growth also emphasizes investment in and revitalization of existing urban and developed areas to provide an alternative to sprawling developments.

By *Natural Resources Defense Council,* 1999

Consider that the Maryland Office of Planning projects that, from 1995 to 2020, more land will be converted to housing in the region surrounding the Chesapeake Bay than in the past three and one-half centuries. In greater Chicago, in the two decades from 1970 to 1990, the consumption of residential land grew an amazing eleven times faster than the region's population. Over the same period, the Chicago region's consumption of land for commercial and industrial uses grew 74 percent, eighteen times faster than population. In greater Cleveland, land consumption has been growing even while population has been *declining:* since 1970, the regional population has declined by 11 percent, while the amount of urbanized land has grown by 33 percent.

In the Sun Belt [southern area of the United States that extends from coast to coast], the prototype "city" of the future, if present trends continue, may well be metropolitan Phoenix, which is reported to be developing open land at the rate of 1.2 acres per hour. Indeed, the geographic reach of Phoenix is now said to be equivalent in size to Delaware. Another Sun-Belt candidate to represent the metropolis of the future is Atlanta, which has grown in population at an average annual rate of 2.9 percent since 1950, with almost all of the growth in the suburbs. Real estate analyst Christopher Leinberger holds that "it is altogether probable that in terms of land area Atlanta is the fastest-growing human settlement in history."

And one of the most daunting aspects of our rapid development of new land is its permanence: every acre of natural or open space paved over for sprawl—every acre claimed by a new subdivision or shopping center on the fringe of a Chicago, Phoenix, or Atlanta—represents an acre lost forever. Barring heroic measures, farmland and other open space cannot be reclaimed.

These patterns of development portend [indicate] serious consequences for our nation's land, air, and water, each of which we discuss below. Although as a society we have made great environmental progress in the last few decades, it is becoming increasingly clear that we cannot continue to do so if we do not change the way we grow.

From Old MacDonald to the New McDonald's

While the United States continues to enjoy the appearance of abundant farmland, the best of that land is being lost at an amazing rate. At the conclusion of exhaustive research on the subject, the American Farmland Trust has reported that from 1982 to 1992 we lost to urban and suburban development an average of 400,000 acres per year of "prime" farmland, the land with the best soils and climate for growing crops. This translates to a loss of 45.7 acres per hour, every single day. During that same period, we lost an additional 26,600 acres per year—three more acres per hour, every single day—of "unique" farmland, used for growing rare and specialty crops. Put another way, for each acre of prime or unique farmland that is being saved by various farmland protection programs across the county, three acres are lost to development.

To make matters worse, there is an unfortunate congruence between that land most suited and productive for farming and that land most in danger of urban encroachment. As Professor Reid Ewing has put it, the "lands most suitable for growing crops also tend to be most suitable for 'growing houses.'" This is because inland urban settlements in the United States have tended to situate in river valleys and other fertile areas that are also highly productive for farming.

Perhaps as a result, most of the country's prime farmland is located within the suburban and exurban counties of metropolitan areas. Such "urban-influenced" counties currently produce more than half the total value of U.S. farm production; their average annual production value per acre is some 2.7 times that of other U.S. counties. Yet, ominously, their population growth is also disproportionately high, over twice the national average. Those counties with prime and unique farmland found to be threatened by particularly high rates of current development collectively produce some 79 percent of our nation's fruit, 69 percent of our vegetables, 52 percent of our dairy products, and over one-fourth of our meat and grains. Among the farming regions most seriously endangered by sprawl are California's Central Valley, the Northern Piedmont near Washington, DC and Baltimore, and the Northern Illinois Drift Plain near Chicago.

Landscapes Lost

To add insult to injury, what is being plopped down on our nation's lost farmland and open space is not pretty, to say the least.

The impacts go beyond the annoyance of visual clutter, of course, to our deeper senses of place and history, and to our ever-more-tenuous connection as human beings to the mysteries of the natural world.

For example, there is plenty of evidence that we place a high value on exactly those benefits that we are losing. Writer Tony Hiss describes research documenting strong human preferences for green landscapes with water, winding paths, long and sweeping vistas, and hidden natural places. Similarly, in a recent public opinion poll, 63 percent of respondents cited "the beauty of nature" as a reason for wanting to protect the environment. A New Jersey survey reported that 78 percent of respondents supported changes in development patterns in order to preserve farmland. In still another study, citizens shown slide images gave the lowest approval rating to images of "cookie cutter" subdivisions and complexes, highway strip development, and shopping plazas with large front parking lots, while they gave the highest rating to natural areas, farmland, woodlots, parks and streams (Figure 20.2).

FIGURE 20.2   People prefer to look at views of undeveloped land, or farmland, that has less visual clutter.

Shutterstock/Philip Bird LRPS CPAGB

In addition to the findings with regard to preference, there is evidence that ugly development is not good for us. Research at Texas A & M and the University of Delaware indicates that humans' reactions to visual clutter may include elevated blood pressure, increased muscle tension, and impacts on mood and work performance. Recovery from stress has been measured as faster and more complete when we are exposed to natural outdoor environments. At least one study indicates that travel along unattractive suburban corridors may make people feel that they are driving for longer periods of time than they actually are.

Moreover, the loss of undeveloped landscapes threatens economic as well as psychological values. Over 130 million Americans enjoy observing, photographing, and feeding wildlife and fish, thus supporting a nature-oriented tourist industry in excess of $14 billion annually. The National Survey of Fishing, Hunting, and Wildlife Associated Recreation found that 77 percent of the U.S. population enjoys some form of wildlife-related recreation, and a 1987 poll sponsored by the President's Commission on Americans Outdoors found that "natural beauty was the single most important criterion for tourists selecting outdoor recreation sites." Independent of recreation and tourism, proximity to open spaces has been found to raise the value of residential property by as much as a third in some cases, raising property tax revenues as well.

Gray Skies and Greenhouses

Current development patterns also bring substantial air pollution, largely because of the increased automobile dependence that is associated with sprawl. As we spread ourselves farther and farther apart, it becomes inevitable that we must travel longer distances to work, shop, enjoy recreation, and visit family and friends.

Shutterstock/Pushish Images

FIGURE 20.3 Traffic backs up the streets of Bangkok, Thailand, a common sight in metropolitan areas around the globe. Idling cars and trucks lead to growing emissions and lower air quality.

The convenience store and even the playground may no longer be within walking distance. The bus stop may be farther away, too, even if we are fortunate enough to have a bus that goes anywhere close to our destination; in some places, there may be no bus service at all. Work may be on the other side of town or even in another town altogether.

The only good choice for most suburbanites is to drive, and to drive a lot. And that is exactly what we are doing. Motor vehicle use in America doubled from one to two trillion miles per year between 1970 and 1990. In the 1980s, vehicle miles traveled grew more than four times faster than the driving-age population and many times faster than the population at large. There are many reasons for this surge in driving, but a growing body of research makes it increasingly clear that sprawl comprises a large portion of the problem: people in spread-out locations drive more.

This translates directly into growing emissions of greenhouse gases and the continued inability of our metropolitan areas to cleanse themselves of unhealthy air. In particular, transportation in the United States already contributes some 450 million metric tons of carbon dioxide—the inevitable by-product of fossil-fuel combustion—each year, around 32 percent of total U.S. carbon emissions, to the atmosphere. The federal Department of Energy projects that total U.S. carbon emissions will continue to grow at an average rate of 1.0 percent per year, with transportation sources growing 20 percent faster than the average. And DOE's projections are conservative; many experts predict the rate of growth could be much greater, precisely because of continued increases in automobile and truck emissions due to sprawl-induced driving.

The news on other air pollutants from sprawl-induced traffic is only slightly more encouraging. A recent government publication summarizes the situation:

Despite considerable progress, the overall goal of clean and healthy air continues to elude much of the country. Unhealthy air pollution levels still plague virtually every major city in the United States. This is largely because development and urban sprawl have created new pollution sources and have contributed to a doubling of vehicle travel since 1970.

In particular, cars and other highway vehicles continue to emit some 60 million tons of carbon monoxide per year, about 62 percent of our national inventory of that pollutant; cars and other highway vehicles continue to emit some seven million tons per year, almost 26 percent, of our volatile organic compounds (VOCs), which constitute a major precursor to ozone smog; and they emit around eight million tons per year, about 32 percent, of our nitrogen oxides, another ozone precursor. Motor vehicles also emit as much as 50 percent of our carcinogenic and toxic air pollutants, such as benzene and formaldehyde. And heavy vehicles, particularly diesel-powered buses and freight trucks, constitute a significant source of soot and other unhealthy fine particles that, when inhaled, lodge in and damage human tissue.

EPA research indicates that, notwithstanding continuing improvements in emission control systems, the total national inventory of hydrocarbon emissions from gasoline vehicles could reverse direction and begin to increase again in the early part of the 21st century, because of increased driving. Total nitrogen oxide emissions from motor vehicles already are at a higher level than they were two decades ago, despite improvements in the emissions performance of individual vehicles. Ozone and particulate pollution are both projected to rise, and some observers believe that, by 2015, carbon monoxide will also be on a rising trend.

Runoff Run Amok

Haphazard sprawl development also brings runoff water pollution to more and more **watersheds**, degrading streams, lakes, and estuaries. Natural landscapes, such as forests, wetlands, and grasslands, are typically varied and porous. They trap rainwater and snowmelt and filter it into the ground slowly. When there is runoff, it tends to reach receiving waterways gradually. Cities and suburbs, by contrast, are characterized by large paved or covered surfaces that are impervious to rain. Instead of percolating slowly into the ground, stormwater becomes trapped above these surfaces, accumulates, and runs off in large amounts into waterways, picking up pollutants as it goes.

It is now thoroughly documented that, as the amount of impervious pavement and rooftops increases in a watershed, the velocity and volume of surface runoff increases; flooding, erosion and pollutant loads in receiving waters increase; groundwater recharge and water tables decline; stream beds and flows are altered; and aquatic habitat is impaired. As a result, there is a strong correlation between the amount of imperviousness in a drainage basin and the health of its receiving stream.

The consequences of watershed degradation from development have been felt across the country. In the Puget Sound region of Washington state, for example, major floods that were 25-year events now occur annually; "the sponge is full," according to King County analyst Tom Kiney. Similarly, in Akron, Ohio, runoff from residential areas has been estimated at up to 10 times that of pre-development conditions, and runoff from commercial development has been estimated at 18 times that before development. In several Maryland, Pennsylvania, and Virginia watersheds that drain into the Chesapeake Bay, pollution from development has been found to exceed—in some cases dramatically— pollution from industry and agriculture. Even in counties that have enacted stormwater-management regulations, the pace of development is causing pollutant loads to increase.

> **watersheds**: land area that delivers water, sediment, and dissolved substances via small streams to a major stream (river)

FIGURE 20.4 **The addition of rooftops and impervious surfaces (pavement) over a landscape funnels water to storm drains as there is a loss of land surface for water infiltration.** When drains cannot handle enough flow, flooded streets are a result.

Partly as a result, runoff pollution is now the nation's leading threat to water quality, affecting about 40 percent of our nation's surveyed rivers, lakes, and estuaries. Among the various pollution categories, urban runoff is the second-most prevalent source of impairments to our estuaries, affecting some 46 percent of the impaired estuaries in EPA's *National Water Quality Inventory*. It is tied for third-most-prevalent among sources of impairments to our lakes.

What to Do?

Fortunately, smart-growth solutions—those that reinvigorate our cities, bring new development that is compact, walkable, and transit-oriented, and preserve the best of our landscape for future generations—work well for the environment. For example, a comprehensive New Jersey study found that, compared to a "current trend" scenario, a plan that channeled job and housing growth to preferred locations (but assumed that a majority of new homes would remain single-family and detached) would consume 28 percent less farmland, 43 percent less open space of all kinds, and an impressive 80 percent less environmentally fragile land, including valuable forests, steep slopes, and sensitive watersheds, all while accommodating a projected growth of 408,000 new households and 654,000 new jobs over 20 years.

Building smart-growth neighborhoods that are more compact can reduce traffic, too. Transportation research indicates that each doubling of average neighborhood density is associated with a decrease in per-household vehicle use of 20–40 percent, with a corresponding decline in emissions. This is one of the reasons that European cities typically exhibit only one-fourth the per-person emissions of carbon dioxide and other pollutants from transportation that are typical of American cities. Similar savings can be had in the effects of runoff water pollution: the New Jersey study, for example, found that, by directing growth to preferred locations and achieving modestly higher average residential densities, the state could lower pollutants in stormwater runoff by some 4,560 tons per year, a 40 percent reduction over that predicted for the "current trend" scenario.

In other words, we don't have to keep doing this to ourselves. Indeed, this is a time of great innovation in the development, architecture, planning, and land preservation communities. There may be no one magic, all-encompassing solution to sprawl, but many promising and successful alternatives are at hand.

For the environment, the task is to apply advocacy to all levels of government, local, state, and national, and to work with the private sector on forging and implementing the alternatives. The rewards can be great because, just as sprawl poses multiple environmental problems, its solutions promise multiple environmental benefits. The task is large, but also exciting. We dedicate ourselves to it.

Adapted from Paving Paradise: Sprawl and the Environment. Natural Resources Defense Council (NRDC). 1999. http://www.nrdc.org/cities/smartgrowth/rpave.asp. Reprinted with permission from the Natural Resources Defense Council.

Looking Ahead Why Smart Growth?

The EPA lays out arguments and strategies for cities to implement as they grow in order to improve health (human and environmental) and quality of life. This report, called "Our Built and Natural Environments: A Technical Review of the Interactions Between Land Use, Transportation, and Environmental Quality (2nd Edition)" was published in 2013 and its detailed findings may be found at the link http://www2.epa.gov/smart-growth/our-built-and-natural-environments-technical-review-interactions-between-land-use. A summary of this report follows.

Decisions about how and where we build our communities have significant impacts on the natural environment and on human health. Cities, regions, states, and the private sector need information about the environmental effects of their land use and transportation decisions to mitigate growth-related environmental impacts and to improve community quality of life and human health. In 2001, EPA published *Our Built and Natural Environments: A Technical Review of the Interactions Between Land Use, Transportation, and Environmental Quality* to show how development patterns affect the environment and human health. Since then, research has continued to clarify and better explain these connections. To capture this research, EPA has revised and updated the report in 2013, incorporating key findings from hundreds of studies.

■ Discusses the status of and trends in land use, development, and transportation and their environmental implications. Findings include:
   ❑ The U.S. population is projected to grow 42 percent between 2010 and 2050, from 310 million to 439 million (Vincent and Velkoff 2010).
   ❑ While the population roughly doubled between 1950 and 2011 (U.S. Census Bureau), vehicle travel during this same period increased nearly sixfold (Federal Highway Administration 2010 and 2012). However, evidence suggests that the growth of vehicle travel might be slowing in recent years.
   ❑ Virtually every metropolitan region in the United States has expanded substantially in land area since 1950—including regions that lost population during that time (U.S. Census Bureau).
■ Articulates the current understanding of the relationship between the built environment and the quality of air, water, land resources, habitat, and human health. Findings include:
   ❑ Biodiversity: For nearly all plants and animals, species diversity declines with increases in the amount of impervious surface, road density, time since development, human population density, and building density (Pickett et al. 2011).

- ❏ Water: Development in watersheds reduces the quantity, quality, and diversity of stream habitat for aquatic life (Booth and Bledsoe 2009). As water is polluted and degraded, it can become unfit for drinking, swimming, fishing, and other uses.
- ❏ Air: More than 38 percent of national carbon monoxide emissions and 38 percent of nitrogen oxide emissions come from highway vehicles. Stationary sources like power plants that provide energy to homes, offices, and industries are also major sources of pollution (EPA 2012).
- ❏ Climate Change: Greenhouse gas emissions from the transportation sector increased 19 percent between 1990 and 2010, due primarily to the increase in vehicle travel but partially offset by a slight increase in average fuel economy as older vehicles were removed from the roads (EPA 2012).
- ❏ Health: While data are lacking to determine whether the built environment determines levels of physical activity and/or obesity, nearly 90 percent of studies found a positive association (Ferdinand et al. 2012), suggesting that the built environment is one of the many factors that could play a role in how much people exercise and levels of obesity.
- ❏ Safety: Car crashes are the third leading cause of death in terms of years of life lost given the young age of so many car crash victims and the number of years they would have been expected to live if they had not died in a car crash. Only cancer and heart disease are responsible for more years of life lost (Subramanian 2011).

- ■ Provides evidence that certain kinds of land use and transportation strategies can reduce the environmental and human health impacts of development. Findings include:

- ❏ Development in and adjacent to already-developed areas can help protect natural resources like wetlands, streams, coastlines, and critical habitat.
- ❏ Residents of transit-oriented developments are two to five times more likely to use transit for commuting and non-work trips than others living in the same region (Arrington and Cervero 2008).
- ❏ In general, the greater the population density of an area, the less the area's residents tend to drive (Transportation Research Board of the National Academies 2003). Doubling residential density across a metropolitan region could reduce household vehicle travel by between 5 and 12 percent (National Research Council of the National Academies, Driving and the Built Environment 2009).

FIGURE 20.5 As part of their "Smart Growth" planning, the city of Phoenix, Arizona is investing in a light rail to help revitalize the downtown area and as a more environmentally friendly commuting option.

- ❏ Communities with streets designed for the safety of all users can encourage walking and biking and help people lead healthier lifestyles (Giles et al. 2011).
- ❏ A review of green building retrofits of commercial buildings around the world found energy savings of 50 to 70 percent (Harvey 2009).
- ❏ Water-efficient household appliances and fixtures can yield significant water savings, and careful selection of construction materials can conserve natural resources and improve indoor air quality. Site-scale green infrastructure can also reduce development's impacts on water quality.

EPA's Smart Growth Program has many resources that give more information on development strategies that reduce environmental and health impacts while improving quality of life, providing more housing and transportation options, and achieving other community goals. See our *publications page* (http://www2.epa.gov/smart-growth/smart-growth-publications) and our *topics pages* (http://www2.epa.gov/smart-growth/smart-growth-topics) for links to these resources.

20.2 Housing Solutions for Developing Countries

Growing Pains in Developing Countries

In developing nations, people are moving to urban centers in unprecedented numbers, seeking work and economic stability. Many of these newcomers cannot afford adequate housing. Up to 80% of the population of some cities in developing countries lives in slums. The UN characterizes an area to be a slum if it meets these criteria:

- "Inadequate access to safe water
- Inadequate access to sanitation and infrastructure
- Poor structural quality of housing
- Overcrowding
- Insecure residential status"

It is estimated that more than 3 billion new people will need housing in these urban centers by 2030. Furthermore, these people need more than just affordable housing. These cities are also rushing to improve their sanitation services to provide clean water, sewage, reliable energy sources and more to this new population. This is an expensive process, and one that takes time.

Access to safe water, and safe wastewater disposal practices, is often a primary issue in slums. In sub-Saharan Africa, more than 40% of the population does not have access to safe water. Worldwide, 1.1 billion people do not have access to safe, clean water. Water sources in slums are often contaminated with sewage waste due to the lack of sanitation systems. Indeed, 2.6 billion people do not have adequate sanitation where they live. The situation is improving. As of 2010, 56% of people in developing countries have access to improved sanitation systems, as compared to 36% of people just 20 years earlier.

Urban Environments and the Growth of Megacities

As more and more of the Earth's human population moves to urban areas, there has been a rapid increase in what are known as megacities—urban regions of more than 10 million people. These megacities put huge strains on the environment for resources such as water and food, and generate mega-quantities of waste in the process. Yet, because they are relatively concentrated in size, megacities also offer the possibility of reducing overall human impact on the environment if they are designed and maintained efficiently. Journalist Harry Bruinius of The Christian Science Monitor examines trends in megacities around the world, with a focus on Mexico City, Beijing, and Tokyo as examples of the perils and promise of this form of development.

For the past few decades many Americans have been moving from cities to suburbs; however, the trend in many developing countries has been just the opposite. The chief reason for this is that there are greater economic opportunities in urban areas, and so people are migrating from rural areas to cities. The net effect of these movements has been a fundamental change in the relative distribution of where people live. From fewer than 30 percent in 1950, cities now hold more than 50 percent of the global population—and this figure is projected to rise to over 70 percent by 2050.

As cities absorb more and more people, they often encounter a series of environmental, resource, and logistical challenges. Providing so many residents with water and food, removing the solid waste and sewage created by such large numbers, and meeting the energy and other resource needs of millions of people pose tremendous challenges to governments and planners. However, this very same concentration of people also creates opportunities for lowering the per capita impacts of development. Highly concentrated populations make the provision of mass transit (such as subway and tram systems) much more viable and can reduce individual automobile use significantly. Water provision and waste removal systems can also benefit from so-called economies of scale—lower per unit costs of service due to a larger scale of operation. The challenge for many urban areas, especially in the poorer developing countries, will be providing these services in a timely fashion to urban residents in cities that are growing so rapidly.

By *Harry Bruinius*

O n a teeming street in Mumbai's Dharavi **slum**, amid a colorful swirl of sweet lime carts and red-clay pottery, Pastor Bala Singh brings an assortment of buckets to retrieve his daily ration of water. The indoor spigot he uses provides water only three hours a day. It is the only source for the six small homes on his street, and each family has 30 minutes to fill its containers.

> **slum**: a densely populated urban area marked by crowding, substandard housing, poverty, and social disorganization

Pastor Singh is not complaining, though. Things are greatly improved from when he first immigrated to Dharavi—the most crowded part of one of the world's most crowded cities. "The roads were muddy," he says from his second-floor office, above the popping sizzle of a man welding, sans protective gear, downstairs. "Now they put down bricks." Singh ministers to a small congregation that meets above the church-sponsored kindergarten where his wife has taught for 17 years. Though relatives have begged him to come home to Tamil Nadu, 700 miles east, he has no plans to leave.

"Three times I tried to go back to my native place," the pastor says, explaining that there were no jobs there. "I don't want to live here but God's plan is different."

Singh's migration to the city, a combination of divine impulsion and the simple need to work, is part of what could be called an epic trend affecting billions of people worldwide. Sometime in 2007, for the first time in human history, more people began to live within the cacophonous [harsh sounding] swirl of cities than in rural hamlets or on countryside farms.

It's a fundamental shift that may be altering the very fabric of human life, from the intimate, intricate structures of individual families to the massive, far-flung infrastructures of human civilizations. In 1950, fewer than 30 percent of the world's 2.5 billion inhabitants lived in urban regions. By 2050, almost 70 percent of the world's estimated 10 billion inhabitants—or more than the number of people living today—will be part of massive urban networks, according to the Population Division of the United Nations' Department of Economic and Social Affairs.

These staggering statistical trends are driving the evolution of the "**megacity**," defined as an urban agglomeration [the process of gathering into a mass] of more than 10 million people. Sixty years ago there were only two: New York/Newark and Tokyo. Today there are 22 such megacities—the majority in the developing countries of Asia, Africa, and Latin America—and by 2025 there will probably be 30 or more.

Consider just India. Though the country is still largely one of villagers—about 70 percent of India's 1.2 billion inhabitants live in rural areas— immigration and internal migrations have transformed it into a country with 25 of the 100 fastest-growing cities worldwide. Two of them, Mumbai (Bombay) and Delhi, already rank among the top five most populous urban areas.

In the "developed" countries of the West, this trend had been building since the Industrial Revolution, which sparked, relatively quickly, the exponential growth of cities seen today. The quest for "efficiency" and the corresponding divisions of labor generated technological innovations that obliterated the need for farm laborers and local artisans. This drove populations from the country to the city over time and transformed the plow and the hoe into mere tools for backyard gardeners.

Randy Olson/National Geographic/SuperStock

FIGURE 20.6   Laundry hangs to dry in a neighborhood in Mumbai, India, an area where an increase in population is met with a lag in modernity.

Today, on average, 3 out of 4 people living in modern industrialized states are already building their lives within an urban area—a ratio that will jump to more than 5 in 6 by 2050. By contrast, today in the least-developed regions of the world, more

> **megacity**: a metropolitan area with a population of more than 10 million people

than 2 out of 3 people still eke out a living in a rural area. For these people, even the slumdog [person who lives in a crowded, poverty-striken urban area] existence in places like Dharavi can offer more opportunities than their villages ever could. And within these developing regions, according to UN-HABITAT, cities are gaining an average of 5 million new residents—per month.

"Most of these [urban immigrants] couldn't earn cash in their rural situations," says Chuck Redman, director of the School of Sustainability at Arizona State University in Tempe. "There's not as much of a cash economy there, but they still want cash to buy radios and mobile phones or TVs—or even send their kids to school, which costs money in many of these countries."

Call it the lag of modernity: The changes wrought by industrialization began slowly 200 years ago, accelerated through the 20th century in the West, and now are spreading exponentially around the globe. Many observers see great promise in this urbanizing trend: The efficiencies of cities can cut energy consumption up to 20 percent, transportation costs for goods and labor can drop significantly, and entertainment industries can thrive when millions live together. In other words, cities are giant cash machines, the primary locus of economic growth.

"Some companies look at this as a huge opportunity," says Fariborz Ghadar, director of Penn State's Center for Global Business Studies and the author of a book on megacities. "We're going to build roads, we're going to build buildings, and [tech companies] love this because you can put the Internet in concentrated cities much more efficiently."

Yet, as megacities evolve in the developing world, many groan under the weight of a sudden, massive, and unprecedented demand for services never seen in the West. The basic necessities of clean water, of sanitation systems to remove megatons of garbage and human waste, of transportation systems to shuttle millions of workers, not to mention the need for electrical networks, health-care facilities, and policing and security, are, simply put, creating one of the greatest logistical challenges ever seen in human history. And this is even before factoring in the challenges of climate change, terrorism, and the preservation of human dignity.

Mexico City: On a Bowl of Pudding

An orange metal elevator heads deep into the bowels of Mexico City, where a crew of technicians and engineers is inspecting a 900-ton machine, longer than a football field, that burrows through a muddy mélange of rock, silt, and water. It's the first stage of the city's plan to build a massive new tunnel that officials hope will relieve the pressures on Mexico City's drainage system.

Nearly 1,000 feet into the passageway, made up of adjacent rings each composed of concrete slabs weighing some 4 tons each, the air is thin. Oxygen roars in through a tube, providing relief for those working on the project's edge. Workers crawl along scaffolding, crouching under an Erector Set of tanks and pipes that pump out water and hurl the deep-earth's rock and mud to the surface.

Their task is to prevent large portions of Mexico City, one of the world's most populous megacities, from catastrophic flooding. The area's growing population has placed demands on water supplies that are simply unsustainable. Its 20 million residents have laid down an urban jungle that obstructs water from naturally filtering into the ground.

FIGURE 20.7 **A worker walks in a drainage tunnel in Mexico City built to prevent catastrophic flooding.** This flood risk exists because Mexico city is located in what some call a giant "saucepan"—because it lacks a natural exit for floodwaters.

Today, the city is sucking up water from the natural aquifers at twice the rate they are being replenished. The result: Mexico City is sinking, in some areas up to 16 inches a year, threatening its entire infrastructure. This includes the city's deteriorating drainage system, whose capacity has diminished by 30 percent since 1975 while the area's population has doubled.

"It's an alarming situation," says Felipe Arreguin, the technical general subdirector at Mexico's National Water Commission (Conagua), which is building the drainage tunnel. "We are taking [out] so much water, the city is sinking. What if an entire block were to go under?"

It nearly has. In 2007, a giant sinkhole swallowed a large swath of a busy street. At Revolution Monument, a water pipe installed over 75 years ago now stands nearly 30 feet above ground. Given Mexico City's history as a "floating city" in the middle of a lake, it's no surprise that water is what vexes most urban planners here. When the Spaniards arrived to conquer the great Aztec Empire, the mode of transportation was not horses but canoes. Today, the city sits essentially on a bowl of pudding. Jose Miguel Guevara, the general coordinator for water supply and drainage projects at Conagua, calls this basin a giant "saucepan," with no natural exit for the torrential rains that fall each year. But these drainage problems and the corresponding threats of catastrophic flooding belie one of the great ironies of its urban plumbing. When it comes to water, the city is also facing the kind of shortages that plague the rest of the globe. Mexico City, which sits at an altitude of over 7,300 feet, must pump water up 3,000 feet to reach residents. Last year it had to ration water after one of the worst droughts in six decades. The drainage program includes plans for treatment plants to turn runoff into clean water for use by farmers.

These problems, and the enormously complex engineering and plumbing challenges they create, reveal a much larger global concern. Like Mexico City, megacities around the world must find ways to control runoff while providing clean water for millions of inhabitants. With 1.1 billion people—or 18 percent of the world's population—now lacking access to safe drinking water, according to the World Health Organization, governments of developing countries need the money and know-how to build massive public works.

Beijing: The Commute That Never Ends

Zhao Ning lives just outside Beijing's Fifth Ring Road, one of the massive concentric expressways that circle the center of China's second-largest city. She wakes up at 5:30 a.m. each workday morning, quickly puts on makeup, and then rushes out to catch her first bus for her interminable commute to work. She barely has time for breakfast.

She transfers to a second bus, which takes her to the subway. Then she transfers twice more, needing three different lines to make her way to another bus that will take her to her office in northwest Beijing, where she is the associate director of an American study-abroad program. The subway system is only two-thirds the size of New York's, but it carries the same number of daily commuters, more than 5 million.

FIGURE 20.8   **Commuting subway trains in Beijing are crowded, as the city struggles to keep up with population growth.**

"Each day I spend four hours on the road," she says. "It is very exhausting and it puts so much pressure on me, especially in the morning."

Despite Beijing's modern, well-kept web of **beltways** and feeder roads into the city, driving is not an option now for Ms. Zhao, even though she and her husband own a car. Like most sprawling megacities, traffic—and the resulting, oft-reported pollution problem—is a constant urban plague. More than 4 million cars jostle along Beijing's roadways, with nearly 1,300 added every day, according to the city's Traffic Management Bureau. In April, the city began to adjust the working hours for nearly 810,000 of these commuters, hoping to alleviate the morning and evening rush.

> **beltways:** a highway encircling an urban area

When Zhao once tried to drive, her car was quickly entombed in traffic. "I was so worried—like an ant dancing in a hot pan," she says, using a classic Chinese expression. "Since then I haven't driven to work."

Indeed, along with water and sanitation, the challenges of mobility virtually define the growth of megacities. At the same time, they reveal the profound social and political upheaval the world's transition to city life can create. Cities bring economic growth and the expansion of the middle class. Members of the middle class want to own property—homes and, increasingly, cars.

A major problem confronting expanding cities is how to graft new subways and sewer systems onto existing neighborhoods. In China, authorities have tried to circumvent that by creating entire cities from scratch. As part of the government's aggressive **urbanization** program, it has poured large resources into building new communities, especially deep within the mainland. In fact, many of the most educated Chinese professionals on the coast have never heard of cities in their own country, some with populations the size of Houston.

> **urbanization**: the movement of a population from rural to urban areas and the resulting growth of the urban area

In 1980, only 51 cities with more than 500,000 people existed in China, according to UN figures. Since then, that number has jumped to 236. By 2025, the UN estimates, China will add 100 more cities to this group, as it pursues moving millions of rural peasants into vast urban networks. And with its robust rate of economic growth, China has the money to pursue the theorem, "If we build it, they will come." Its centralized political system also makes it easier to plan new urban networks without significant resistance.

Tokyo: The Megacity That Works

Zhao's friends raise a basic question: As the world tilts inexorably urban, will the megacities of tomorrow even be livable? Experts point to cities like Lagos, Nigeria, as the kind of urban beehive that doesn't work—traffic, untold pollution, the lack of even the most basic services.

Yet other megacities have certainly found the right blend of concrete and urban cachet. Most notable is the world's largest urban conglomeration—Tokyo. Though the multitudes in Tokyo proper are shoehorned into a relatively small area, the city consistently ranks near the top in surveys of the world's most livable places. It boasts high-quality goods and services, a wealth of world-class restaurants, and an enviable choice of museums, galleries, and architectural wonders.

But its near-faultless transportation system may be the most impressive and efficient means of public mobility ever built. Many residents cite the ease with which they can explore their city as a primary reason Tokyo is a desirable place to live.

"You can be anywhere in the city within an hour, easily," says Mami Ishikawa, a university student.

Outside Shinjuku Station, the busiest train station in the world, a swarm of 3.64 million commuters per day spill out onto the streets, seemingly in unison, via countless exits and well-designed traffic lights. Innovative "cycle trees," multilevel mechanized parking lots for cyclists, make it simple to get around without a car.

Yet Tokyo's urban efficiency is due as much to social factors as it is to its transportation system and technological prowess. "Cultural aspects, such as the Japanese penchant for order, respect for social rules and norms, and

FIGURE 20.9 **Residents of Tokyo, Japan often use bicycles for commuting to alleviate congestion and air pollution.**

Shutterstock/longtaildog

reluctance to intrude on others' private realms is also very important to minimizing friction," says Julian Worrall, an expert at Waseda University.

Undeniably, Tokyo has its challenges: high costs, dense living, patience-sapping gridlock for those brave enough to drive. Mr. Worrall points to aesthetic deficiencies, too—the spread of high-rise condos, the lack of urban space devoted to something other than consumption and production.

Maybe so. But to someone like Pastor Singh, who has to line up each day in Mumbai just to get water, those might seem like petty annoyances.

Adapted from Bruinius, H. 2010. Megacities of the World: A Glimpse of How We'll Live Tomorrow. The Christian Science Monitor, May 5. Used with permission of the author. http://www.csmonitor.com/layout/set/print/content/view/print/298332